TOYOTA/LEXUS
HIGHLANDER 2001-2006
RX 300/330 1999-2006
REPAIR MANUAL

CHILTON'S

Deleted

Covers U.S. and Canadian models of Toyota Highlander and Lexus RX 300/330 models

Does not include information specific to hybrid models

by Joe L. Hamilton

CHILTON *Automotive Books*

PUBLISHED BY **HAYNES NORTH AMERICA. Inc.**

Manufactured in USA
©2006 Haynes North America, Inc.
ISBN-13: 978-1-56392-624-2
ISBN-10: 1-56392-624-5
Library of Congress Control Number 2006933551

Haynes Publishing Group
Sparkford Nr Yeovil
Somerset BA22 7JJ England

Haynes North America, Inc
861 Lawrence Drive
Newbury Park
California 91320 USA

ABCDE
FGHIJ
KLMNO
PQRST

D1088050

Contents

Mechanic, author and photographer with a 2001 Highlander

ACKNOWLEDGEMENTS

Technical writers who contributed to this project include John Wegmann, Rob Maddox and Mike Stubblefield. Wiring diagrams provided exclusively for the publisher by Solution Builders.

About this manual

ITS PURPOSE

The purpose of this manual is to help you get the best value from your vehicle. It can do so in several ways. It can help you decide what work must be done, even if you choose to have it done by a dealer service department or a repair shop; it provides information and procedures for routine maintenance and servicing; and it offers diagnostic and repair procedures to follow when trouble occurs.

We hope you use the manual to tackle the work yourself. For many simpler jobs, doing it yourself may be quicker than arranging an appointment to get the vehicle into a shop and making the trips to leave it and pick it up. More importantly, a lot of money can be saved by avoiding the expense the shop must pass on to you to cover its labor and overhead costs. An added benefit is the sense of satisfaction and accomplishment that you feel after doing the job yourself.

USING THE MANUAL

The manual is divided into Chapters. Each Chapter is divided into numbered Sections. Each Section consists of consecutively numbered paragraphs.

At the beginning of each numbered Section you will be referred to any illustrations which apply to the procedures in that Section. The reference numbers used in illustration captions pinpoint the pertinent Section and the Step within that Section. That is, illustration 3.2 means the illustration refers to Section 3 and Step (or paragraph) 2 within that Section.

Procedures, once described in the text, are not normally repeated. When it's necessary to refer to another Chapter, the reference will be given as Chapter and Section number. Cross references given without use of the word "Chapter" apply to Sections and/or paragraphs in the same Chapter. For example, "see Section 8" means in the same Chapter.

References to the left or right side of the vehicle assume you are sitting in the driver's seat, facing forward.

Even though we have prepared this manual with extreme care, neither the publisher nor the author can accept responsibility for any errors in, or omissions from, the information given.

➡ **NOTE**

A *Note* provides information necessary to properly complete a procedure or information which will make the procedure easier to understand.

❊❊ **CAUTION**

A *Caution* provides a special procedure or special steps which must be taken while completing the procedure where the Caution is found. Not heeding a Caution can result in damage to the assembly being worked on.

❊❊ **WARNING**

A *Warning* provides a special procedure or special steps which must be taken while completing the procedure where the Warning is found. Not heeding a Warning can result in personal injury.

Introduction

This manual covers the Toyota Highlander and Lexus RX 300/330. Highlander models feature either an in-line four-cylinder engine or a V6 engine, while Lexus RX 300/330 models are only equipped with a V6 engine.

The engine drives the front wheels through either a four or five-speed automatic transaxle via independent driveaxles. On 4WD models the rear wheels are also propelled, via a driveshaft, rear differential, and two rear driveaxles.

Suspension is independent at all four wheels, MacPherson struts being used at the front end and at the rear. The rack-and-pinion steering unit is mounted on the suspension crossmember.

The brakes are disc at the front and rear, with power assist standard. All models are equipped with an Anti-lock Brake System (ABS).

Vehicle Identification Numbers

Modifications are a continuing and unpublicized process in vehicle manufacturing. Since spare parts manuals and lists are compiled on a numerical basis, the individual vehicle numbers are essential to correctly identify the component required.

VEHICLE IDENTIFICATION NUMBER (VIN)

This very important identification number is stamped on a plate attached to the dashboard inside the windshield on the driver's side of the vehicle (see illustration). It can also be found on the certification label located on the driver's side door post. The VIN also appears on the Vehicle Certificate of Title and Registration. It contains information such as where and when the vehicle was manufactured, the model year and the body style.

CERTIFICATION LABEL

The certification label is attached to the end of the driver's door post (see illustration). The plate contains the name of the manufacturer, the month and year of production, the Gross Vehicle Weight Rating (GVWR), the Gross Axle Weight Rating (GAWR) and the certification statement.

ENGINE NUMBER

On four-cylinder models, the engine identification number is stamped into a machined pad on the front-side of the engine, near the exhaust manifold (see illustration).

On V6 models, the engine identification number is stamped into a machined pad on the left end (driver's side) of the engine block (see illustration).

The Vehicle Identification Number (VIN) is located on a plate (arrow) on top of the dash (visible through the windshield)

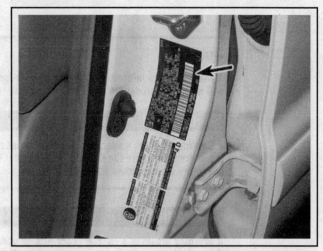

The vehicle certification label is located at the driver's door pillar

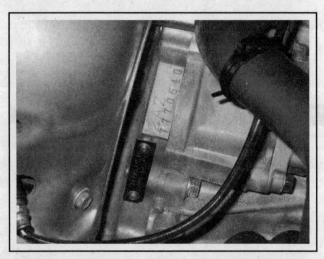

The four-cylinder engine number is located at the rear of the block on the front side, near the exhaust manifold

The V6 engine serial number is located on the front side of the block, adjacent to the transaxle

Recall information

Vehicle recalls are carried out by the manufacturer in the rare event of a possible safety-related defect. The vehicle's registered owner is contacted at the address on file at the Department of Motor Vehicles and given the details of the recall. Remedial work is carried out free of charge at a dealer service department.

If you are the new owner of a used vehicle which was subject to a recall and you want to be sure that the work has been carried out, it's best to contact a dealer service department and ask about your individual vehicle - you'll need to furnish them your Vehicle Identification Number (VIN).

The table below is based on information provided by the National Highway Traffic Safety Administration (NHTSA), the body which oversees vehicle recalls in the United States. The recall database is updated constantly. For the latest information on vehicle recalls, check the NHTSA website at www.nhtsa.gov, or call the NHTSA hotline at 1-888-327-4236.

Recall date	Recall campaign no.	Model(s) affected	Concern
Jan 19, 1999	99V011000	1999 RX300	The headlights and taillights may not automatically illuminate in low ambient light when the headlight switch is placed in the "AUTO" position.
Jan 19, 1999	99V012000	1999 RX300	Passenger vehicles equipped with the optional traction control system. Due to improper design of the rear brake tube distribution, if one of the dual brake lines fails when the driver applies the brake, the vehicle will feel unstable compared to vehicles with proper rear brake tube distribution.
Jul 03, 2001	01V228000	2001 Highlander	The reservoir filler cap of the brake master cylinder may induce a vacuum, introducing some air into the brake master cylinder reservoir. This could cause abnormal brake noise and increased pedal stroke when the brakes are applied.
Jul 30, 2002	02V208000	2001, 2002 Highlander	There is a breather hose that attaches to the nozzle of the on-board refueling vapor recovery (ORVR) valve, which is located on the topside of the fuel tank. The breather hose is attached with a clamp, whose "tabs" face in the upward direction. When the vehicle was crashed under the New Car Assessment Program (NCAP) test (i.e., 35 mph frontal barrier crash), the clamp "tabs" contacted the underside of the body due to movement of the fuel tank, causing the nozzle to break.
Dec 12, 2002	02V339000	2003 Highlander	On certain sport utility vehicles equipped with five factory alloy wheels (model 6934 and 6936), with factory LLAT, and with port installed WR4, four alloy wheel upgrade processed at the Jacksonville, Florida port and distributed by Southeast Toyota Distributors in the States of Alabama, Florida, Georgia, North and South Carolina, when the vehicles were processed at the port, an alloy tire/wheel upgrade was installed. The upgrade included 4 alloy wheels. The original spare tire/alloy wheel was not changed. The spare tire requires the use of a different style of wheel nut to attach it to the vehicle.
Apr 07, 2004	04V181000	2001, 2002, 2003 and 2004 Highlander	Certain sport utility vehicles have a Child Protection Lock (CPL) System on both of the rear side doors. When an operator of the vehicle closes the door very hard with the CPL lever set to the lock position, there is a possibility that the lever may contact the body panel, causing the CPL lever to move into the unlock position.
Nov 18, 2004	04V558000	2004 RX330	Certain passenger vehicles may be equipped with an improperly designed brake light switch. A silicon oxide build-up occurs on the contacts inside the brake light switch, which can make it inoperable.
Mar 21, 2006	06E026000	1999, 2000, 2001, 2002 and 2003 RX300 2004 RX330	Certain Pro-A motors corner lamps, turn signals, and headlights sold as replacement lamps for use on certain passenger vehicles. Some combination lamps that are not equipped with amber side reflectors fail to conform to Federal Motor Vehicle Safety Standard No. 108, lamps, reflective devices, and associated equipment. This recall only pertains to Pro-A Motors aftermarket lamps and has no relation to any original equipment installed on the listed passenger vehicles.

Buying parts

Replacement parts are available from many sources, which generally fall into one of two categories - authorized dealer parts departments and independent retail auto parts stores. Our advice concerning these parts is as follows:

Retail auto parts stores: Good auto parts stores will stock frequently needed components which wear out relatively fast, such as clutch components, exhaust systems, brake parts, tune-up parts, etc. These stores often supply new or reconditioned parts on an exchange basis, which can save a considerable amount of money. Discount auto parts stores are often very good places to buy materials and parts needed for general vehicle maintenance such as oil, grease, filters, spark plugs, belts, touch-up paint, bulbs, etc. They also usually sell

tools and general accessories, have convenient hours, charge lower prices and can often be found not far from home.

Authorized dealer parts department: This is the best source for parts which are unique to the vehicle and not generally available elsewhere (such as major engine parts, transmission parts, trim pieces, etc.).

Warranty information: If the vehicle is still covered under warranty, be sure that any replacement parts purchased - regardless of the source - do not invalidate the warranty!

To be sure of obtaining the correct parts, have engine and chassis numbers available and, if possible, take the old parts along for positive identification.

Maintenance techniques, tools and working facilities

MAINTENANCE TECHNIQUES

There are a number of techniques involved in maintenance and repair that will be referred to throughout this manual. Application of these techniques will enable the home mechanic to be more efficient, better organized and capable of performing the various tasks properly, which will ensure that the repair job is thorough and complete.

Fasteners

Fasteners are nuts, bolts, studs and screws used to hold two or more parts together. There are a few things to keep in mind when working with fasteners. Almost all of them use a locking device of some type, either a lockwasher, locknut, locking tab or thread adhesive. All threaded fasteners should be clean and straight, with undamaged threads and undamaged corners on the hex head where the wrench fits. Develop the habit of replacing all damaged nuts and bolts with new ones. Special locknuts with nylon or fiber inserts can only be used once. If they are removed, they lose their locking ability and must be replaced with new ones.

Rusted nuts and bolts should be treated with a penetrating fluid to ease removal and prevent breakage. Some mechanics use turpentine in a spout-type oil can, which works quite well. After applying the rust penetrant, let it work for a few minutes before trying to loosen the nut or bolt. Badly rusted fasteners may have to be chiseled or sawed off or removed with a special nut breaker, available at tool stores.

If a bolt or stud breaks off in an assembly, it can be drilled and removed with a special tool commonly available for this purpose. Most automotive machine shops can perform this task, as well as other repair procedures, such as the repair of threaded holes that have been stripped out.

Flat washers and lockwashers, when removed from an assembly, should always be replaced exactly as removed. Replace any damaged washers with new ones. Never use a lockwasher on any soft metal surface (such as aluminum), thin sheet metal or plastic.

Fastener sizes

For a number of reasons, automobile manufacturers are making wider and wider use of metric fasteners. Therefore, it is important to be able to tell the difference between standard (sometimes called U.S.

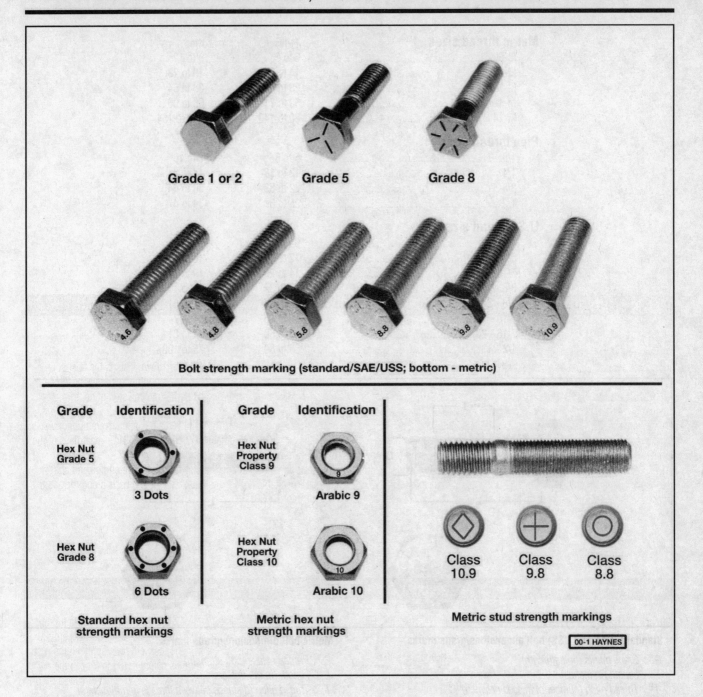

Bolt strength marking (standard/SAE/USS; bottom - metric)

Grade 1 or 2   Grade 5   Grade 8

Grade	Identification
Hex Nut Grade 5	3 Dots
Hex Nut Grade 8	6 Dots

Standard hex nut strength markings

Grade	Identification
Hex Nut Property Class 9	Arabic 9
Hex Nut Property Class 10	Arabic 10

Metric hex nut strength markings

Class 10.9   Class 9.8   Class 8.8

Metric stud strength markings

00-1 HAYNES

or SAE) and metric hardware, since they cannot be interchanged.

All bolts, whether standard or metric, are sized according to diameter, thread pitch and length. For example, a standard 1/2 - 13 x 1 bolt is 1/2 inch in diameter, has 13 threads per inch and is 1 inch long. An M12 - 1.75 x 25 metric bolt is 12 mm in diameter, has a thread pitch of 1.75 mm (the distance between threads) and is 25 mm long. The two bolts are nearly identical, and easily confused, but they are not interchangeable.

In addition to the differences in diameter, thread pitch and length, metric and standard bolts can also be distinguished by examining the bolt heads. To begin with, the distance across the flats on a standard bolt head is measured in inches, while the same dimension on a metric bolt is sized in millimeters (the same is true for nuts). As a result, a standard wrench should not be used on a metric bolt and a metric wrench should not be used on a standard bolt. Also, most standard bolts have slashes

radiating out from the center of the head to denote the grade or strength of the bolt, which is an indication of the amount of torque that can be applied to it. The greater the number of slashes, the greater the strength of the bolt. Grades 0 through 5 are commonly used on automobiles. Metric bolts have a property class (grade) number, rather than a slash, molded into their heads to indicate bolt strength. In this case, the higher the number, the stronger the bolt. Property class numbers 8.8, 9.8 and 10.9 are commonly used on automobiles.

Strength markings can also be used to distinguish standard hex nuts from metric hex nuts. Many standard nuts have dots stamped into one side, while metric nuts are marked with a number. The greater the number of dots, or the higher the number, the greater the strength of the nut.

Metric studs are also marked on their ends according to property class (grade). Larger studs are numbered (the same as metric bolts), while smaller studs carry a geometric code to denote grade.

Metric thread sizes	Ft-lbs	Nm
M-6	6 to 9	9 to 12
M-8	14 to 21	19 to 28
M-10	28 to 40	38 to 54
M-12	50 to 71	68 to 96
M-14	80 to 140	109 to 154

Pipe thread sizes		
1/8	5 to 8	7 to 10
1/4	12 to 18	17 to 24
3/8	22 to 33	30 to 44
1/2	25 to 35	34 to 47

U.S. thread sizes		
1/4 - 20	6 to 9	9 to 12
5/16 - 18	12 to 18	17 to 24
5/16 - 24	14 to 20	19 to 27
3/8 - 16	22 to 32	30 to 43
3/8 - 24	27 to 38	37 to 51
7/16 - 14	40 to 55	55 to 74
7/16 - 20	40 to 60	55 to 81
1/2 - 13	55 to 80	75 to 108

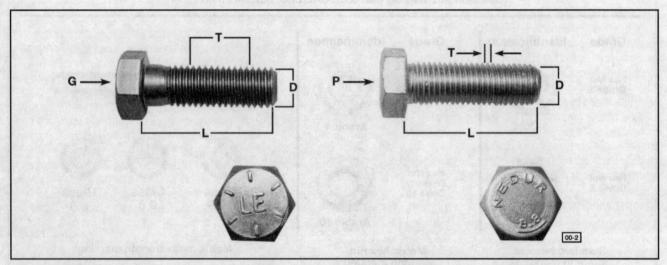

Standard (SAE and USS) bolt dimensions/grade marks

G   Grade marks (bolt strength)
L   Length (in inches)
T   Thread pitch (number of threads per inch)
D   Nominal diameter (in inches)

Metric bolt dimensions/grade marks

P   Property class (bolt strength)
L   Length (in millimeters)
T   Thread pitch (distance between threads in millimeters)
D   Diameter

It should be noted that many fasteners, especially Grades 0 through 2, have no distinguishing marks on them. When such is the case, the only way to determine whether it is standard or metric is to measure the thread pitch or compare it to a known fastener of the same size.

Standard fasteners are often referred to as SAE, as opposed to metric. However, it should be noted that SAE technically refers to a non-metric fine thread fastener only. Coarse thread non-metric fasteners are referred to as USS sizes.

Since fasteners of the same size (both standard and metric) may have different strength ratings, be sure to reinstall any bolts, studs or nuts removed from your vehicle in their original locations. Also, when replacing a fastener with a new one, make sure that the new one has a strength rating equal to or greater than the original.

Tightening sequences and procedures

Most threaded fasteners should be tightened to a specific torque value (torque is the twisting force applied to a threaded component such as a nut or bolt). Overtightening the fastener can weaken it and cause it to break, while undertightening can cause it to eventually come loose. Bolts, screws and studs, depending on the material they are made of and their thread diameters, have specific torque values, many of which are noted in the Specifications at the end of each Chapter. Be sure to follow the torque recommendations closely. For fasteners not assigned a specific torque, a general torque value chart is presented here as a guide. These torque values are for dry (unlubricated) fasteners threaded into steel or cast iron (not aluminum). As was previously mentioned, the size and grade of a fastener determine the amount of torque that can

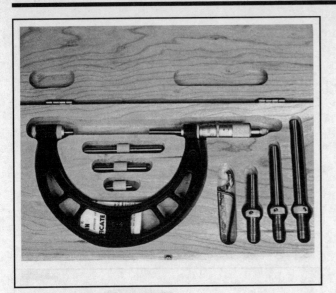

Micrometer set

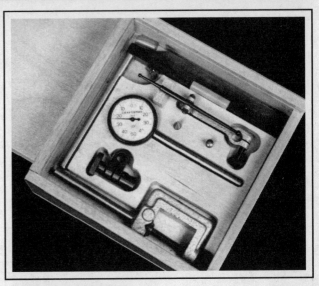

Dial indicator set

safely be applied to it. The figures listed here are approximate for Grade 2 and Grade 3 fasteners. Higher grades can tolerate higher torque values.

Fasteners laid out in a pattern, such as cylinder head bolts, oil pan bolts, differential cover bolts, etc., must be loosened or tightened in sequence to avoid warping the component. This sequence will normally be shown in the appropriate Chapter. If a specific pattern is not given, the following procedures can be used to prevent warping.

Initially, the bolts or nuts should be assembled finger-tight only. Next, they should be tightened one full turn each, in a criss-cross or diagonal pattern. After each one has been tightened one full turn, return to the first one and tighten them all one-half turn, following the same pattern. Finally, tighten each of them one-quarter turn at a time until each fastener has been tightened to the proper torque. To loosen and remove the fasteners, the procedure would be reversed.

Component disassembly

Component disassembly should be done with care and purpose to help ensure that the parts go back together properly. Always keep track of the sequence in which parts are removed. Make note of special characteristics or marks on parts that can be installed more than one way, such as a grooved thrust washer on a shaft. It is a good idea to lay the disassembled parts out on a clean surface in the order that they were removed. It may also be helpful to make sketches or take instant photos of components before removal.

When removing fasteners from a component, keep track of their locations. Sometimes threading a bolt back in a part, or putting the washers and nut back on a stud, can prevent mix-ups later. If nuts and bolts cannot be returned to their original locations, they should be kept in a compartmented box or a series of small boxes. A cupcake or muffin tin is ideal for this purpose, since each cavity can hold the bolts and nuts from a particular area (i.e. oil pan bolts, valve cover bolts, engine mount bolts, etc.). A pan of this type is especially helpful when working on assemblies with very small parts, such as the carburetor, alternator, valve train or interior dash and trim pieces. The cavities can be marked with paint or tape to identify the contents.

Whenever wiring looms, harnesses or connectors are separated, it is a good idea to identify the two halves with numbered pieces of masking tape so they can be easily reconnected.

Gasket sealing surfaces

Throughout any vehicle, gaskets are used to seal the mating surfaces between two parts and keep lubricants, fluids, vacuum or pressure contained in an assembly.

Many times these gaskets are coated with a liquid or paste-type gasket sealing compound before assembly. Age, heat and pressure can sometimes cause the two parts to stick together so tightly that they are very difficult to separate. Often, the assembly can be loosened by striking it with a soft-face hammer near the mating surfaces. A regular hammer can be used if a block of wood is placed between the hammer and the part. Do not hammer on cast parts or parts that could be easily damaged. With any particularly stubborn part, always recheck to make sure that every fastener has been removed.

Avoid using a screwdriver or bar to pry apart an assembly, as they can easily mar the gasket sealing surfaces of the parts, which must remain smooth. If prying is absolutely necessary, use an old broom handle, but keep in mind that extra clean up will be necessary if the wood splinters.

After the parts are separated, the old gasket must be carefully scraped off and the gasket surfaces cleaned. Stubborn gasket material can be soaked with rust penetrant or treated with a special chemical to soften it so it can be easily scraped off.

❄❄ CAUTION:

Never use gasket removal solutions or caustic chemicals on plastic or other composite components.

A scraper can be fashioned from a piece of copper tubing by flattening and sharpening one end. Copper is recommended because it is usually softer than the surfaces to be scraped, which reduces the chance of gouging the part. Some gaskets can be removed with a wire brush, but regardless of the method used, the mating surfaces must be left clean and smooth. If for some reason the gasket surface is gouged, then a gasket sealer thick enough to fill scratches will have to be used during reassembly of the components. For most applications, a non-drying (or semi-drying) gasket sealer should be used.

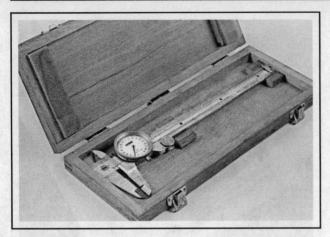

Dial caliper

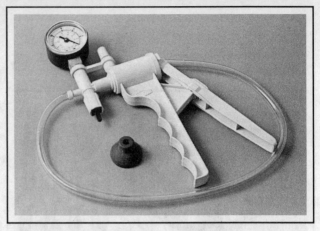

Hand-operated vacuum pump

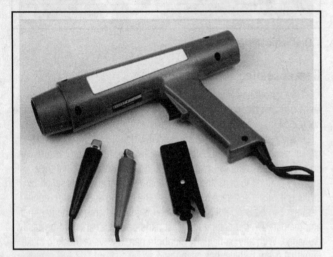

Timing light

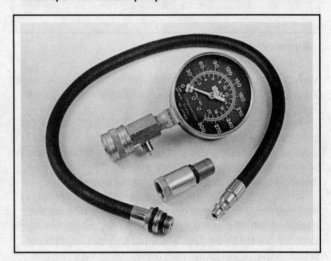

Compression gauge with spark plug hole adapter

Hose removal tips

✳✳ WARNING:

If the vehicle is equipped with air conditioning, do not disconnect any of the A/C hoses without first having the system depressurized by a dealer service department or a service station.

Hose removal precautions closely parallel gasket removal precautions. Avoid scratching or gouging the surface that the hose mates against or the connection may leak. This is especially true for radiator hoses. Because of various chemical reactions, the rubber in hoses can bond itself to the metal spigot that the hose fits over. To remove a hose, first loosen the hose clamps that secure it to the spigot. Then, with slip-joint pliers, grab the hose at the clamp and rotate it around the spigot. Work it back and forth until it is completely free, then pull it off. Silicone or other lubricants will ease removal if they can be applied between the hose and the outside of the spigot. Apply the same lubricant to the inside of the hose and the outside of the spigot to simplify installation.

As a last resort (and if the hose is to be replaced with a new one anyway), the rubber can be slit with a knife and the hose peeled from the spigot. If this must be done, be careful that the metal connection is not damaged.

If a hose clamp is broken or damaged, do not reuse it. Wire-type clamps usually weaken with age, so it is a good idea to replace them with screw-type clamps whenever a hose is removed.

TOOLS

A selection of good tools is a basic requirement for anyone who plans to maintain and repair his or her own vehicle. For the owner who has few tools, the initial investment might seem high, but when compared to the spiraling costs of professional auto maintenance and repair, it is a wise one.

To help the owner decide which tools are needed to perform the tasks detailed in this manual, the following tool lists are offered: *Maintenance and minor repair, Repair/overhaul and Special.*

The newcomer to practical mechanics should start off with the *maintenance and minor repair* tool kit, which is adequate for the simpler jobs performed on a vehicle. Then, as confidence and experience grow, the owner can tackle more difficult tasks, buying additional tools as they are needed. Eventually the basic kit will be expanded into the *repair and overhaul* tool set. Over a period of time, the experienced do-it-yourselfer will assemble a tool set complete enough for most repair and overhaul procedures and will add tools from the special category when it is felt that the expense is justified by the frequency of use.

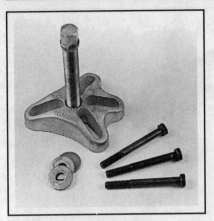

Damper/steering wheel puller

General purpose puller

Hydraulic lifter removal tool

Valve spring compressor

Valve spring compressor

Ridge reamer

Maintenance and minor repair tool kit

The tools in this list should be considered the minimum required for performance of routine maintenance, servicing and minor repair work. We recommend the purchase of combination wrenches (box-end and open-end combined in one wrench). While more expensive than open end wrenches, they offer the advantages of both types of wrench.

> *Combination wrench set (1/4-inch to 1 inch or 6 mm to 19 mm)*
> *Adjustable wrench, 8 inch*
> *Spark plug wrench with rubber insert*
> *Spark plug gap adjusting tool*
> *Feeler gauge set*
> *Brake bleeder wrench*
> *Standard screwdriver (5/16-inch x 6 inch)*
> *Phillips screwdriver (No. 2 x 6 inch)*
> *Combination pliers - 6 inch*
> *Hacksaw and assortment of blades*
> *Tire pressure gauge*
> *Grease gun*
> *Oil can*
> *Fine emery cloth*
> *Wire brush*
> *Battery post and cable cleaning tool*
> *Oil filter wrench*
> *Funnel (medium size)*
> *Safety goggles*
> *Jackstands (2)*
> *Drain pan*

➡ **Note: If basic tune-ups are going to be part of routine maintenance, it will be necessary to purchase a good quality stroboscopic timing light and combination tachometer/dwell meter. Although they are included in the list of special tools, it is mentioned here because they are absolutely necessary for tuning most vehicles properly.**

Repair and overhaul tool set

These tools are essential for anyone who plans to perform major repairs and are in addition to those in the maintenance and minor repair tool kit. Included is a comprehensive set of sockets which, though expensive, are invaluable because of their versatility, especially when various extensions and drives are available. We recommend the 1/2-inch drive over the 3/8-inch drive. Although the larger drive is bulky and more expensive, it has the capacity of accepting a very wide range of large sockets. Ideally, however, the mechanic should have a 3/8-inch drive set and a 1/2-inch drive set.

> *Socket set(s)*
> *Reversible ratchet*
> *Extension - 10 inch*
> *Universal joint*
> *Torque wrench (same size drive as sockets)*
> *Ball peen hammer - 8 ounce*
> *Soft-face hammer (plastic/rubber)*
> *Standard screwdriver (1/4-inch x 6 inch)*
> *Standard screwdriver (stubby - 5/16-inch)*
> *Phillips screwdriver (No. 3 x 8 inch)*
> *Phillips screwdriver (stubby - No. 2)*
> *Pliers - vise grip*

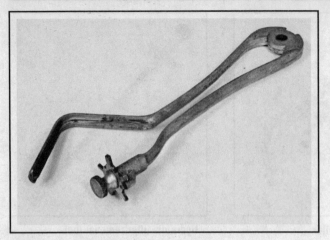

Piston ring groove cleaning tool

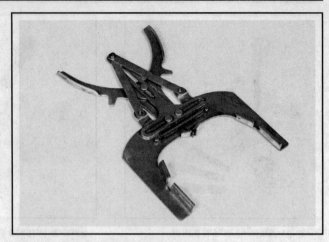

Ring removal/installation tool

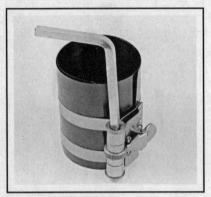

Ring compressor

Cylinder hone

Brake hold-down spring tool

Pliers - lineman's
Pliers - needle nose
Pliers - snap-ring (internal and external)
Cold chisel - 1/2-inch
Scribe
Scraper (made from flattened copper tubing)
Centerpunch
Pin punches (1/16, 1/8, 3/16-inch)
Steel rule/straightedge - 12 inch
Allen wrench set (1/8 to 3/8-inch or 4 mm to 10 mm)
A selection of files
Wire brush (large)
Jackstands (second set)
Jack (scissor or hydraulic type)

➡**Note: Another tool which is often useful is an electric drill with a chuck capacity of 3/8-inch and a set of good quality drill bits.**

Special tools

The tools in this list include those which are not used regularly, are expensive to buy, or which need to be used in accordance with their manufacturer's instructions. Unless these tools will be used frequently, it is not very economical to purchase many of them. A consideration would be to split the cost and use between yourself and a friend or friends. In addition, most of these tools can be obtained from a tool rental shop on a temporary basis.

This list primarily contains only those tools and instruments widely available to the public, and not those special tools produced by the vehicle manufacturer for distribution to dealer service depart-ments. Occasionally, references to the manufacturer's special tools are included in the text of this manual. Generally, an alternative method of doing the job without the special tool is offered. However, sometimes there is no alternative to their use. Where this is the case, and the tool cannot be purchased or borrowed, the work should be turned over to the dealer service department or an automotive repair shop.

Valve spring compressor
Piston ring groove cleaning tool
Piston ring compressor
Piston ring installation tool
Cylinder compression gauge
Cylinder ridge reamer
Cylinder surfacing hone
Cylinder bore gauge
Micrometers and/or dial calipers
Hydraulic lifter removal tool
Balljoint separator
Universal-type puller
Impact screwdriver
Dial indicator set
Stroboscopic timing light (inductive pick-up)
Hand operated vacuum/pressure pump
Tachometer/dwell meter
Universal electrical multimeter
Cable hoist
Brake spring removal and installation tools
Floor jack

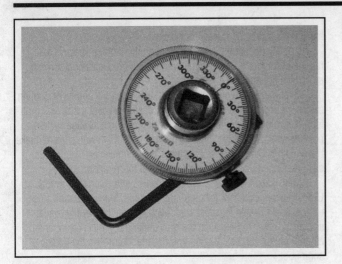

Torque angle gauge

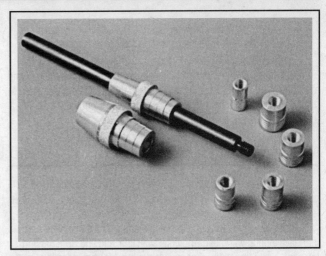

Clutch plate alignment tool

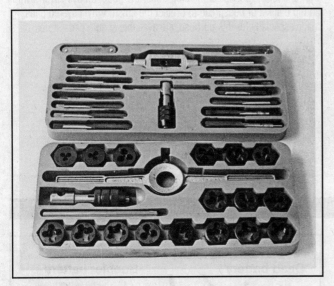

Tap and die set

Buying tools

For the do-it-yourselfer who is just starting to get involved in vehicle maintenance and repair, there are a number of options available when purchasing tools. If maintenance and minor repair is the extent of the work to be done, the purchase of individual tools is satisfactory. If, on the other hand, extensive work is planned, it would be a good idea to purchase a modest tool set from one of the large retail chain stores. A set can usually be bought at a substantial savings over the individual tool prices, and they often come with a tool box. As additional tools are needed, add-on sets, individual tools and a larger tool box can be purchased to expand the tool selection. Building a tool set gradually allows the cost of the tools to be spread over a longer period of time and gives the mechanic the freedom to choose only those tools that will actually be used.

Tool stores will often be the only source of some of the special tools that are needed, but regardless of where tools are bought, try to avoid cheap ones, especially when buying screwdrivers and sockets, because they won't last very long. The expense involved in replacing cheap tools will eventually be greater than the initial cost of quality tools.

Care and maintenance of tools

Good tools are expensive, so it makes sense to treat them with respect. Keep them clean and in usable condition and store them properly when not in use. Always wipe off any dirt, grease or metal chips before putting them away. Never leave tools lying around in the work area. Upon completion of a job, always check closely under the hood for tools that may have been left there so they won't get lost during a test drive.

Some tools, such as screwdrivers, pliers, wrenches and sockets, can be hung on a panel mounted on the garage or workshop wall, while others should be kept in a tool box or tray. Measuring instruments, gauges, meters, etc. must be carefully stored where they cannot be damaged by weather or impact from other tools.

When tools are used with care and stored properly, they will last a very long time. Even with the best of care, though, tools will wear out if used frequently. When a tool is damaged or worn out, replace it. Subsequent jobs will be safer and more enjoyable if you do.

HOW TO REPAIR DAMAGED THREADS

Sometimes, the internal threads of a nut or bolt hole can become stripped, usually from overtightening. Stripping threads is an all-too-common occurrence, especially when working with aluminum parts, because aluminum is so soft that it easily strips out.

Usually, external or internal threads are only partially stripped. After they've been cleaned up with a tap or die, they'll still work. Sometimes, however, threads are badly damaged. When this happens, you've got three choices:

1) *Drill and tap the hole to the next suitable oversize and install a larger diameter bolt, screw or stud.*
2) *Drill and tap the hole to accept a threaded plug, then drill and tap the plug to the original screw size. You can also buy a plug already threaded to the original size. Then you simply drill a hole to the specified size, then run the threaded plug into the hole with a bolt and jam nut. Once the plug is fully seated, remove the jam nut and bolt.*
3) *The third method uses a patented thread repair kit like Heli-Coil or Slimsert. These easy-to-use kits are designed to repair damaged threads in straight-through holes and blind holes. Both are available as kits which can handle a variety of sizes and thread*

patterns. Drill the hole, then tap it with the special included tap. Install the Heli-Coil and the hole is back to its original diameter and thread pitch.

Regardless of which method you use, be sure to proceed calmly and carefully. A little impatience or carelessness during one of these relatively simple procedures can ruin your whole day's work and cost you a bundle if you wreck an expensive part.

WORKING FACILITIES

Not to be overlooked when discussing tools is the workshop. If anything more than routine maintenance is to be carried out, some sort of suitable work area is essential.

It is understood, and appreciated, that many home mechanics do not have a good workshop or garage available, and end up removing an engine or doing major repairs outside. It is recommended, however, that the overhaul or repair be completed under the cover of a roof.

A clean, flat workbench or table of comfortable working height is an absolute necessity. The workbench should be equipped with a vise that has a jaw opening of at least four inches.

As mentioned previously, some clean, dry storage space is also required for tools, as well as the lubricants, fluids, cleaning solvents, etc. which soon become necessary.

Sometimes waste oil and fluids, drained from the engine or cooling system during normal maintenance or repairs, present a disposal problem. To avoid pouring them on the ground or into a sewage system, pour the used fluids into large containers, seal them with caps and take them to an authorized disposal site or recycling center. Plastic jugs, such as old antifreeze containers, are ideal for this purpose.

Always keep a supply of old newspapers and clean rags available. Old towels are excellent for mopping up spills. Many mechanics use rolls of paper towels for most work because they are readily available and disposable. To help keep the area under the vehicle clean, a large cardboard box can be cut open and flattened to protect the garage or shop floor.

Whenever working over a painted surface, such as when leaning over a fender to service something under the hood, always cover it with an old blanket or bedspread to protect the finish. Vinyl covered pads, made especially for this purpose, are available at auto parts stores.

Booster battery (jump) starting

Observe these precautions when using a booster battery to start a vehicle:

a) *Before connecting the booster battery, make sure the ignition switch is in the Off position.*
b) *Turn off the lights, heater and other electrical loads.*
c) *Your eyes should be shielded. Safety goggles are a good idea.*
d) *Make sure the booster battery is the same voltage as the dead one in the vehicle.*
e) *The two vehicles MUST NOT TOUCH each other!*
f) *Make sure the transaxle is in Neutral (manual) or Park (automatic).*
g) *If the booster battery is not a maintenance-free type, remove the vent caps and lay a cloth over the vent holes.*

Connect the red jumper cable to the positive (+) terminals of each battery (see illustration).

Connect one end of the black jumper cable to the negative (-) terminal of the booster battery. The other end of this cable should be connected to a good ground on the vehicle to be started, such as a bolt or bracket on the body.

Start the engine using the booster battery, then, with the engine running at idle speed, disconnect the jumper cables in the reverse order of connection.

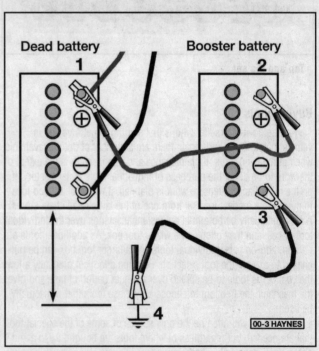

Make the booster battery cable connections in the numerical order shown (note that the negative cable of the booster battery is NOT attached to the negative terminal of the dead battery)

Jacking and towing

JACKING

The jack supplied with the vehicle should only be used for raising the vehicle for changing a tire or placing jackstands under the frame.

✳✳ WARNING:

Never crawl under the vehicle or start the engine when the jack is being used as the only means of support.

All vehicles are supplied with a scissors-type jack. When jacking the vehicle, it should be engaged with the seam notch, between the two

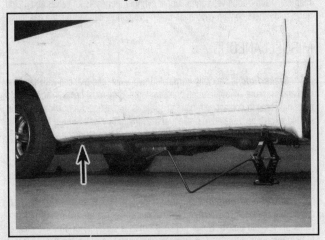

The jack fits over the rocker panel flange, between the two notches (there are two jacking points on each side of the vehicle)

dimples (see illustration).

The vehicle should be on level ground with the wheels blocked and the transmission in Park (automatic). Pry off the hub cap (if equipped) using the tapered end of the lug wrench. Loosen the lug nuts one-half turn and leave them in place until the wheel is raised off the ground.

Place the jack under the side of the vehicle in the indicated position. Use the supplied wrench to turn the jackscrew clockwise until the wheel is raised off the ground. Remove the lug nuts, pull off the wheel and replace it with the spare.

With the beveled side in, reinstall the lug nuts and tighten them until snug. Lower the vehicle by turning the jackscrew counterclockwise. Remove the jack and tighten the nuts in a diagonal pattern to the torque listed in the Chapter 1 Specifications. If a torque wrench is not available, have the torque checked by a service station as soon as possible. Replace the hubcap by placing it in position and using the heel of your hand or a rubber mallet to seat it.

TOWING

Two-wheel drive models can be towed from the front with the front wheels off the ground, using a wheel lift type tow truck. If towed from the rear, the front wheels must be placed on a dolly. Four wheel drive models must be towed with all four wheels off the ground. A sling-type tow truck cannot be used, as body damage will result. The best way to tow the vehicle is with a flat-bed car carrier.

In an emergency the vehicle can be towed a short distance with a cable or chain attached to one of the towing eyelets located under the front or rear bumpers. The driver must remain in the vehicle to operate the steering and brakes (remember that power steering and power brakes will not work with the engine off).

Automotive chemicals and lubricants

A number of automotive chemicals and lubricants are available for use during vehicle maintenance and repair. They include a wide variety of products ranging from cleaning solvents and degreasers to lubricants and protective sprays for rubber, plastic and vinyl.

CLEANERS

Carburetor cleaner and choke cleaner is a strong solvent for gum, varnish and carbon. Most carburetor cleaners leave a dry-type lubricant film which will not harden or gum up. Because of this film it is not recommended for use on electrical components.

Brake system cleaner is used to remove brake dust, grease and brake fluid from the brake system, where clean surfaces are absolutely necessary. It leaves no residue and often eliminates brake squeal caused by contaminants.

Electrical cleaner removes oxidation, corrosion and carbon deposits from electrical contacts, restoring full current flow. It can also be used to clean spark plugs, carburetor jets, voltage regulators and other parts where an oil-free surface is desired.

Demoisturants remove water and moisture from electrical components such as alternators, voltage regulators, electrical connectors and fuse blocks. They are non-conductive and non-corrosive.

Degreasers are heavy-duty solvents used to remove grease from the outside of the engine and from chassis components. They can be sprayed or brushed on and, depending on the type, are rinsed off either with water or solvent.

LUBRICANTS

Motor oil is the lubricant formulated for use in engines. It normally contains a wide variety of additives to prevent corrosion and reduce foaming and wear. Motor oil comes in various weights (viscosity ratings) from 0 to 50. The recommended weight of the oil depends on the season, temperature and the demands on the engine. Light oil is used in cold climates and under light load conditions. Heavy oil is used in hot climates and where high loads are encountered. Multi-viscosity oils are designed to have characteristics of both light and heavy oils and are available in a number of weights from 5W-20 to 20W-50.

Gear oil is designed to be used in differentials, manual transmissions and other areas where high-temperature lubrication is required.

Chassis and wheel bearing grease is a heavy grease used where increased loads and friction are encountered, such as for wheel bearings, balljoints, tie-rod ends and universal joints.

High-temperature wheel bearing grease is designed to withstand the extreme temperatures encountered by wheel bearings in disc brake equipped vehicles. It usually contains molybdenum disulfide (moly), which is a dry-type lubricant.

White grease is a heavy grease for metal-to-metal applications where water is a problem. White grease stays soft under both low and high temperatures (usually from -100 to +190-degrees F), and will not wash off or dilute in the presence of water.

Assembly lube is a special extreme pressure lubricant, usually containing moly, used to lubricate high-load parts (such as main and rod bearings and cam lobes) for initial start-up of a new engine. The assembly lube lubricates the parts without being squeezed out or washed away until the engine oiling system begins to function.

Silicone lubricants are used to protect rubber, plastic, vinyl and nylon parts.

Graphite lubricants are used where oils cannot be used due to contamination problems, such as in locks. The dry graphite will lubricate metal parts while remaining uncontaminated by dirt, water, oil or acids. It is electrically conductive and will not foul electrical contacts in locks such as the ignition switch.

Moly penetrants loosen and lubricate frozen, rusted and corroded fasteners and prevent future rusting or freezing.

Heat-sink grease is a special electrically non-conductive grease that is used for mounting electronic ignition modules where it is essential that heat is transferred away from the module.

SEALANTS

RTV sealant is one of the most widely used gasket compounds. Made from silicone, RTV is air curing, it seals, bonds, waterproofs, fills surface irregularities, remains flexible, doesn't shrink, is relatively easy to remove, and is used as a supplementary sealer with almost all low and medium temperature gaskets.

Anaerobic sealant is much like RTV in that it can be used either to seal gaskets or to form gaskets by itself. It remains flexible, is solvent resistant and fills surface imperfections. The difference between an anaerobic sealant and an RTV-type sealant is in the curing. RTV cures when exposed to air, while an anaerobic sealant cures only in the absence of air. This means that an anaerobic sealant cures only after the assembly of parts, sealing them together.

Thread and pipe sealant is used for sealing hydraulic and pneumatic fittings and vacuum lines. It is usually made from a Teflon compound, and comes in a spray, a paint-on liquid and as a wrap-around tape.

CHEMICALS

Anti-seize compound prevents seizing, galling, cold welding, rust and corrosion in fasteners. High-temperature anti-seize, usually made with copper and graphite lubricants, is used for exhaust system and exhaust manifold bolts.

Anaerobic locking compounds are used to keep fasteners from vibrating or working loose and cure only after installation, in the absence of air. Medium strength locking compound is used for small nuts, bolts and screws that may be removed later. High-strength locking compound is for large nuts, bolts and studs which aren't removed on a regular basis.

Oil additives range from viscosity index improvers to chemical treatments that claim to reduce internal engine friction. It should be noted that most oil manufacturers caution against using additives with their oils.

Gas additives perform several functions, depending on their chemical makeup. They usually contain solvents that help dissolve gum and varnish that build up on carburetor, fuel injection and intake parts. They also serve to break down carbon deposits that form on the inside surfaces of the combustion chambers. Some additives contain upper cylinder lubricants for valves and piston rings, and others contain chemicals to remove condensation from the gas tank.

MISCELLANEOUS

Brake fluid is specially formulated hydraulic fluid that can withstand the heat and pressure encountered in brake systems. Care must be taken so this fluid does not come in contact with painted surfaces or plastics. An opened container should always be resealed to prevent contamination by water or dirt.

Weatherstrip adhesive is used to bond weatherstripping around doors, windows and trunk lids. It is sometimes used to attach trim pieces.

Undercoating is a petroleum-based, tar-like substance that is designed to protect metal surfaces on the underside of the vehicle from corrosion. It also acts as a sound-deadening agent by insulating the bottom of the vehicle.

Waxes and polishes are used to help protect painted and plated surfaces from the weather. Different types of paint may require the use of different types of wax and polish. Some polishes utilize a chemical or abrasive cleaner to help remove the top layer of oxidized (dull) paint on older vehicles. In recent years many non-wax polishes that contain a wide variety of chemicals such as polymers and silicones have been introduced. These non-wax polishes are usually easier to apply and last longer than conventional waxes and polishes.

CONVERSION FACTORS

LENGTH (distance)

Inches (in)	X	25.4	= Millimeters (mm)	X 0.0394	= Inches (in)
Feet (ft)	X	0.305	= Meters (m)	X 3.281	= Feet (ft)
Miles	X	1.609	= Kilometers (km)	X 0.621	= Miles

VOLUME (capacity)

Cubic inches (cu in; in³)	X	16.387	= Cubic centimeters (cc; cm³)	X 0.061	= Cubic inches (cu in; in³)
Imperial pints (Imp pt)	X	0.568	= Liters (l)	X 1.76	= Imperial pints (Imp pt)
Imperial quarts (Imp qt)	X	1.137	= Liters (l)	X 0.88	= Imperial quarts (Imp qt)
Imperial quarts (Imp qt)	X	1.201	= US quarts (US qt)	X 0.833	= Imperial quarts (Imp qt)
US quarts (US qt)	X	0.946	= Liters (l)	X 1.057	= US quarts (US qt)
Imperial gallons (Imp gal)	X	4.546	= Liters (l)	X 0.22	= Imperial gallons (Imp gal)
Imperial gallons (Imp gal)	X	1.201	= US gallons (US gal)	X 0.833	= Imperial gallons (Imp gal)
US gallons (US gal)	X	3.785	= Liters (l)	X 0.264	= US gallons (US gal)

MASS (weight)

Ounces (oz)	X	28.35	= Grams (g)	X 0.035	= Ounces (oz)
Pounds (lb)	X	0.454	= Kilograms (kg)	X 2.205	= Pounds (lb)

FORCE

Ounces-force (ozf; oz)	X	0.278	= Newtons (N)	X 3.6	= Ounces-force (ozf; oz)
Pounds-force (lbf; lb)	X	4.448	= Newtons (N)	X 0.225	= Pounds-force (lbf; lb)
Newtons (N)	X	0.1	= Kilograms-force (kgf; kg)	X 9.81	= Newtons (N)

PRESSURE

Pounds-force per square inch (psi; lbf/in²; lb/in²)	X	0.070	= Kilograms-force per square centimeter (kgf/cm²; kg/cm²)	X 14.223	= Pounds-force per square inch (psi; lbf/in²; lb/in²)
Pounds-force per square inch (psi; lbf/in²; lb/in²)	X	0.068	= Atmospheres (atm)	X 14.696	= Pounds-force per square inch (psi; lbf/in²; lb/in²)
Pounds-force per square inch (psi; lbf/in²; lb/in²)	X	0.069	= Bars	X 14.5	= Pounds-force per square inch (psi; lbf/in²; lb/in²)
Pounds-force per square inch (psi; lbf/in²; lb/in²)	X	6.895	= Kilopascals (kPa)	X 0.145	= Pounds-force per square inch (psi; lbf/in²; lb/in²)
Kilopascals (kPa)	X	0.01	= Kilograms-force per square centimeter (kgf/cm²; kg/cm²)	X 98.1	= Kilopascals (kPa)

TORQUE (moment of force)

Pounds-force inches (lbf in; lb in)	X	1.152	= Kilograms-force centimeter (kgf cm; kg cm)	X 0.868	= Pounds-force inches (lbf in; lb in)
Pounds-force inches (lbf in; lb in)	X	0.113	= Newton meters (Nm)	X 8.85	= Pounds-force inches (lbf in; lb in)
Pounds-force inches (lbf in; lb in)	X	0.083	= Pounds-force feet (lbf ft; lb ft)	X 12	= Pounds-force inches (lbf in; lb in)
Pounds-force feet (lbf ft; lb ft)	X	0.138	= Kilograms-force meters (kgf m; kg m)	X 7.233	= Pounds-force feet (lbf ft; lb ft)
Pounds-force feet (lbf ft; lb ft)	X	1.356	= Newton meters (Nm)	X 0.738	= Pounds-force feet (lbf ft; lb ft)
Newton meters (Nm)	X	0.102	= Kilograms-force meters (kgf m; kg m)	X 9.804	= Newton meters (Nm)

VACUUM

Inches mercury (in. Hg)	X	3.377	= Kilopascals (kPa)	X 0.2961	= Inches mercury
Inches mercury (in. Hg)	X	25.4	= Millimeters mercury (mm Hg)	X 0.0394	= Inches mercury

POWER

Horsepower (hp)	X	745.7	= Watts (W)	X 0.0013	= Horsepower (hp)

VELOCITY (speed)

Miles per hour (miles/hr; mph)	X	1.609	= Kilometers per hour (km/hr; kph)	X 0.621	= Miles per hour (miles/hr; mph)

FUEL CONSUMPTION *

Miles per gallon, Imperial (mpg)	X	0.354	= Kilometers per liter (km/l)	X 2.825	= Miles per gallon, Imperial (mpg)
Miles per gallon, US (mpg)	X	0.425	= Kilometers per liter (km/l)	X 2.352	= Miles per gallon, US (mpg)

TEMPERATURE

Degrees Fahrenheit = (°C x 1.8) + 32

Degrees Celsius (Degrees Centigrade; °C) = (°F - 32) x 0.56

It is common practice to convert from miles per gallon (mpg) to liters/100 kilometers (l/100km), where mpg (Imperial) x l/100 km = 282 and mpg (US) x l/100 km = 235

FRACTION/DECIMAL/MILLIMETER EQUIVALENTS

DECIMALS to MILLIMETERS

Decimal	mm	Decimal	mm
0.001	0.0254	0.500	12.7000
0.002	0.0508	0.510	12.9540
0.003	0.0762	0.520	13.2080
0.004	0.1016	0.530	13.4620
0.005	0.1270	0.540	13.7160
0.006	0.1524	0.550	13.9700
0.007	0.1778	0.560	14.2240
0.008	0.2032	0.570	14.4780
0.009	0.2286	0.580	14.7320
		0.590	14.9860
0.010	0.2540		
0.020	0.5080		
0.030	0.7620		
0.040	1.0160	0.600	15.2400
0.050	1.2700	0.610	15.4940
0.060	1.5240	0.620	15.7480
0.070	1.7780	0.630	16.0020
0.080	2.0320	0.640	16.2560
0.090	2.2860	0.650	16.5100
		0.660	16.7640
0.100	2.5400	0.670	17.0180
0.110	2.7940	0.680	17.2720
0.120	3.0480	0.690	17.5260
0.130	3.3020		
0.140	3.5560		
0.150	3.8100		
0.160	4.0640	0.700	17.7800
0.170	4.3180	0.710	18.0340
0.180	4.5720	0.720	18.2880
0.190	4.8260	0.730	18.5420
		0.740	18.7960
0.200	5.0800	0.750	19.0500
0.210	5.3340	0.760	19.3040
0.220	5.5880	0.770	19.5580
0.230	5.8420	0.780	19.8120
0.240	6.0960	0.790	20.0660
0.250	6.3500		
0.260	6.6040		
0.270	6.8580	0.800	20.3200
0.280	7.1120	0.810	20.5740
0.290	7.3660	0.820	21.8280
		0.830	21.0820
0.300	7.6200	0.840	21.3360
0.310	7.8740	0.850	21.5900
0.320	8.1280	0.860	21.8440
0.330	8.3820	0.870	22.0980
0.340	8.6360	0.880	22.3520
0.350	8.8900	0.890	22.6060
0.360	9.1440		
0.370	9.3980		
0.380	9.6520		
0.390	9.9060		
		0.900	22.8600
0.400	10.1600	0.910	23.1140
0.410	10.4140	0.920	23.3680
0.420	10.6680	0.930	23.6220
0.430	10.9220	0.940	23.8760
0.440	11.1760	0.950	24.1300
0.450	11.4300	0.960	24.3840
0.460	11.6840	0.970	24.6380
0.470	11.9380	0.980	24.8920
0.480	12.1920	0.990	25.1460
0.490	12.4460	1.000	25.4000

FRACTIONS to DECIMALS to MILLIMETERS

Fraction	Decimal	mm	Fraction	Decimal	mm
1/64	0.0156	0.3969	33/64	0.5156	13.0969
1/32	0.0312	0.7938	17/32	0.5312	13.4938
3/64	0.0469	1.1906	35/64	0.5469	13.8906
1/16	0.0625	1.5875	9/16	0.5625	14.2875
5/64	0.0781	1.9844	37/64	0.5781	14.6844
3/32	0.0938	2.3812	19/32	0.5938	15.0812
7/64	0.1094	2.7781	39/64	0.6094	15.4781
1/8	0.1250	3.1750	5/8	0.6250	15.8750
9/64	0.1406	3.5719	41/64	0.6406	16.2719
5/32	0.1562	3.9688	21/32	0.6562	16.6688
11/64	0.1719	4.3656	43/64	0.6719	17.0656
3/16	0.1875	4.7625	11/16	0.6875	17.4625
13/64	0.2031	5.1594	45/64	0.7031	17.8594
7/32	0.2188	5.5562	23/32	0.7188	18.2562
15/64	0.2344	5.9531	47/64	0.7344	18.6531
1/4	0.2500	6.3500	3/4	0.7500	19.0500
17/64	0.2656	6.7469	49/64	0.7656	19.4469
9/32	0.2812	7.1438	25/32	0.7812	19.8438
19/64	0.2969	7.5406	51/64	0.7969	20.2406
5/16	0.3125	7.9375	13/16	0.8125	20.6375
21/64	0.3281	8.3344	53/64	0.8281	21.0344
11/32	0.3438	8.7312	27/32	0.8438	21.4312
23/64	0.3594	9.1281	55/64	0.8594	21.8281
3/8	0.3750	9.5250	7/8	0.8750	22.2250
25/64	0.3906	9.9219	57/64	0.8906	22.6219
13/32	0.4062	10.3188	29/32	0.9062	23.0188
27/64	0.4219	10.7156	59/64	0.9219	23.4156
7/16	0.4375	11.1125	15/16	0.9375	23.8125
29/64	0.4531	11.5094	61/64	0.9531	24.2094
15/32	0.4688	11.9062	31/32	0.9688	24.6062
31/64	0.4844	12.3031	63/64	0.9844	25.0031
1/2	0.5000	12.7000	1	1.0000	25.4000

Safety first!

Regardless of how enthusiastic you may be about getting on with the job at hand, take the time to ensure that your safety is not jeopardized. A moment's lack of attention can result in an accident, as can failure to observe certain simple safety precautions. The possibility of an accident will always exist, and the following points should not be considered a comprehensive list of all dangers. Rather, they are intended to make you aware of the risks and to encourage a safety conscious approach to all work you carry out on your vehicle.

ESSENTIAL DOS AND DON'TS

DON'T rely on a jack when working under the vehicle. Always use approved jackstands to support the weight of the vehicle and place them under the recommended lift or support points.

DON'T attempt to loosen extremely tight fasteners (i.e. wheel lug nuts) while the vehicle is on a jack - it may fall.

DON'T start the engine without first making sure that the transmission is in Neutral (or Park where applicable) and the parking brake is set.

DON'T remove the radiator cap from a hot cooling system - let it cool or cover it with a cloth and release the pressure gradually.

DON'T attempt to drain the engine oil until you are sure it has cooled to the point that it will not burn you.

DON'T touch any part of the engine or exhaust system until it has cooled sufficiently to avoid burns.

DON'T siphon toxic liquids such as gasoline, antifreeze and brake fluid by mouth, or allow them to remain on your skin.

DON'T inhale brake lining dust - it is potentially hazardous (see *Asbestos* below).

DON'T allow spilled oil or grease to remain on the floor - wipe it up before someone slips on it.

DON'T use loose fitting wrenches or other tools which may slip and cause injury.

DON'T push on wrenches when loosening or tightening nuts or bolts. Always try to pull the wrench toward you. If the situation calls for pushing the wrench away, push with an open hand to avoid scraped knuckles if the wrench should slip.

DON'T attempt to lift a heavy component alone - get someone to help you.

DON'T rush or take unsafe shortcuts to finish a job.

DON'T allow children or animals in or around the vehicle while you are working on it.

DO wear eye protection when using power tools such as a drill, sander, bench grinder, etc. and when working under a vehicle.

DO keep loose clothing and long hair well out of the way of moving parts.

DO make sure that any hoist used has a safe working load rating adequate for the job.

DO get someone to check on you periodically when working alone on a vehicle.

DO carry out work in a logical sequence and make sure that everything is correctly assembled and tightened.

DO keep chemicals and fluids tightly capped and out of the reach of children and pets.

DO remember that your vehicle's safety affects that of yourself and others. If in doubt on any point, get professional advice.

ASBESTOS

Certain friction, insulating, sealing, and other products - such as brake linings, brake bands, clutch linings, torque converters, gaskets, etc. - may contain asbestos. Extreme care must be taken to avoid inhalation of dust from such products, since it is hazardous to health. If in doubt, assume that they do contain asbestos.

FIRE

Remember at all times that gasoline is highly flammable. Never smoke or have any kind of open flame around when working on a vehicle. But the risk does not end there. A spark caused by an electrical short circuit, by two metal surfaces contacting each other, or even by static electricity built up in your body under certain conditions, can ignite gasoline vapors, which in a confined space are highly explosive. Do not, under any circumstances, use gasoline for cleaning parts. Use an approved safety solvent.

Always disconnect the battery ground (-) cable at the battery before working on any part of the fuel system or electrical system. Never risk spilling fuel on a hot engine or exhaust component. It is strongly recommended that a fire extinguisher suitable for use on fuel and electrical fires be kept handy in the garage or workshop at all times. Never try to extinguish a fuel or electrical fire with water.

FUMES

Certain fumes are highly toxic and can quickly cause unconsciousness and even death if inhaled to any extent. Gasoline vapor falls into this category, as do the vapors from some cleaning solvents. Any draining or pouring of such volatile fluids should be done in a well ventilated area.

When using cleaning fluids and solvents, read the instructions on the container carefully. Never use materials from unmarked containers.

Never run the engine in an enclosed space, such as a garage. Exhaust fumes contain carbon monoxide, which is extremely poisonous. If you need to run the engine, always do so in the open air, or at least have the rear of the vehicle outside the work area.

If you are fortunate enough to have the use of an inspection pit, never drain or pour gasoline and never run the engine while the vehicle is over the pit. The fumes, being heavier than air, will concentrate in the pit with possibly lethal results.

THE BATTERY

Never create a spark or allow a bare light bulb near a battery. They normally give off a certain amount of hydrogen gas, which is highly explosive.

Always disconnect the battery ground (-) cable at the battery before working on the fuel or electrical systems.

If possible, loosen the filler caps or cover when charging the battery from an external source (this does not apply to sealed or maintenance-free batteries). Do not charge at an excessive rate or the battery may burst.

Take care when adding water to a non maintenance-free battery and when carrying a battery. The electrolyte, even when diluted, is very corrosive and should not be allowed to contact clothing or skin.

Always wear eye protection when cleaning the battery to prevent the caustic deposits from entering your eyes.

HOUSEHOLD CURRENT

When using an electric power tool, inspection light, etc., which operates on household current, always make sure that the tool is correctly connected to its plug and that, where necessary, it is properly grounded. Do not use such items in damp conditions and, again, do not create a spark or apply excessive heat in the vicinity of fuel or fuel vapor.

SECONDARY IGNITION SYSTEM VOLTAGE

A severe electric shock can result from touching certain parts of the ignition system (such as the spark plug wires) when the engine is running or being cranked, particularly if components are damp or the insulation is defective. In the case of an electronic ignition system, the secondary system voltage is much higher and could prove fatal.

Troubleshooting

CONTENTS

This section provides an easy reference guide to the more common problems which may occur during the operation of your vehicle. These problems and their possible causes are grouped under headings denoting various components or systems, such as Engine, Cooling system, etc. They also refer you to the chapter and/or section which deals with the problem.

Remember that successful troubleshooting is not a mysterious art practiced only by professional mechanics. It is simply the result of the right knowledge combined with an intelligent, systematic approach to the problem. Always work by a process of elimination, starting with the simplest solution and working through to the most complex - and

never overlook the obvious. Anyone can run the gas tank dry or leave the lights on overnight, so don't assume that you are exempt from such oversights.

Finally, always establish a clear idea of why a problem has occurred and take steps to ensure that it doesn't happen again. If the electrical system fails because of a poor connection, check the other connections in the system to make sure that they don't fail as well. If a particular fuse continues to blow, find out why - don't just replace one fuse after another. Remember, failure of a small component can often be indicative of potential failure or incorrect functioning of a more important component or system.

ENGINE

1 Engine will not rotate when attempting to start

1 Battery terminal connections loose or corroded (Chapter 1).
2 Battery discharged or faulty (Chapter 1).
3 Automatic transaxle not completely engaged in Park (Chapter 7).
4 Broken, loose or disconnected wiring in the starting circuit (Chapters 5 and 12).
5 Starter motor pinion jammed in flywheel ring gear (Chapter 5).
6 Starter solenoid faulty (Chapter 5).
7 Starter motor faulty (Chapter 5).
8 Ignition switch faulty (Chapter 12).
9 Starter pinion or flywheel teeth worn or broken (Chapter 5).

2 Engine rotates but will not start

1 Fuel tank empty.
2 Battery discharged (engine rotates slowly) (Chapter 5).
3 Battery terminal connections loose or corroded (Chapter 1).
4 Leaking fuel injector(s), faulty fuel pump, pressure regulator, etc. (Chapter 4).
5 Fuel not reaching fuel rail (Chapter 4).
6 Ignition components damp or damaged (Chapter 5).
7 Worn, faulty or incorrectly gapped spark plugs (Chapter 1).
8 Broken, loose or disconnected wiring in the starting circuit (Chapter 5).
9 Broken, loose or disconnected wires at the ignition coil(s) or faulty coil(s) (Chapter 5).
10 Faulty camshaft or crankshaft position sensor (Chapter 6).

3 Engine hard to start when cold

1 Battery discharged or low (Chapter 1).
2 Malfunctioning fuel system (Chapter 4).
3 Faulty cold start injector (Chapter 4).
4 Injector(s) leaking (Chapter 4).
5 Faulty coolant temperature sensor (Chapter 6).

4 Engine hard to start when hot

1 Air filter clogged (Chapter 1).
2 Fuel not reaching the fuel injection system (Chapter 4).
3 Corroded battery connections, especially ground (Chapter 1).

5 Starter motor noisy or excessively rough in engagement

1 Pinion or flywheel gear teeth worn or broken (Chapter 5).
2 Starter motor mounting bolts loose or missing (Chapter 5).

6 Engine starts but stops immediately

1 Loose or faulty electrical connections at coil(s) or alternator (Chapter 5).
2 Insufficient fuel reaching the fuel injector(s) (Chapters 1 and 4).
3 Vacuum leak at the gasket between the intake manifold/plenum and throttle body (Chapters 1 and 4).

7 Oil puddle under engine

1 Oil pan gasket and/or oil pan drain bolt washer leaking (Chapter 2).
2 Oil pressure sending unit leaking (Chapter 2).
3 Valve covers leaking (Chapter 2).
4 Engine oil seals leaking (Chapter 2).
5 Oil pump housing leaking (Chapter 2).

8 Engine lopes while idling or idles erratically

1 Vacuum leakage (Chapters 2 and 4).
2 Leaking EGR valve (Chapter 6).
3 Air filter clogged (Chapter 1).
4 Fuel pump not delivering sufficient fuel to the fuel injection system (Chapter 4).
5 Leaking head gasket (Chapter 2).
6 Timing belt and/or sprockets worn (Chap-ter 2).
7 Camshaft lobes worn (Chapter 2).

9 Engine misses at idle speed

1 Spark plugs worn or faulty (Chapter 1).
2 Faulty spark plug wires (Chapter 1).
3 Vacuum leaks (Chapter 1).
4 Fault in engine management system (Chapter 6).
5 Uneven or low compression (Chapter 2).

10 · Engine misses throughout driving speed range

1 Fuel filter clogged and/or impurities in the fuel system (Chapter 1).
2 Low fuel output at the injector(s) (Chapter 4).
3 Faulty or worn spark plugs (Chapter 1).
4 Fault in engine management system (Chapter 6).
5 Faulty emission system components (Chapter 6).
6 Low or uneven cylinder compression pressures (Chapter 2).
7 Weak or faulty ignition coil(s) (Chapter 5).
8 Vacuum leak in fuel injection system, intake manifold/plenum, air control valve or vacuum hoses (Chapter 4).

11 Engine stumbles on acceleration

1 Spark plugs fouled (Chapter 1).
2 Fuel injection system faulty (Chapter 4).
3 Fuel filter clogged (Chapters 1 and 4).
4 Fault in engine management system (Chapter 6).
5 Intake manifold or plenum air leak (Chapters 2 and 4).

12 Engine surges while holding accelerator steady

1 Intake air leak (Chapter 4).
2 Fuel pump faulty (Chapter 4).
3 Loose fuel injector wire harness connectors (Chapter 4).
4 Defective ECM or information sensor (Chapter 6).

13 Engine stalls

1 Fuel filter clogged and/or water and impurities in the fuel system (Chapters 1 and 4).

2 Ignition components damp or damaged (Chapter 5).
3 Faulty emissions system components (Chapter 6).
4 Faulty or incorrectly gapped spark plugs (Chapter 1).
5 Vacuum leak in the fuel injection system, intake manifold or vacuum hoses (Chapters 2 and 4).
6 Valve clearances incorrectly set (Chapter 1).

14 Engine lacks power

1 Fault in engine management system (Chapter 6).
2 Faulty or worn spark plugs (Chapter 1).
3 Fuel injection system malfunction (Chapter 4).
4 Faulty coil(s) (Chapter 5).
5 Brakes binding (Chapter 9).
6 Automatic transaxle fluid level incorrect (Chapter 1).
7 Fuel filter clogged and/or impurities in the fuel system (Chapters 1 and 4).
8 Emissions control systems not functioning properly (Chapter 6).
9 Low or uneven cylinder compression pressures (Chapter 2).
10 Obstructed exhaust system (Chapter 4).

15 Engine backfires

1 Emission control system not functioning properly (Chapter 6).
2 Fault in engine management system (Chapter 6).
3 Faulty spark plug insulator (Chapter 1).
4 Fuel injection system malfunction (Chapter 4).
5 Vacuum leak at fuel injector(s), intake manifold, air control valve or vacuum hoses (Chapters 2 and 4).
6 Valve clearances incorrectly set and/or valves sticking (Chapter 1).

16 Pinging or knocking engine sounds during acceleration or uphill

1 Incorrect grade of fuel.
2 Fault in engine management system (Chapter 6).
3 Fuel injection system faulty (Chapter 4).
4 Improper or damaged spark plug(s) (Chapter 1).
5 Vacuum leak (Chapters 2 and 4).
6 Defective knock sensor (Chapter 6).

17 Engine runs with oil pressure light on

1 Low oil level (Chapter 1).
2 Short in wiring circuit (Chapter 12).
3 Faulty oil pressure sender (Chapter 2).
4 Worn engine bearings and/or oil pump (Chapter 2).

18 Engine diesels (continues to run) after switching off

1 Excessive engine operating temperature (Chapter 3).
2 Fault in engine management system (Chapter 6).

ENGINE ELECTRICAL SYSTEM

19 Battery will not hold a charge

1 Alternator drivebelt defective or not adjusted properly (Chapter 1).

2 Battery electrolyte level low (Chapter 1).
3 Battery terminals loose or corroded (Chapter 1).
4 Alternator not charging properly (Chapter 5).
5 Loose, broken or faulty wiring in the charging circuit (Chapter 5).
6 Short in vehicle wiring (Chapter 12).
7 Internally defective battery (Chapters 1 and 5).

20 Alternator light fails to go out

1 Faulty alternator or charging circuit (Chapter 5).
2 Alternator drivebelt defective or out of adjustment (Chapter 1).
3 Alternator voltage regulator inoperative (Chapter 5).

21 Alternator light fails to come on when key is turned on

1 Warning light bulb defective (Chapter 12).
2 Fault in the printed circuit, dash wiring or bulb holder (Chapter 12).

FUEL SYSTEM

22 Excessive fuel consumption

1 Dirty or clogged air filter element (Chapter 1).
2 Fault in engine management system (Chapter 6).
3 Emissions systems not functioning properly (Chapter 6).
4 Fuel injection system not functioning properly (Chapter 4).
5 Low tire pressure or incorrect tire size (Chapter 1).

23 Fuel leakage and/or fuel odor

1 Leaking fuel feed or return line (Chapters 1 and 4).
2 Tank overfilled.
3 Evaporative canister filter clogged (Chapters 1 and 6).
4 Fuel injection system not functioning properly (Chapter 4).

COOLING SYSTEM

24 Overheating

1 Insufficient coolant in system (Chapter 1).
2 Water pump defective (Chapter 3).
3 Radiator core blocked or grille restricted (Chapter 3).
4 Thermostat faulty (Chapter 3).
5 Electric coolant fan blades broken or cracked (Chapter 3).
6 Radiator cap not maintaining proper pressure (Chapter 3).
7 Fault in engine management system (Chapter 6).

25 Overcooling

1 Faulty thermostat (Chapter 3).
2 Inaccurate temperature gauge sending unit (Chapter 3)

26 External coolant leakage

1 Deteriorated/damaged hoses; loose clamps (Chapters 1 and 3).
2 Water pump defective (Chapter 3).

3 Leakage from radiator core or coolant reservoir bottle (Chapter 3).
4 Engine drain or water jacket core plugs leaking (Chapter 2).

27 Internal coolant leakage

1 Leaking cylinder head gasket (Chapter 2).
2 Cracked cylinder bore or cylinder head (Chapter 2).

28 Coolant loss

1 Too much coolant in system (Chapter 1).
2 Coolant boiling away because of overheating (Chapter 3).
3 Internal or external leakage (Chapter 3).
4 Faulty radiator cap (Chapter 3).

29 Poor coolant circulation

1 Inoperative water pump (Chapter 3).
2 Restriction in cooling system (Chapters 1 and 3).
3 Water pump drivebelt defective/out of adjustment (Chapter 1).
4 Thermostat sticking (Chapter 3).

AUTOMATIC TRANSAXLE

➡Note: Due to the complexity of the automatic transaxle, it is difficult for the home mechanic to properly diagnose and service this component. For problems other than the following, the vehicle should be taken to a dealer or transaxle shop.

30 Fluid leakage

1 Automatic transaxle fluid is a deep red color. Fluid leaks should not be confused with engine oil, which can easily be blown onto the transaxle by air flow.
2 To pinpoint a leak, first remove all built-up dirt and grime from the transaxle housing with degreasing agents and/or steam cleaning. Then drive the vehicle at low speeds so air flow will not blow the leak far from its source. Raise the vehicle and determine where the leak is coming from. Common areas of leakage are:

   a) Pan (Chapters 1 and 7)
   b) Dipstick tube (Chapters 1 and 7)
   c) Transaxle oil lines (Chapter 7)
   d) Speed sensor (Chapter 7)
   e) Differential drain plug (Chapters 1 and 7)

31 Transaxle fluid brown or has a burned smell

Transaxle fluid overheated (Chapter 1).

32 General shift mechanism problems

1 Chapter 7 deals with checking and adjusting the shift linkage on automatic transaxles. Common problems which may be attributed to poorly adjusted linkage are:

   a) Engine starting in gears other than Park or Neutral.
   b) Indicator on shifter pointing to a gear other than the one actually being used.
   c) Vehicle moves when in Park.
2 Refer to Chapter 7 for the shift linkage adjustment procedure.

33 Transaxle will not downshift with accelerator pedal pressed to the floor

These transaxles are electronically controlled. Check for trouble codes stored in the PCM (see Chapter 6).

34 Engine will start in gears other than Park or Neutral

Transmission Range (TR) sensor malfunctioning (Chapter 7).

35 Transaxle slips, shifts roughly, is noisy or has no drive in forward or reverse gears

There are many probable causes for the above problems, but the home mechanic should be concerned with only one possibility - fluid level. Before taking the vehicle to a repair shop, check the level and condition of the fluid as described in Chapter 1. Correct the fluid level as necessary or change the fluid and filter if needed. If the problem persists, have a professional diagnose the cause.

DRIVEAXLES

36 Clicking noise in turns

Worn or damaged outboard CV joint (Chapter 8).

37 Shudder or vibration during acceleration

1 Excessive toe-in (Chapter 10).
2 Worn or damaged inboard or outboard CV joints (Chapter 8).
3 Sticking inboard CV joint assembly (Chapter 8).

38 Vibration at highway speeds

1 Out-of-balance front wheels and/or tires (Chapters 1 and 10).
2 Out-of-round front tires (Chapters 1 and 10).
3 Worn CV joint(s) (Chapter 8).

BRAKES

➡Note: Before assuming that a brake problem exists, make sure that:

   a) The tires are in good condition and properly inflated (Chapter 1).
   b) The front end alignment is correct.
   c) The vehicle is not loaded with weight in an unequal manner.

39 Vehicle pulls to one side during braking

1 Incorrect tire pressures (Chapter 1).
2 Front end out of alignment (have the front end aligned).
3 Front, or rear, tires not matched to one another.
4 Restricted brake lines or hoses (Chapter 9).
5 Malfunctioning caliper assembly (Chapter 9).
6 Loose suspension parts (Chapter 10).
7 Loose calipers (Chapter 9).
8 Excessive wear of brake pad material or disc on one side.

40 Noise (high-pitched squeal when the brakes are applied)

Front and/or rear disc brake pads worn out. The noise comes from the wear sensor rubbing against the disc. Replace pads with new ones immediately (Chapter 9).

41 Brake roughness or chatter (pedal pulsates)

1 Excessive lateral runout (Chapter 9).
2 Uneven pad wear (Chapter 9).
3 Defective disc (Chapter 9).

42 Excessive brake pedal effort required to stop vehicle

1 Malfunctioning power brake booster (Chapter 9).
2 Partial system failure (Chapter 9).
3 Excessively worn pads (Chapter 9).
4 Piston in caliper stuck or sluggish (Chapter 9).
5 Brake pads contaminated with oil, grease or brake fluid (Chapter 9).
6 New pads installed and not yet seated. It will take a while for the new material to seat against the disc.

43 Excessive brake pedal travel

1 Partial brake system failure (Chapter 9).
2 Insufficient fluid in master cylinder (Chapters 1 and 9).
3 Air trapped in system (Chapters 1 and 9).

44 Dragging brakes

1 Incorrect adjustment of brake light switch (Chapter 9).
2 Master cylinder pistons not returning correctly (Chapter 9).
3 Restricted brakes lines or hoses (Chapters 1 and 9).
4 Incorrect parking brake adjustment (Chapter 9).

45 Grabbing or uneven braking action

1 Malfunction of proportioning valve (Chapter 9).
2 Malfunction of power brake booster unit (Chapter 9).
3 Binding brake pedal mechanism (Chapter 9).

46 Brake pedal feels spongy when depressed

1 Air in hydraulic lines (Chapter 9).
2 Master cylinder mounting bolts loose (Chapter 9).
3 Master cylinder defective (Chapter 9).

47 Brake pedal travels to the floor with little resistance

1 Little or no fluid in the master cylinder reservoir caused by leaking caliper piston(s) (Chapter 9).
2 Loose, damaged or disconnected brake lines (Chapter 9).
3 Defective master cylinder (Chapter 9).

48 Parking brake does not hold

Parking brake improperly adjusted (Chapters 1 and 9).

SUSPENSION AND STEERING SYSTEMS

➡Note: Before attempting to diagnose the suspension and steering systems, perform the following preliminary checks:

a) Tires for wrong pressure and uneven wear.
b) Steering universal joints from the column to the rack and pinion for loose connectors or wear.
c) Front and rear suspension and the steering gear assembly for loose or damaged parts.
d) Out-of-round or out-of-balance tires, bent rims and loose and/or rough wheel bearings.

49 Vehicle pulls to one side

1 Mismatched or uneven tires (Chapter 10).
2 Broken or sagging springs (Chapter 10).
3 Wheel alignment (Chapter 10).
4 Front brake dragging (Chapter 9).

50 Abnormal or excessive tire wear

1 Wheel alignment (Chapter 10).
2 Sagging or broken springs (Chapter 10).
3 Tire out of balance (Chapter 10).
4 Worn strut damper (Chapter 10).
5 Overloaded vehicle.
6 Tires not rotated regularly.

51 Wheel makes a thumping noise

1 Blister or bump on tire (Chapter 10).
2 Improper strut damper action (Chapter 10).

52 Shimmy, shake or vibration

1 Tire or wheel out-of-balance or out-of-round (Chapter 10).
2 Loose or worn wheel bearings (Chapters 1, 8 and 10).
3 Worn tie-rod ends (Chapter 10).
4 Worn lower balljoints (Chapters 1 and 10).
5 Excessive wheel runout (Chapter 10).
6 Blister or bump on tire (Chapter 10).

53 Hard steering

1 Lack of lubrication at balljoints, tie-rod ends and steering gear assembly (Chapter 10).
2 Front wheel alignment (Chapter 10).
3 Low tire pressure(s) (Chapters 1 and 10).

54 Poor returnability of steering to center

1 Lack of lubrication at balljoints and tie-rod ends (Chapter 10).
2 Binding in balljoints (Chapter 10).
3 Binding in steering column (Chapter 10).
4 Lack of lubricant in steering gear assembly (Chapter 10).
5 Front wheel alignment (Chapter 10).

55 Abnormal noise at the front end

1 Lack of lubrication at balljoints and tie-rod ends (Chapters 1 and 10).

2 Damaged strut mounting (Chapter 10).
3 Worn control arm bushings or tie-rod ends (Chapter 10).
4 Loose stabilizer bar (Chapter 10).
5 Loose wheel nuts (Chapters 1 and 10).
6 Loose suspension bolts (Chapter 10)

56 Wander or poor steering stability

1 Mismatched or uneven tires (Chapter 10).
2 Lack of lubrication at balljoints and tie-rod ends (Chapters 1 and 10).
3 Worn strut assemblies (Chapter 10).
4 Loose stabilizer bar (Chapter 10).
5 Broken or sagging springs (Chapter 10).
6 Wheels out of alignment (Chapter 10).

57 Erratic steering when braking

1 Wheel bearings worn (Chapter 10).
2 Broken or sagging springs (Chapter 10).
3 Leaking wheel cylinder or caliper (Chapter 9).
4 Warped brake discs (Chapter 9).

58 Excessive pitching and/or rolling around corners or during braking

1 Loose stabilizer bar (Chapter 10).
2 Worn strut dampers or mountings (Chapter 10).
3 Broken or sagging springs (Chapter 10).
4 Overloaded vehicle.

59 Suspension bottoms

1 Overloaded vehicle.
2 Worn strut dampers (Chapter 10).
3 Incorrect, broken or sagging springs (Chapter 10).

60 Cupped tires

1 Front wheel or rear wheel alignment (Chapter 10).

2 Worn strut dampers (Chapter 10).
3 Wheel bearings worn (Chapter 10).
4 Excessive tire or wheel runout (Chapter 10).
5 Worn balljoints (Chapter 10).

61 Excessive tire wear on outside edge

1 Inflation pressures incorrect (Chapter 1).
2 Excessive speed in turns.
3 Front end alignment incorrect (excessive toe-in). Have professionally aligned.
4 Suspension arm bent or twisted (Chapter 10).

62 Excessive tire wear on inside edge

1 Inflation pressures incorrect (Chapter 1).
2 Front end alignment incorrect (toe-out). Have professionally aligned.
3 Loose or damaged steering components (Chapter 10).

63 Tire tread worn in one place

1 Tires out of balance.
2 Damaged wheel. Inspect and replace if necessary.
3 Defective tire (Chapter 1).

64 Excessive play or looseness in steering system

1 Wheel bearing(s) worn (Chapter 10).
2 Tie-rod end loose (Chapter 10).
3 Steering gear loose (Chapter 10).
4 Worn or loose steering intermediate shaft (Chapter 10).

65 Rattling or clicking noise in steering gear

1 Steering gear loose (Chapter 10).
2 Steering gear defective.

Notes

TUNE-UP
AND ROUTINE
MAINTENANCE

Section

Typical four-cylinder engine compartment

1	Relay box	5	Automatic transaxle dipstick	9	Engine oil filler cap
2	Brake fluid reservoir	6	Upper radiator hose	10	Windshield washer fluid reservoir
3	Air filter housing	7	Radiator cap	11	Engine coolant reservoir
4	Battery	8	Engine oil dipstick	12	Power steering fluid reservoir

Typical V6 engine compartment

1 Battery	6 Windshield washer fluid reservoir	10 Engine oil filler cap
2 Automatic transaxle dipstick	7 Engine coolant reservoir	11 Relay box
3 Radiator cap	8 Power steering fluid reservoir	12 Brake fluid reservoir
4 Engine oil dipstick	9 Engine cover (remove for access to	13 Air filter housing
5 Upper radiator hose	spark plugs)	14 Main fuse/relay box

Typical front underside components

1	Front brake calipers	3	Engine oil drain plug (V6 engine shown)
2	Brake hoses	4	Automatic transaxle drain plug

5 Balljoints
6 Driveaxle boots

Typical rear underside components

| 1 Muffler | 2 Outer driveaxle boot | 3 Inner driveaxle boot | 4 Rear differential drain plug |

1 Toyota Highlander and Lexus RX 300/330 Maintenance Schedule

The maintenance intervals in this manual are provided with the assumption that you, not the dealer, will be doing the work. These are the minimum maintenance intervals recommended by the factory for vehicles that are driven daily. If you wish to keep your vehicle in peak condition at all times, you may wish to perform some of these procedures even more often. Because frequent maintenance enhances the efficiency, performance and resale value of your car, we encourage you to do so. If you drive in dusty areas, tow a trailer, idle or drive at low speeds for extended periods or drive for short distances (less than four miles) in below freezing temperatures, shorter intervals are also recommended.

When your vehicle is new, it should be serviced by a factory authorized dealer service department to protect the factory warranty. In many cases, the initial maintenance check is done at no cost to the owner.

EVERY 250 MILES (400 KM) OR WEEKLY, WHICHEVER COMES FIRST

Check the engine oil level (Section 4)
Check the engine coolant level (Section 4)
Check the windshield washer fluid level (Section 4)
Check the brake fluid level (Section 4)
Check the power steering fluid level (Section 4)
Check the automatic transaxle fluid level (Section 4)
Check the tires and tire pressures (Section 5)

EVERY 3000 MILES (4,800 KM) OR 3 MONTHS, WHICHEVER COMES FIRST

All items listed above plus:
Change the engine oil and oil filter (Section 6)

EVERY 5000 MILES (8000 KM) OR 6 MONTHS, WHICHEVER COMES FIRST

All items listed above plus:
Check and service the battery (Section 7)
Rotate the tires (Section 8)
Inspect and replace if necessary the windshield wiper blades (Section 9)
Inspect and replace if necessary all underhood hoses (Section 10)
Check the cooling system (Section 11)
Inspect the brake system (Section 12)

EVERY 15,000 MILES (24,000 KM) OR 18 MONTHS, WHICHEVER COMES FIRST

All items listed above plus:
Inspect the suspension, steering components and driveaxle boots (Section 13)*
Inspect the exhaust system (Section 14)

EVERY 30,000 MILES (48,000 KM) OR 36 MONTHS, WHICHEVER COMES FIRST

All items listed above plus:
Replace the air filter (Section 15)*
Inspect the fuel system (Section 16)
Replace interior ventilation filter (Section 17)*
Replace the fuel tank cap gasket (Section 18)
Check and replace if necessary the PCV valve (Section 19)
Change the brake fluid (Section 20)

EVERY 50,000 MILES (80,500KM) OR 60 MONTHS, WHICHEVER COMES FIRST

Service the cooling system (drain, flush and refill) (Section 25) (after the initial 100,000-mile [160,000 km] or 120-month service)

EVERY 60,000 MILES (96,000 KM) OR 72 MONTHS, WHICHEVER COMES FIRST

All items listed above plus:
Check and adjust if necessary the engine drivebelts (Section 21) (after the initial 60,000-mile [96,000 km] or 72-month check)
Inspect and if necessary adjust the valve clearance (Section 22)
Change the automatic transaxle lubricant (Section 23)**
Change the transfer case lubricant (4WD models) (Section 24)**
Change the rear differential lubricant (4WD models) (Section 25)**

100,000 MILES (160,000 KM) OR 120 MONTHS, WHICHEVER COMES FIRST - THEREAFTER EVERY 50,000 MILES (80,500 KM) OR 60 MONTHS, WHICHEVER COMES FIRST

Service the cooling system (drain, flush and refill) (Section 26)

EVERY 120,000 MILES (193,000 KM) OR 144 MONTHS, WHICHEVER COMES FIRST

Replace the spark plugs (Section 27)

  * *This item is affected by "severe" operating conditions as described below. If your vehicle is operated under "severe" conditions, inspect all maintenance indicated with an asterisk (*) at 3000 mile/3 month intervals and perform maintenance or replace parts as necessary. Severe conditions are indicated if you mainly operate your vehicle under one or more of the following conditions:*

Operating in dusty areas
Idling for extended periods and/or low speed operation
Operating when outside temperatures remain below freezing and when most trips are less than 4 miles

  ** *If used for trailer towing, change the automatic transaxle, transfer case or rear differential fluid lubricant every 30,000 miles (Section 23, Section 24 or Section 25)*

2   Introduction

This Chapter is designed to help the home mechanic maintain his vehicle for peak performance, economy, safety and long life.

Included is a master maintenance schedule, followed by sections dealing specifically with each item on the schedule. Visual checks, adjustments, component replacement and other helpful items are included. Refer to the accompanying illustrations of the engine compartment and the underside of the vehicle for the location of various components.

Servicing your vehicle in accordance with the mileage/time maintenance schedule and the following Sections will provide it with a planned maintenance program that should result in a long and reliable service life. This is a comprehensive plan, so maintaining some items but not others at the specified service intervals won't produce the same results.

As you service your vehicle, you will discover that many of the procedures can - and should - be grouped together because of the nature of the particular procedure you're performing or because of the close proximity of two otherwise unrelated components to one another.

For example, if the vehicle is raised for any reason, you should inspect the exhaust, suspension, steering and fuel systems while you're under the vehicle. When you're rotating the tires, it makes good sense to check the brakes and wheel bearings since the wheels are already removed.

Finally, let's suppose you have to borrow or rent a torque wrench. Even if you only need to tighten the spark plugs, you might as well check the torque of as many critical fasteners as time allows.

The first step of this maintenance program is to prepare yourself before the actual work begins. Read through all sections pertinent to the procedures you're planning to do, then make a list of and gather together all the parts and tools you will need to do the job. If it looks as if you might run into problems during a particular segment of some procedure, seek advice from your local parts man or dealer service department.

3   Tune-up general information

The term tune-up is used in this manual to represent a combination of individual operations rather than one specific procedure.

If, from the time the vehicle is new, the routine maintenance schedule is followed closely and frequent checks are made of fluid levels and high wear items, as suggested throughout this manual, the engine will be kept in relatively good running condition and the need for additional work will be minimized.

More likely than not, however, there will be times when the engine is running poorly due to lack of regular maintenance. This is even more likely if a used vehicle, which has not received regular and frequent maintenance checks, is purchased. In such cases, an engine tune-up will be needed outside of the regular routine maintenance intervals.

The first step in any tune-up or engine diagnosis to help correct a poor running engine would be a cylinder compression check. A check of the engine compression (Chapter 2 Part C) will give valuable information regarding the overall performance of many internal components and should be used as a basis for tune-up and repair procedures. If, for instance, a compression check indicates serious internal engine wear, a conventional tune-up will not help the running condition of the engine and would be a waste of time and money. Also in Chapter 2, Part C is information on checking engine vacuum, which also gives information on the engine's state-of-tune and condition.

The following series of operations are those most often needed to bring a generally poor-running engine back into a proper state of tune.

MINOR TUNE-UP

Check all engine related fluids (Section 4)
Clean, inspect and test the battery (Section 7)
Check all underhood hoses (Section 10)
Check the cooling system (Section 11)
Check the air filter (Section 15)
Check and adjust the drivebelts (Section 21)

MAJOR TUNE-UP

All items listed under Minor tune-up, plus . . .
Replace the air filter (Section 15)
Check the fuel system (Section 16)
Replace the spark plugs (Section 27)
Check the charging system (Chapter 5)

4   Fluid level checks (every 250 miles [400km] or weekly)

1   Fluids are an essential part of the lubrication, cooling, brake, clutch and other systems. Because these fluids gradually become depleted and/or contaminated during normal operation of the vehicle, they must be periodically replenished. See *Recommended lubricants and fluids* and *Capacities* at the end of this Chapter before adding fluid to any of the following components.

➡Note: **The vehicle must be on level ground before fluid levels can be checked.**

ENGINE OIL

▶ **Refer to illustrations 4.2, 4.4 and 4.6**

2   The engine oil level is checked with a dipstick located at the front side of the engine (see illustration). The dipstick extends through a

4.2 The engine oil dipstick is mounted on the front (radiator) side of the engine

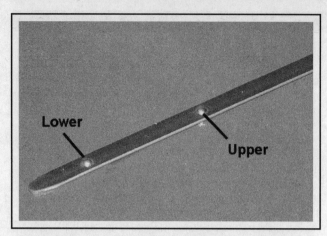

4.4 The oil level should be at or near the upper mark on the dipstick - if it isn't, add enough oil to bring the level to or near the upper mark (it takes one quart to raise the level from the lower mark to the upper mark)

4.6 The threaded oil filler cap is located on the valve cover - to prevent dirt from contaminating the engine, always make sure the area around this opening is clean before removing the cap

metal tube from which it protrudes down into the engine oil pan.

3   The oil level should be checked before the vehicle has been driven, or about 5 minutes after the engine has been shut off. If the oil is checked immediately after driving the vehicle, some of the oil will remain in the upper engine components, producing an inaccurate reading on the dipstick.

4   Pull the dipstick from the tube and wipe all the oil from the end with a clean rag or paper towel. Insert the clean dipstick all the way back into its metal tube and pull it out again. Observe the oil at the end of the dipstick. At its highest point, the level should be between the lower and upper marks (see illustration).

5   It takes one quart of oil to raise the level from the lower mark to the upper mark on the dipstick. Do not allow the level to drop below the lower mark or oil starvation may cause engine damage. Conversely, overfilling the engine (adding oil above the upper mark) may cause oil-fouled spark plugs, oil leaks or oil seal failures.

6   Remove the threaded cap from the valve cover to add oil (see illustration). Use a funnel to prevent spills. After adding the oil, install the filler cap hand tight. Start the engine and look carefully for any small leaks around the oil filter or drain plug. Stop the engine and check the oil level again after it has had sufficient time to drain from the upper block and cylinder head galleys.

7   Checking the oil level is an important preventive maintenance step. A continually dropping oil level indicates oil leakage through damaged seals, from loose connections, or past worn rings or valve

guides. If the oil looks milky in color or has water droplets in it, a cylinder head gasket may be blown. The engine should be checked immediately. The condition of the oil should also be checked. Each time you check the oil level, slide your thumb and index finger up the dipstick before wiping off the oil. If you see small dirt or metal particles clinging to the dipstick, the oil should be changed (see Section 8).

ENGINE COOLANT

▶ Refer to illustrations 4.8a and 4.8b

❋❋ WARNING:

Do not allow antifreeze to come in contact with your skin or painted surfaces of the vehicle. Flush contaminated areas immediately with plenty of water. Don't store new coolant or leave old coolant lying around where it's accessible to children or pets - they're attracted by its sweet smell. Ingestion of even a small amount of coolant can be fatal! Wipe up garage floor and drip pan spills immediately. Keep antifreeze containers covered and repair cooling system leaks as soon as they're noticed.

8   All models covered by this manual are equipped with a coolant

4.8a The coolant reservoir is located at the ride side of the engine compartment on early models...

4.8b ...and on later models it's located on the radiator

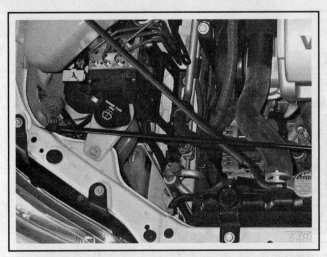

4.14 The windshield washer fluid reservoir is located at the right front corner of the engine compartment

4.16 The brake fluid should be kept between the Min and Max marks on the reservoir

recovery system. The coolant reservoir is connected by a hose to the base of the coolant filler cap (see illustrations). If the coolant heats up during engine operation, coolant can escape through the pressurized filler cap and connecting hose into the reservoir. As the engine cools, the coolant is automatically drawn back into the cooling system to maintain the correct level.

9   The coolant level should be checked regularly. It must be between the Full and Low lines on the tank. The level will vary with the temperature of the engine. When the engine is cold, the coolant level should be at or slightly above the Low mark on the tank. Once the engine has warmed up, the level should be at or near the Full mark. If it isn't, allow the fluid in the tank to cool, then remove the cap from the reservoir and add coolant to bring the level up to the Full line. Use only the type of coolant listed in this Chapter's Specifications or in your owner's manual. Do not use supplemental inhibitors or additives. If only a small amount of coolant is required to bring the system up to the proper level, water can be used. However, repeated additions of water will dilute the recommended antifreeze and water solution. In order to maintain the proper ratio of antifreeze and water, it is advisable to top up the coolant level with the correct mixture.

→Note: The coolant recommended by the manufacturer is pre-mixed to the correct ratio, so water should not be mixed with it when adding coolant. If you're using a non-Toyota coolant, check the container carefully.

10 If the coolant level drops within a short time after replenishment, there may be a leak in the system. Inspect the radiator, hoses, engine coolant filler cap, drain plugs, air bleeder plugs and water pump. If no leak is evident, have the radiator cap pressure tested.

❋❋ WARNING:

Never remove the radiator pressure cap when the engine is running or has just been shut down, because the cooling system is hot. Escaping steam and scalding liquid could cause serious injury.

11 If it is necessary to open the radiator cap, wait until the system has cooled completely, then wrap a thick cloth around the cap and turn it to the first stop. If any steam escapes, wait until the system has cooled further, then remove the cap.

12 When checking the coolant level, always note its condition. It should be relatively clear. If it is brown or rust colored, the system should be drained, flushed and refilled. Even if the coolant appears to be normal, the corrosion inhibitors wear out with use, so it must be replaced at the specified intervals.

13 Do not allow antifreeze to come in contact with your skin or painted surfaces of the vehicle. Flush contacted areas immediately with plenty of water.

WINDSHIELD WASHER FLUID

▶ Refer to illustration 4.14

14 Fluid for the windshield washer system is stored in a plastic reservoir which is located at the right front corner of the engine compartment (see illustration). In milder climates, plain water can be used to top up the reservoir, but the reservoir should be kept no more than two-thirds full to allow for expansion should the water freeze. In colder climates, the use of a specially designed windshield washer fluid, available at your dealer and any auto parts store, will help lower the freezing point of the fluid. Mix the solution with water in accordance with the manufacturer's directions on the container. Do not use regular antifreeze. It will damage the vehicle's paint.

BRAKE FLUID

▶ Refer to illustration 4.16

15 The brake master cylinder is mounted on the front of the power booster unit in the engine compartment.

16 To check the fluid level of the brake master cylinder reservoir, simply look at the MAX and MIN marks on the reservoir (see illustration). The level should be within the specified distance from the maximum fill line.

17 If the level is low, wipe the top of the reservoir cover with a clean rag to prevent contamination of the brake system before lifting the cover.

18 Add only the specified brake fluid to the brake reservoir (refer to *Recommended lubricants and fluids* at the end of this Chapter or your owner's manual). Mixing different types of brake fluid can damage the system. Fill the brake master cylinder reservoir only to the dotted line - this brings the fluid to the correct level when you put the cover back on.

4.26 The power steering fluid reservoir is located on the right side of the engine compartment - the reservoir is translucent so the fluid level can be checked either hot or cold without removing the cap

4.32a The automatic transaxle dipstick is located next to the battery

4.32b If the automatic transaxle fluid is cold, the level should be between the lower two notches; if it's at normal operating temperature, the level should be between the upper notches on the dipstick

✳✳ WARNING:

Use caution when filling the reservoir - brake fluid can harm your eyes and damage painted surfaces. Do not use brake fluid that has been opened for more than one year or has been left open. Brake fluid absorbs moisture from the air. Excess moisture can cause a dangerous loss of braking.

19 While the reservoir cap is removed, inspect the master cylinder reservoir for contamination. If deposits, dirt particles or water droplets are present, the system should be drained and refilled (see Chapter 9).

20 After filling the reservoir to the proper level, make sure the lid is properly seated to prevent fluid leakage and/or system pressure loss.

21 The brake fluid in the master cylinder will drop slightly as the brake pads at each wheel wear down during normal operation. If the master cylinder requires repeated replenishing to keep it at the proper level, this is an indication of leakage in the brake system, which should be corrected immediately. Check all brake lines and connections, along with the wheel cylinders and booster (see Section 12 for more information).

22 If, upon checking the master cylinder fluid level, you discover the reservoir empty or nearly empty, the brake system should be thouroughly inspected for leaks (see Chapter 9 for more information on the brake system).

POWER STEERING FLUID

▶ **Refer to illustrations 4.26**

23 Unlike manual steering, the power steering system relies on fluid which may, over a period of time, require replenishing.

24 The fluid reservoir for the power steering pump is located on the right (passenger side) inner fender panel near the front of the engine.

25 For the check, the front wheels should be pointed straight ahead and the engine should be off.

26 The reservoir is translucent plastic and the fluid level can be checked visually (see illustration).

27 If additional fluid is required, pour the specified type directly into the reservoir, using a funnel to prevent spills.

28 If the reservoir requires frequent fluid additions, all

power steering hoses, hose connections, the power steering pump and the rack and pinion assembly should be carefully checked for leaks.

AUTOMATIC TRANSAXLE FLUID

▶ **Refer to illustrations 4.32a and 4.32b**

29 The level of the automatic transaxle fluid should be carefully maintained. Low fluid level can lead to slipping or loss of drive, while overfilling can cause foaming, loss of fluid and transaxle damage.

30 The transaxle fluid level should only be checked when the transaxle is hot (at its normal operating temperature). If the vehicle has just been driven over 10 miles (15 miles in a frigid climate), and the fluid temperature is 160 to 175-degrees F, the transaxle is hot.

✳✳ CAUTION:

If the vehicle has just been driven for a long time at high speed or in city traffic in hot weather, or if it has been pulling a trailer, an accurate fluid level reading cannot be obtained. Allow the fluid to cool down for about 30 minutes.

31 If the vehicle has not just been driven, park the vehicle on level ground, set the parking brake and start the engine. While the engine is idling, depress the brake pedal and move the selector lever through all the gear ranges, beginning and ending in Park.

32 With the engine still idling, remove the dipstick from its tube (see illustration). Check the level of the fluid on the dipstick (see illustration) and note its condition.

33 Wipe the fluid from the dipstick with a clean rag and reinsert it back into the filler tube until the cap seats.

34 Pull the dipstick out again and note the fluid level. If the transaxle is cold, the level should be in the COLD or COOL range on the dipstick. If it is hot, the fluid level should be in the HOT range. If the level is at the low side of either range, add the specified automatic transaxle fluid through the dipstick tube with a funnel.

35 Add just enough of the recommended fluid to fill the transaxle to the proper level. It takes about one pint to raise the level from the low mark to the high mark when the fluid is hot, so add the fluid a little at a time and keep checking the level until it is correct.

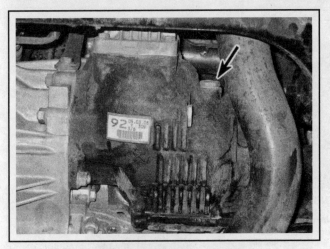

4.37 Location of the transfer case check/fill plug

4.41 Location of the rear differential check/fill plug

36 The condition of the fluid should also be checked along with the level. If the fluid at the end of the dipstick is black or a dark reddish brown color, or if it emits a burned smell, the fluid should be changed (see Section 23). If you are in doubt about the condition of the fluid, purchase some new fluid and compare the two for color and smell.

TRANSFER CASE LUBRICANT (4WD MODELS)

▶ **Refer to illustration 4.37**

❋❋ WARNING:

If the vehicle is equipped with electronically modulated air suspension, make sure that the height control switch is turned off.

37 The transfer case does not have a dipstick. To check the fluid level, raise the vehicle and support it securely on jackstands. On the back side of the transfer case housing, you will see a plug (see illustration) - remove it. If the lubricant level is correct, it should be up to the lower edge of the hole.

38 If the transaxle needs more lubricant (if the level is not up to the hole), use a syringe or a gear oil pump to add more. Stop filling the transaxle when the lubricant begins to run out the hole.

39 Install the plug and tighten it securely. Drive the vehicle a short distance, then check for leaks.

REAR DIFFERENTIAL LUBRICANT LEVEL (4WD MODELS)

▶ **Refer to illustration 4.41**

40 Raise the vehicle and support it securely on jackstands.

❋❋ WARNING:

If the vehicle is equipped with electronically modulated air suspension, make sure that the height control switch is turned off.

41 Using the appropriate wrench, unscrew the plug from the rear differential (see illustration).

42 Use your little finger to reach inside the housing to feel the lubricant level. The level should be at or near the bottom of the plug hole. If it isn't, add the recommended lubricant through the plug hole with a syringe or squeeze bottle.

43 Install the plug and tighten it securely. Check for leaks after the first few miles of driving.

5   Tire and tire pressure checks (every 250 miles [400 km] or weekly)

▶ **Refer to illustrations 5.2, 5.3, 5.4a, 5.4b and 5.8**

1   Periodic inspection of the tires may spare you from the inconvenience of being stranded with a flat tire. It can also provide you with vital information regarding possible problems in the steering and suspension systems before major damage occurs.

2   Normal tread wear can be monitored with a simple, inexpensive device known as a tread depth indicator (see illustration). When the tread depth reaches approximately 1/16-inch, replace the tire(s) (preferably long before that).

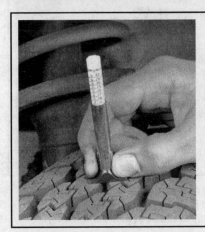

5.2 Use a tire tread depth gauge to monitor tire wear - they are available at auto parts stores and service stations and cost very little

UNDERINFLATION

CUPPING

OVERINFLATION

Cupping may be caused by:
- Underinflation and/or mechanical irregularities such as out-of-balance condition of wheel and/or tire, and bent or damaged wheel.
- Loose or worn steering tie-rod or steering idler arm.
- Loose, damaged or worn front suspension parts.

INCORRECT TOE-IN OR EXTREME CAMBER

FEATHERING DUE TO MISALIGNMENT

5.3 This chart will help you determine the condition of your tires, the probable cause(s) of abnormal wear and the corrective action necessary

3 Note any abnormal tread wear (see illustration). Tread pattern irregularities such as cupping, flat spots and more wear on one side than the other are indications of front end alignment and/or balance problems. If any of these conditions are noted, take the vehicle to a tire shop or service station to correct the problem.

4 Look closely for cuts, punctures and embedded nails or tacks. Sometimes a tire will hold its air pressure for a short time or leak down very slowly even after a nail has embedded itself into the tread. If a slow leak persists, check the valve stem core to make sure it is tight (see illustration). Examine the tread for an object that may have embedded itself into the tire or for a "plug" that may have begun to leak (radial

tire punctures are repaired with a plug that is fitted in a puncture). If a puncture is suspected, it can be easily verified by spraying a solution of soapy water onto the puncture area (see illustration). The soapy solution will bubble if there is a leak. Unless the puncture is inordinately large, a tire shop or gas station can usually repair the punctured tire.

5 Carefully inspect the inner sidewall of each tire for evidence of brake fluid leakage. If you see any, inspect the brakes immediately.

6 Correct tire air pressure adds miles to the lifespan of the tires, improves mileage and enhances overall ride quality. Tire pressure cannot be accurately estimated by looking at a tire, particularly if it is a radial. A tire pressure gauge is therefore essential. Keep an accurate

5.4a If a tire loses air on a steady basis, check the valve core first to make sure it's snug (special inexpensive wrenches are commonly available at auto parts stores)

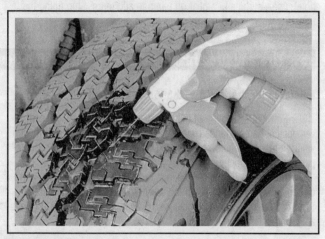

5.4b If the valve core is tight, raise the corner of the vehicle with the low tire and spray a soapy water solution onto the tread as the tire is turned slowly - slow leaks will cause small bubbles to appear

gauge in the glovebox. The pressure gauges fitted to the nozzles of air hoses at gas stations are often inaccurate.

7   Always check tire pressure when the tires are cold. "Cold," in this case, means the vehicle has not been driven over a mile in the three hours preceding a tire pressure check. A pressure rise of four to eight pounds is not uncommon once the tires are warm.

8   Unscrew the valve cap protruding from the wheel or hubcap and push the gauge firmly onto the valve (see illustration). Note the reading on the gauge and compare this figure to the recommended tire pressure shown on the tire placard in the glovebox. Be sure to reinstall the valve cap to keep dirt and moisture out of the valve stem mechanism. Check all four tires and, if necessary, add enough air to bring them up to the recommended pressure levels.

9   Don't forget to keep the spare tire inflated to the specified pressure (consult your owner's manual). Note that the air pressure specified for a compact spare is significantly higher than the pressure of the regular tires.

5.8 To extend the life of your tires, check the air pressure at least once a week with an accurate gauge (don't forget the spare!)

6   Engine oil and oil filter change (every 3000 miles [4,800 km] or 3 months)

▶ Refer to illustrations 6.2, 6.7, 6.13a, 6.13b, and 6.15

❋❋ WARNING:

If the vehicle is equipped with electronically modulated air suspension, make sure that the height control switch is turned off.

1   Frequent oil changes are the best preventive maintenance the home mechanic can give the engine, because aging oil becomes diluted and contaminated, which leads to premature engine wear.

2   Make sure that you have all the necessary tools before you begin this procedure (see illustration). You should also have plenty of rags or newspapers handy for mopping up any spills.

3   Access to the underside of the vehicle is greatly improved if the vehicle can be lifted on a hoist, driven onto ramps or supported by jackstands.

❋❋ WARNING:

Do not work under a vehicle which is supported only by a bumper, hydraulic or scissors-type jack.

4   If this is your first oil change, get under the vehicle and familiarize yourself with the location of the oil drain plug. The engine and exhaust components will be warm during the actual work, so try to anticipate any potential problems before the engine and accessories are hot.

5   Park the vehicle on a level spot. Start the engine and allow it to reach its normal operating temperature (the needle on the temperature gauge should be at least above the bottom mark). Warm oil and sludge will flow out more easily. Turn off the engine when it's warmed up. Remove the filler cap.

6   Raise the vehicle and support it securely on jackstands.

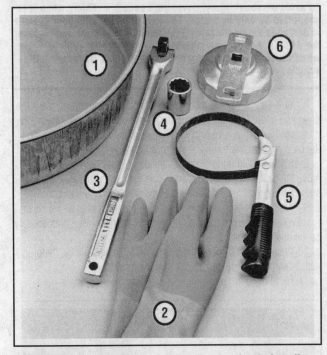

6.2 These tools are required when changing the engine oil and filter

1   *Drain pan* - It should be fairly shallow in depth, but wide in order to prevent spills

2   *Rubber gloves* - When removing the drain plug and filter, it is inevitable that you will get oil on your hands (the gloves will prevent burns)

3   *Breaker bar* - Sometimes the oil drain plug is pretty tight and a long breaker bar is needed to loosen it

4   *Socket* - To be used with the breaker bar or a ratchet (must be the correct size to fit the drain plug)

5   *Filter wrench* - This is a metal band-type wrench, which requires clearance around the filter to be effective

6   *Filter wrench* - This type fits on the bottom of the filter and can be turned with a ratchet or beaker bar (different size wrenches are available for different types of filters)

6.7 Use the proper size box-end wrench or socket to remove the oil drain plug without rounding off the corners

6.13a Location of the oil filter on four-cylinder models

✳✳ WARNING:

To avoid personal injury, never get beneath a vehicle when it is supported by only by a jack. The jack provided with your vehicle is designed solely for raising the vehicle to remove and replace the wheels. Always use jackstands to support the vehicle when it becomes necessary to place your body underneath the vehicle.

---

7 Being careful not to touch the hot exhaust components, place the drain pan under the drain plug in the bottom of the pan and remove the plug (see illustration). You may want to wear gloves while unscrewing the plug the final few turns if the engine is really hot.

8 Allow the old oil to drain into the pan. It may be necessary to move the pan farther under the engine as the oil flow slows to a trickle. Inspect the old oil for the presence of metal shavings and chips.

9 After all the oil has drained, wipe off the drain plug with a clean rag. Even minute metal particles clinging to the plug would immediately contaminate the new oil.

10 Clean the area around the drain plug opening, reinstall the plug and tighten it securely, but do not strip the threads.

11 Move the drain pan into position under the oil filter.

12 Remove all tools, rags, etc. from under the vehicle, being careful not to spill the oil in the drain pan, then lower the vehicle.

13 Loosen the oil filter (see illustrations) by turning it counterclockwise with an oil filter wrench. Once the filter is loose, use your hands to unscrew it from the block. Just as the filter is detached from the block, immediately tilt the open end up to prevent the oil inside the filter from spilling out.

✳✳ WARNING:

The engine exhaust manifold may still be hot, so be careful.

---

14 With a clean rag, wipe off the mounting surface on the block. If a residue of old oil is allowed to remain, it will smoke when the block is heated up. It will also prevent the new filter from seating properly. Also make sure that the none of the old gasket remains stuck to the mounting surface. It can be removed with a scraper if necessary.

15 Compare the old filter with the new one to make sure they are the same type. Smear some engine oil on the rubber gasket of the new filter and screw it into place (see illustration). Overtightening the filter will damage the gasket, so don't use a filter wrench. Most filter manufactur-

6.13b On V6 models, the oil filter is mounted on the side of the engine block - since it is usually on very tight, you'll need a special wrench for removal - DO NOT use the wrench to tighten the new filter

6.15 Lubricate the oil filter gasket with clean engine oil before installing the filter on the engine

ers recommend tightening the filter by hand only. Normally they should be tightened 3/4-turn after the gasket contacts the block, but be sure to follow the directions on the filter or container.

16 Add new oil to the engine through the oil filler cap in the valve cover. Use a funnel to prevent oil from spilling onto the top of the engine. Pour three quarts of fresh oil into the engine. Wait a few minutes to allow the oil to drain into the pan, then check the level on the oil dipstick (see Section 4 if necessary). If the oil level is at or near the F mark, install the filler cap hand tight, start the engine and allow the new oil to circulate.

17 Allow the engine to run for about a minute. While the engine is running, look under the vehicle and check for leaks at the oil pan drain plug and around the oil filter. If either is leaking, stop the engine and

tighten the plug or filter slightly.

18 Wait a few minutes to allow the oil to trickle down into the pan, then recheck the level on the dipstick and, if necessary, add enough oil to bring the level to the F mark.

19 During the first few trips after an oil change, make it a point to check frequently for leaks and proper oil level.

20 The old oil drained from the engine cannot be reused in its present state and should be disposed of. Check with your local auto parts store, disposal facility or environmental agency to see if they will accept the oil for recycling. After the oil has cooled it can be drained into a container (capped plastic jugs, topped bottles, milk cartons, etc.) for transport to one of these disposal sites. Don't dispose of the oil by pouring it on the ground or down a drain!

7   Battery check, maintenance and charging (every 5000 miles [8000 km] or 6 months)

▶ Refer to illustrations 7.1, 7.6a, 7.6b, 7.7a and 7.7b

✳✳ WARNING:

Certain precautions must be followed when checking and servicing the battery. Hydrogen gas, which is highly flammable, is always present in the battery cells, so keep lighted tobacco and all other open flames and sparks away from the battery. The electrolyte inside the battery is actually dilute sulfuric acid, which will cause injury if splashed on your skin or in your eyes. It will also ruin clothes and painted surfaces. When removing the battery cables, always detach the negative cable first and hook it up last!

MAINTENANCE

1   A routine preventive maintenance program for the battery in your vehicle is the only way to ensure quick and reliable starts. But before performing any battery maintenance, make sure that you have the proper equipment necessary to work safely around the battery (see illustration).

2   There are also several precautions that should be taken whenever battery maintenance is performed. Before servicing the battery, always turn the engine and all accessories off and disconnect the cable from the negative terminal of the battery.

3   The battery produces hydrogen gas, which is both flammable and explosive. Never create a spark, smoke or light a match around the battery. Always charge the battery in a ventilated area.

4   Electrolyte contains poisonous and corrosive sulfuric acid. Do not allow it to get in your eyes, on your skin on your clothes. Never ingest it. Wear protective safety glasses when working near the battery. Keep children away from the battery.

5   Note the external condition of the battery. If the positive terminal and cable clamp on your vehicle's battery is equipped with a rubber protector, make sure that it's not torn or damaged. It should completely cover the terminal. Look for any corroded or loose connections, cracks in the case or cover or loose hold-down clamps. Also check the entire length of each cable for cracks and frayed conductors.

7.1 Tools and materials required for battery maintenance

1   **Face shield/safety goggles** - When removing corrosion with a brush, the acidic particles can easily fly up into your eyes

2   **Baking soda** - A solution of baking soda and water can be used to neutralize corrosion

3   **Petroleum jelly** - A layer of this on the battery posts will help prevent corrosion

4   **Battery post/cable cleaner** - This wire brush cleaning tool will remove all traces of corrosion from the battery posts and cable clamps

5   **Treated felt washers** - Placing one of these on each post, directly under the cable clamps, will help prevent corrosion

6   **Puller** - Sometimes the cable clamps are very difficult to pull off the posts, even after the nut/bolt has been completely loosened. This tool pulls the clamp straight up and off the post without damage

7   **Battery post/cable cleaner** - Here is another cleaning tool which is a slightly different version of number 4 above, but it does the same thing

8   **Rubber gloves** - Another safety item to consider when servicing the battery; remember that's acid inside the battery

6   If corrosion, which looks like white, fluffy deposits (see illustration) is evident, particularly around the terminals, the battery should be removed for cleaning. Loosen the cable clamp bolts with a wrench, being careful to remove the ground cable first, and slide them off the terminals (see illustration). Then disconnect the hold-down clamp bolt and nut, remove the clamp and lift the battery from the engine compartment.

7   Clean the cable clamps thoroughly with a battery brush or a terminal cleaner and a solution of warm water and baking soda (see illustration). Wash the terminals and the top of the battery case with the same solution but make sure that the solution doesn't get into the battery. When cleaning the cables, terminals and battery top, wear safety goggles and rubber gloves to prevent any solution from coming in contact with your eyes or hands. Wear old clothes too - even diluted, sulfuric acid splashed onto clothes will burn holes in them. If the terminals have been extensively corroded, clean them up with a terminal cleaner (see illustration). Thoroughly wash all cleaned areas with plain water.

8   Make sure that the battery tray is in good condition and the hold-down clamp bolts are tight. If the battery is removed from the tray, make sure no parts remain in the bottom of the tray when the battery is reinstalled. When reinstalling the hold-down clamp bolts, do not overtighten them.

9   Any metal parts of the vehicle damaged by corrosion should be covered with a zinc-based primer, then painted.

10   Information on removing and installing the battery can be found in Chapter 5. Information on jump starting can be found at the front of this manual.

CHARGING

✳✳ WARNING:

When batteries are being charged, hydrogen gas, which is very explosive and flammable, is produced. Do not smoke or allow open flames near a battery. Wear eye protection when near the battery during charging. Also, make sure the charger is unplugged before connecting or disconnecting the battery from the charger.

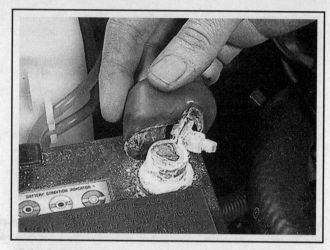

7.6a   Battery terminal corrosion usually appears as light, fluffy powder

➡Note: The manufacturer recommends the battery be removed from the vehicle for charging because the gas that escapes during this procedure can damage the paint. Fast charging with the battery cables connected can result in damage to the electrical system.

11   Slow-rate charging is the best way to restore a battery that's discharged to the point where it will not start the engine. It's also a good way to maintain the battery charge in a vehicle that's only driven a few miles between starts. Maintaining the battery charge is particularly important in the winter when the battery must work harder to start the engine and electrical accessories that drain the battery are in greater use.

12   It's best to use a one or two-amp battery charger (sometimes called a "trickle" charger). They are the safest and put the least strain on the battery. They are also the least expensive. For a faster charge, you can use a higher amperage charger, but don't use one rated more than 1/10th the amp/hour rating of the battery. Rapid boost charges that claim to restore the power of the battery in one to two hours are hardest on the battery and can damage batteries not in good condition. This

7.6b   Removing a cable from the battery post with a wrench - sometimes a pair of special battery pliers are required for this procedure if corrosion has caused deterioration of the nut hex (always remove the ground (-) cable first and hook it up last!)

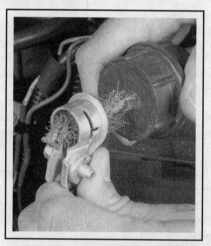

7.7a   When cleaning the cable clamps, all corrosion must be removed (the inside of the clamp is tapered to match the taper on the post, so don't remove too much material)

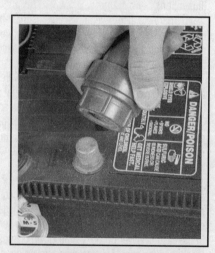

7.7b   Regardless of the type of tool used to clean the battery posts, a clean, shiny surface should be the result

type of charging should only be used in emergency situations.

13 The average time necessary to charge a battery should be listed in the instructions that come with the charger. As a general rule, a trickle charger will charge a battery in 12 to 16 hours.

14 Remove all the cell caps (if equipped) and cover the holes with a clean cloth to prevent spattering electrolyte. Disconnect the negative battery cable and hook the battery charger cable clamps up to the battery posts (positive to positive, negative to negative), then plug in the charger. Make sure it is set at 12-volts if it has a selector switch.

15 If you're using a charger with a rate higher than two amps, check the battery regularly during charging to make sure it doesn't overheat. If you're using a trickle charger, you can safely let the battery charge overnight after you've checked it regularly for the first couple of hours.

16 If the battery has removable cell caps, measure the specific gravity with a hydrometer every hour during the last few hours of the charging cycle. Hydrometers are available inexpensively from auto parts

stores - follow the instructions that come with the hydrometer. Consider the battery charged when there's no change in the specific gravity reading for two hours and the electrolyte in the cells is gassing (bubbling) freely. The specific gravity reading from each cell should be very close to the others. If not, the battery probably has a bad cell(s).

17 Some batteries with sealed tops have built-in hydrometers on the top that indicate the state of charge by the color displayed in the hydrometer window. Normally, a bright-colored hydrometer indicates a full charge and a dark hydrometer indicates the battery still needs charging.

18 If the battery has a sealed top and no built-in hydrometer, you can hook up a voltmeter across the battery terminals to check the charge. A fully charged battery should read 12.6 volts or higher after the surface charge has been removed.

19 Further information on the battery and jump starting can be found in Chapter 5 and at the front of this manual.

8   Tire rotation (every 5000 miles [8000 km] or 6 months)

▶ **Refer to illustrations 8.2a and 8.2b**

❋❋ **WARNING:**

If the vehicle is equipped with electronically modulated air suspension, make sure that the height control switch is turned off.

1   The tires should be rotated at the specified intervals and whenever uneven wear is noticed. Since the vehicle will be raised and the tires removed anyway, check the brakes (see Section 12) at this time.

2   Radial tires must be rotated in a specific pattern (see illustrations).

3   Refer to the information in *Jacking and towing* at the front of this manual for the proper procedures to follow when raising the vehicle

and changing a tire. If the brakes are to be checked, do not apply the parking brake as stated. Make sure the tires are blocked to prevent the vehicle from rolling.

4   Preferably, the entire vehicle should be raised at the same time. This can be done on a hoist or by jacking up each corner and then lowering the vehicle onto jackstands placed under the frame rails. Always use four jackstands and make sure the vehicle is firmly supported.

5   After rotation, check and adjust the tire pressures as necessary and be sure to check the lug nut tightness.

6   For further information on the wheels and tires, refer to Chapter 10.

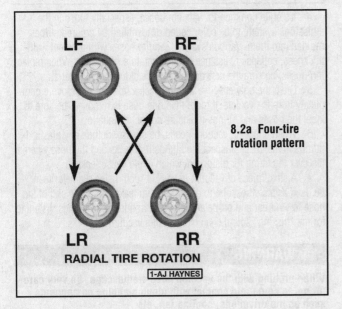

8.2a Four-tire rotation pattern

RADIAL TIRE ROTATION

1-AJ HAYNES

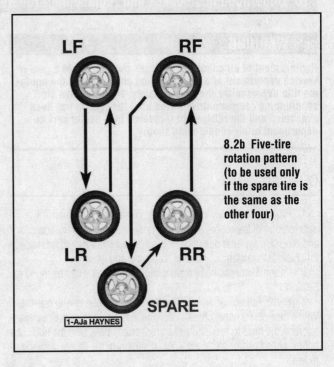

8.2b Five-tire rotation pattern (to be used only if the spare tire is the same as the other four)

1-AJa HAYNES

9 Windshield wiper blade inspection and replacement (every 5000 miles [8000 km] or 6 months)

♦ **Refer to illustrations 9.5a and 9.5b**

1 The windshield wiper and blade assembly should be inspected periodically for damage, loose components and cracked or worn blade elements.

2 Road film can build up on the wiper blades and affect their efficiency, so they should be washed regularly with a mild detergent solution.

3 The action of the wiping mechanism can loosen bolts, nuts and fasteners, so they should be checked and tightened, as necessary, at the same time the wiper blades are checked.

4 If the wiper blade elements are cracked, worn or warped, or no longer clean adequately, they should be replaced with new ones.

5 Lift the arm assembly away from the glass for clearance, press the release lever, then slide the wiper blade assembly out of the hook at the end of the arm (see illustrations).

6 Attach the new wiper to the arm. Connection can be confirmed by an audible click.

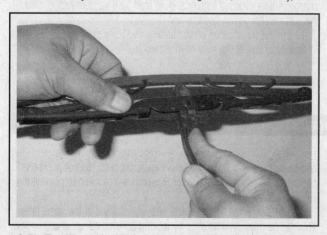

9.5a To release the blade holder, push the release pin . . .

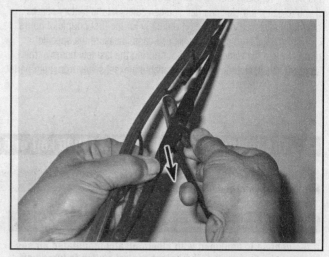

9.5b . . . and pull the wiper blade in the direction of the arrow to separate it from the arm

10 Underhood hose check and replacement (every 5000 miles [8000 km] or 6 months)

✳✳ WARNING:

Replacement of air conditioning hoses must be left to a dealer service department or air conditioning shop that has the equipment to depressurize the system safely. Never remove air conditioning components or hoses until the system has been evacuated and the refrigerant recovered by a dealer service department or air-conditioning shop.

GENERAL

1 High temperatures in the engine compartment can cause the deterioration of the rubber and plastic hoses used for engine, accessory and emission systems operation. Periodic inspection should be made for cracks, loose clamps, material hardening and leaks.

2 Information specific to the cooling system hoses can be found in Section 11.

3 Some, but not all, hoses are secured to the fittings with clamps. Where clamps are used, check to be sure they haven't lost their tension, allowing the hose to leak. If clamps aren't used, make sure the hose has not expanded and/or hardened where it slips over the fitting, allowing it to leak.

VACUUM HOSES

4 It's quite common for vacuum hoses, especially those in the emissions system, to be color coded or identified by colored stripes molded into them. Various systems require hoses with different wall thickness, collapse resistance and temperature resistance. When replacing hoses, be sure the new ones are made of the same material.

5 Often the only effective way to check a hose is to remove it completely from the vehicle. If more than one hose is removed, be sure to label the hoses and fittings to ensure correct installation.

6 When checking vacuum hoses, be sure to include any plastic T-fittings in the check. Inspect the fittings for cracks and the hose where it fits over the fitting for distortion, which could cause leakage.

7 A small piece of vacuum hose (1/4-inch inside diameter) can be used as a stethoscope to detect vacuum leaks. Hold one end of the hose to your ear and probe around vacuum hoses and fittings, listening for the "hissing" sound characteristic of a vacuum leak.

✳✳ WARNING:

When probing with the vacuum hose stethoscope, be very careful not to come into contact with moving engine components such as the drivebelts, cooling fan, etc.

FUEL HOSE

Gasoline is extremely flammable, so take extra precautions when you work on any part of the fuel system. Don't smoke or allow open flames or bare light bulbs near the work area, and don't work in a garage where a gas-type appliance (such as a water heater or a clothes dryer) is present. Since gasoline is carcinogenic, wear fuel resistant gloves when there's a possibility of being exposed to fuel, and, if you spill any fuel on your skin, rinse it off immediately with soap and water. Mop up any spills immediately and do not store fuel-soaked rags where they could ignite. The fuel system is under constant pressure, so, if any fuel lines are to be disconnected, the fuel pressure in the system must be relieved first. When you perform any kind of work on the fuel system, wear safety glasses and have a Class B type fire extinguisher on hand.

8   Check all rubber fuel lines for deterioration and chafing. Check especially for cracks in areas where the hose bends and just before fittings, such as where a hose attaches to the fuel filter.

9   High quality fuel line should be used for fuel line replacement. Never, under any circumstances, use unreinforced vacuum line, clear plastic tubing or water hose for fuel lines.

10   Spring-type clamps are commonly used on fuel lines. These clamps often lose their tension over a period of time, and can be "sprung" during removal. Replace all spring-type clamps with screw clamps whenever a hose is replaced.

METAL LINES

11   Sections of metal line are often used for fuel line between the fuel pump and fuel injection unit. Check carefully to be sure the line has not been bent or crimped and that cracks have not started in the line.

12   If a section of metal fuel line must be replaced, only seamless steel tubing should be used, since copper and aluminum tubing don't have the strength necessary to withstand normal engine vibration.

13   Check the metal brake lines where they enter the master cylinder and brake proportioning unit (if used) for cracks in the lines or loose fittings. Any sign of brake fluid leakage calls for an immediate thorough inspection of the brake system.

11   Cooling system check (every 5000 miles [8000 km] or 6 months)

▶ **Refer to illustration 11.4**

1   Many major engine failures can be attributed to a faulty cooling system. If the vehicle is equipped with an automatic transaxle, the cooling system also cools the transaxle fluid and thus plays an important role in prolonging transaxle life.

2   The cooling system should be checked with the engine cold. Do this before the vehicle is driven for the day or after the engine has been shut off for at least three hours.

Never remove the radiator pressure cap when the engine is running or has just been shut down, because the cooling system is hot. Escaping steam and scalding liquid could cause serious injury.

3   Remove the radiator pressure cap by turning it to the left until it reaches a stop. If you hear a hissing sound (indicating there is still pressure in the system), wait until it stops. Now press down on the cap with the palm of your hand and continue turning to the left until the cap can be removed. Thoroughly clean the cap, inside and out, with clean water. Also clean the filler neck on the radiator. All traces of corrosion should be removed. The coolant inside the radiator should be relatively transparent. If it's rust colored, the system should be drained and refilled (see Section 26). If the coolant level isn't up to the top, add additional antifreeze/coolant mixture (see Section 4).

4   Carefully check the large upper and lower radiator hoses along with the smaller diameter heater hoses which run from the engine to the bulkhead. Inspect each hose along its entire length, replacing any hose which is cracked, swollen or shows signs of deterioration. Cracks may become more apparent if the hose is squeezed (see illustration). Regardless of condition, it's a good idea to replace hoses with new ones every two years.

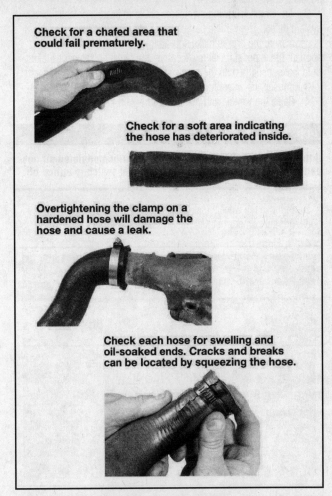

Check for a chafed area that could fail prematurely.

Check for a soft area indicating the hose has deteriorated inside.

Overtightening the clamp on a hardened hose will damage the hose and cause a leak.

Check each hose for swelling and oil-soaked ends. Cracks and breaks can be located by squeezing the hose.

11.4 Hoses, like drivebelts, have a habit of failing at the worst possible time - to prevent the inconvenience of a blown radiator or heater hose, inspect them carefully as shown here

5   Make sure that all hose connections are tight. A leak in the cooling system will usually show up as white or rust colored deposits on the areas adjoining the leak. If wire-type clamps are used at the ends of the hoses, it may be a good idea to replace them with more secure screw-type clamps.

6   Use compressed air or a soft brush to remove bugs, leaves, etc.

from the front of the radiator or air conditioning condenser. Be careful not to damage the delicate cooling fins or cut yourself on them.

7   Every other inspection, or at the first indication of cooling system problems, have the cap and system pressure tested. If you don't have a pressure tester, most gas stations and garages will do this for a minimal charge.

12 Brake check (every 5000 miles [8000 km] or 6 months)

✳✳ WARNING:

The dust created by the brake system is harmful to your health. Never blow it out with compressed air and don't inhale any of it. An approved filtering mask should be worn when working on the brakes. Do not, under any circumstances, use petroleum-based solvents to clean brake parts. Use brake system cleaner only!

➡**Note: For detailed photographs of the brake system, refer to Chapter 9.**

1   In addition to the specified intervals, the brakes should be inspected every time the wheels are removed or whenever a defect is suspected.

2   Any of the following symptoms could indicate a potential brake system defect: The vehicle pulls to one side when the brake pedal is depressed; the brakes make squealing or dragging noises when applied; brake pedal travel is excessive; the pedal pulsates; or brake fluid leaks, usually onto the inside of the tire or wheel.

3   Loosen the wheel lug nuts.

4   Raise the vehicle and place it securely on jackstands.

✳✳ WARNING:

If the vehicle is equipped with electronically modulated air suspension, make sure that the height control switch is turned off.

5   Remove the wheels (see *Jacking and towing* at the front of this book, or your owner's manual, if necessary).

DISC BRAKES

▸ **Refer to illustrations 12.7a and 12.7b**

6   There are two pads (an outer and an inner) in each caliper. The pads are visible with the wheels removed. The vehicles covered by this manual have disc brakes front and rear, with a mechanical, drum-type parking brake mechanism inside the rear discs.

7   Check the pad thickness by looking at each end of the caliper and through the inspection window in the caliper body (see illustrations). If the lining material is less than the thickness listed in this Chapter's Specifications, replace the pads.

➡**Note: Keep in mind that the lining material is riveted or bonded to a metal backing plate and the metal portion is not included in this measurement.**

8   If it is difficult to determine the exact thickness of the remaining pad material by the above method, or if you are at all concerned about the condition of the pads, remove the caliper(s), then remove the pads from the calipers for further inspection (refer to Chapter 9).

9   Once the pads are removed from the calipers, clean them with brake cleaner and re-measure them with a ruler or a vernier caliper.

10 Measure the disc thickness with a micrometer to make sure that it still has service life remaining. If any disc is thinner than the specified minimum thickness, replace it (refer to Chapter 9). Even if the disc has service life remaining, check its condition. Look for scoring, gouging and burned spots. If these conditions exist, remove the disc and have it resurfaced (see Chapter 9).

11 Before installing the wheels, check all brake lines and hoses

12.7a You'll find an inspection hole like this in each caliper through which you can view the inner brake pad lining

12.7b The outer pad is more easily checked at the edge of the caliper

for damage, wear, deformation, cracks, corrosion, leakage, bends and twists, particularly in the vicinity of the rubber hoses at the calipers. Check the clamps for tightness and the connections for leakage. Make sure that all hoses and lines are clear of sharp edges, moving parts and the exhaust system. If any of the above conditions are noted, repair, reroute or replace the lines and/or fittings as necessary (see Chapter 9).

BRAKE BOOSTER CHECK

12 Sit in the driver's seat and perform the following sequence of tests.

13 With the brake fully depressed, start the engine - the pedal should move down a little when the engine starts.

14 With the engine running, depress the brake pedal several times - the travel distance should not change.

15 Depress the brake, stop the engine and hold the pedal in for about 30 seconds - the pedal should neither sink nor rise.

16 Restart the engine, run it for about a minute and turn it off. Then firmly depress the brake several times - the pedal travel should decrease with each application.

17 If your brakes do not operate as described, the brake booster has failed. Refer to Chapter 9 for the replacement procedure.

PARKING BRAKE

18 One method of checking the parking brake is to park the vehicle on a steep hill with the parking brake set and the transmission in Neutral (be sure to stay in the vehicle for this check). If the parking brake cannot prevent the vehicle from rolling, it's in need of attention (see Chapter 9).

---

13 Steering, suspension and driveaxle boot check (every 15,000 miles [24,000 km] or 18 months)

STEERING CHECK

➥Note: For detailed illustrations of the steering and suspension components, refer to Chapter 10.

1 With the vehicle on the ground and the front wheels pointed straight ahead, rock the steering wheel gently back and forth. If freeplay is excessive, a front wheel bearing, main shaft yoke, intermediate shaft yoke, lower arm balljoint or steering system joint is worn or the steering gear is out of adjustment or broken. Steering wheel freeplay is the amount of travel (measured at the rim of the steering wheel) between the initial steering input and the point at which the front wheels begin to turn (indicated by slight resistance). Refer to Chapter 10 for the appropriate repair procedure.

2 Other symptoms, such as excessive vehicle body movement over rough roads, swaying (leaning) around corners and binding as the steering wheel is turned, may indicate faulty steering and/or suspension components.

SUSPENSION CHECK

▸ Refer to illustrations 13.7 and 13.8

3 Check the shock absorbers by pushing down and releasing the vehicle several times at each corner. If the vehicle does not come back to a level position within one or two bounces, the shocks/struts are worn and must be replaced. When bouncing the vehicle up and down, listen for squeaks and noises from the suspension components. Additional

information on suspension components can be found in Chapter 10.

4 Raise the vehicle with a floor jack and support it securely on jackstands. See *Jacking and towing* at the front of this book for the proper jacking points.

5 Check the tires for irregular wear patterns and proper inflation. See Section 5 in this Chapter for information regarding tire wear and Chapter 10 for the wheel bearing replacement procedures.

6 Inspect the universal joint between the steering shaft and the steering gear housing. Check the steering gear housing for lubricant leakage or oozing. Make sure that the dust seals and boots are not damaged and that the boot clamps are not loose. Check the steering linkage for looseness or damage. Check the track rod ends for excessive play. Look for loose bolts, broken or disconnected parts and deteriorated rubber bushings on all suspension and steering components. While an assistant turns the steering wheel from side to side, check the steering components for free movement, chafing and binding. If the steering components do not seem to be reacting with the movement of the steering wheel, try to determine where the slack is located.

7 Check the balljoints for wear by trying to move each lower arm up and down with a prybar (see illustration) to ensure that its balljoint has no play. If any balljoint does have play, replace it. See Chapter 10 for the front balljoint replacement procedure.

8 Inspect the balljoint boots for damage and leaking grease (see illustration). Replace the balljoints with new ones if they are damaged (see Chapter 10).

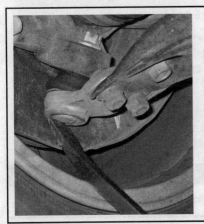

13.7 To check the balljoints attempt to move the lower arm up and down with a prybar to make sure here is no play in the balljoint (if there is, replace it)

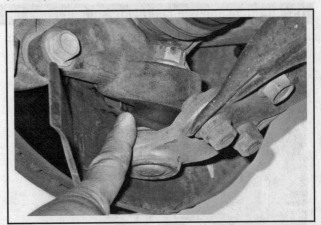

13.8 Push on the balljoint boot to check for tears and grease leaks

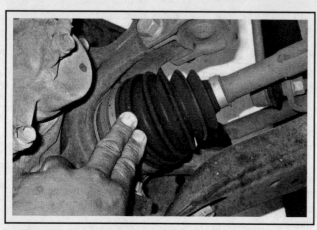

13.10 Flex the driveaxle boots by hand to check for tears, cracks and leaking grease

DRIVEAXLE BOOT CHECK

▶ **Refer to illustration 13.10**

9 The driveaxle boots are very important because they prevent dirt, water and foreign material from entering and damaging the constant velocity (CV) joints. Oil and grease can cause the boot material to deteriorate prematurely, so it's a good idea to wash the boots with soap and water.

10 Inspect the boots for tears and cracks as well as loose clamps (see illustration). If there is any evidence of cracks or leaking lubricant, they must be replaced as described in Chapter 8.

14 Exhaust system check (every 15,000 miles [24,000 km] or 18 months)

▶ **Refer to illustration 14.2**

1 With the engine cold (at least three hours after the vehicle has been driven), check the complete exhaust system from its starting point at the engine to the end of the tailpipe. This should be done on a hoist where unrestricted access is available.

2 Check the pipes and connections for evidence of leaks, severe corrosion or damage. Make sure that all brackets and hangers are in good condition and tight (see illustration).

3 At the same time, inspect the underside of the body for holes, corrosion, open seams, etc. which may allow exhaust gases to enter the passenger compartment. Seal all body openings with silicone or body putty.

4 Rattles and other noises can often be traced to the exhaust system, especially the mounts and hangers. Try to move the pipes, silencer and catalytic converter. If the components can come in contact with the body or suspension parts, secure the exhaust system with new mounts.

5 Check the running condition of the engine by inspecting inside the end of the tailpipe. The exhaust deposits here are an indication of engine state-of-tune. If the pipe is black and sooty or coated with white deposits, the engine is in need of a tune-up, including a thorough fuel system inspection.

14.2 Check the exhaust system for damage, or worn rubber hangers

15 Air filter replacement (every 30,000 miles [48,000 km] or 36 months)

15.1 Release the clamps securing the air filter housing lid

▶ **Refer to illustrations 15.1 and 15.2**

1 The air filter is located inside a housing at the left (driver's) side of the engine compartment. To remove the air filter, release the clamps retaining the two halves of the air filter housing (see illustration).

2 Lift the cover up and remove the air filter element (see illustration).

3 Inspect the outer surface of the filter element. If it is dirty, replace it. If it is only moderately dusty, it can be reused by blowing it clean from the back to the front surface with compressed air.

✳✳ WARNING:

Always wear eye protection when using compressed air!

Because it is a pleated paper type filter, it cannot be washed or oiled. If it cannot be cleaned satisfactorily with compressed air, discard and replace it.

✳✳ CAUTION:

Never drive the vehicle with the air cleaner removed. Excessive engine wear could result and backfiring could even cause a fire under the hood.

4   Installation is the reverse of removal. Make sure the hinge tabs on the housing cover engage properly with the lower part of the housing.

15.2 Lift the cover up and remove the filter

16 Fuel system check (every 30,000 miles [48,000 km] or 36 months)

✳✳ WARNING:

Gasoline is extremely flammable, so take extra precautions when you work on any part of the fuel system. Don't smoke or allow open flames or bare light bulbs near the work area, and don't work in a garage where a gas-type appliance (such as a water heater or a clothes dryer) is present. Since gasoline is carcinogenic, wear fuel resistant gloves when there's a possibility of being exposed to fuel, and, if you spill any fuel on your skin, rinse it off immediately with soap and water. Mop up any spills immediately and do not store fuel-soaked rags where they could ignite. The fuel system is under constant pressure, so, if any fuel lines are to be disconnected, the fuel pressure in the system must be relieved first. When you perform any kind of work on the fuel system, wear safety glasses and have a Class B type fire extinguisher on hand.

1   If you smell fuel while driving or after the vehicle has been sitting in the sun, inspect the fuel system immediately.

2   Remove the fuel filler cap and inspect it for damage and corrosion. The gasket should have an unbroken sealing imprint. If the gasket is damaged or corroded, remove it and install a new one (see Section 18).

3   Inspect the fuel feed and return lines for cracks. Make sure that the threaded flare nut type connectors which secure the metal fuel lines to the fuel injection system and the banjo bolts which secure the banjo

fittings to the in-line fuel filter are tight.

4   Since some components of the fuel system - the fuel tank and part of the fuel feed and return lines, for example - are underneath the vehicle, they can be inspected more easily with the vehicle raised on a hoist. If that's not possible, raise the vehicle and support it securely on jackstands.

5   With the vehicle raised and safely supported, inspect the fuel tank and filler neck for punctures, cracks and other damage. The connection between the filler neck and the tank is particularly critical. Sometimes a rubber filler neck will leak because of loose clamps or deteriorated rubber. These are problems a home mechanic can usually rectify.

✳✳ WARNING:

Do not, under any circumstances, try to repair a fuel tank (except rubber components). A welding torch or any open flame can easily cause fuel vapors inside the tank to explode.

6   Carefully check all rubber hoses and metal lines leading away from the fuel tank. Check for loose connections, deteriorated hoses, crimped lines and other damage. Carefully inspect the lines from the tank to the fuel injection system. Repair or replace damaged sections as necessary (see Chapter 4).

17 Interior ventilation filter replacement (every 30,000 miles [48,000 km] or 36 months)

♦ **Refer to illustration 17.3**

1   There is an air filter in the blower housing that cleans the air before it enters the passenger's compartment.

2   To remove the air filter, remove the glove box (see Chapter 11).

3   Depress the filter cover mounting tabs (see illustration).

4   Lift the cover off and remove the air filter element.

5   Installation is the reverse of removal.

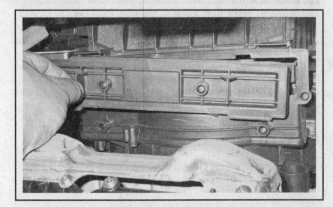

17.3 Release the ventilation filter mounting tabs

18 Fuel tank cap gasket inspection and replacement (every 30,000 miles [48,000 km] or 36 months)

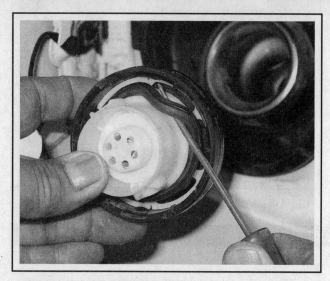

18.2 Use a small screwdriver to carefully pry out the old gasket - take care not to damage the cap

▶ **Refer to illustration 18.2**

1   Remove the tank cap and inspect the rubber gasket for cracks or tears.

2   If replacement is necessary, carefully pry the old gasket out of the recess (see illustration). Be very careful not to damage the sealing surface inside the cap.

3   Work the new gasket into the cap recess.

4   Install the cap, then remove it and make sure the gasket seals all the way around.

19 Positive Crankcase Ventilation (PCV) valve check and replacement (every 30,000 miles [48,000 km] or 36 months)

1   The PCV valve and hose is located in the valve cover.

2   Disconnect the hose, pull the PCV valve from the cover, then reconnect the hose.

3   With the engine idling at normal operating temperature, place your finger over the valve opening. If there's no vacuum at the valve, check for a plugged hose or valve. Replace any plugged or deteriorated hoses.

4   Turn off the engine. Remove the PCV valve from the hose. Blow through the valve from the valve cover (cylinder head) end. If air will not pass through the valve in this direction, replace it with a new one.

5   When purchasing a replacement PCV valve, make sure it's for your particular vehicle and engine size. Compare the old valve with the new one to make sure they're the same.

20 Brake fluid change (every 30,000 miles [48,000 km] or 36 months)

✳✳ **WARNING:**

Brake fluid can harm your eyes and damage painted surfaces, so use extreme caution when handling or pouring it. Do not use brake fluid that has been standing open or is more than one year old. Brake fluid absorbs moisture from the air. Excess moisture can cause a dangerous loss of braking effectiveness.

1   At the specified intervals, the brake fluid should be drained and replaced. Since the brake fluid may drip or splash when pouring it, place plenty of rags around the master cylinder to protect any surrounding painted surfaces.

2   Before beginning work, purchase the specified brake fluid (see *Recommended lubricants and fluids* at the end of this Chapter).

3   Remove the cap from the master cylinder reservoir.

4   Using a hand suction pump or similar device, withdraw the fluid from the master cylinder reservoir.

5   Add new fluid to the master cylinder until it rises to the base of the filler neck.

6   Bleed the brake system as described in Chapter 9 at all four brakes until new and uncontaminated fluid is expelled from the bleeder screw. Be sure to maintain the fluid level in the master cylinder as you perform the bleeding process. If you allow the master cylinder to run dry, air will enter the system.

7   Refill the master cylinder with fluid and check the operation of the brakes. The pedal should feel solid when depressed, with no sponginess.

✳✳ **WARNING:**

Do not operate the vehicle if you are in doubt about the effectiveness of the brake system.

21 Drivebelt check, adjustment and replacement (60,000 miles [96,000 km] or 72 months and every 15,000 miles [24,000 km] and 18 months thereafter)

CHECK

▶ **Refer to illustration 21.3**

1 The drivebelt(s) are located at the right side of the engine compartment. The good condition and proper adjustment of the belts is critical to the operation of the engine. Because of their composition and the high stresses to which they are subjected, drivebelts stretch and deteriorate as they get older. They must therefore be periodically inspected.

2 On four-cylinder models, one belt drives the alternator, air conditioning and power steering. On V6 models, one belt transmits power from the crankshaft to the alternator and air conditioning, and one belt drives the power steering pump.

3 With the engine off, open the hood and locate the drivebelts at the right side of the engine compartment. With a flashlight, check each belt for separation of the adhesive rubber on both sides of the core, core separation from the belt side, a severed core, separation of the ribs from the adhesive rubber, cracking or separation of the ribs, and torn or worn ribs or cracks in the inner ridges of the ribs (see illustration). Also check for fraying and glazing, which gives the belt a shiny appearance. Both sides of the belt should be inspected, which means you will have to twist the belt to check the underside. Use your fingers to feel the belt where you can't see it. If any of the above conditions are evident, replace the belt (go to Step 8 or Step 16).

4 The tension of each belt is checked by pushing on the belt at a distance halfway between the pulleys. Push firmly with your thumb and see how much the belt moves (deflects) (see illustration). As rule of thumb, the belt should deflect approximately 1/4-inch.

ADJUSTMENT (V6 ENGINE ONLY)

▶ **Refer to illustration 21.5**

5 If the alternator/air conditioner compressor belt must be adjusted, loosen the alternator pivot bolt located on the front left corner of the block. Loosen the locking bolt and turn the adjusting bolt (see illustration). Measure the belt tension in accordance with the above method. Repeat this step until the air conditioning compressor drivebelt is adjusted.

6 Adjust the power steering pump belt by loosening the adjustment bolt that secures the pump to the slotted bracket and pivot the pump (away from the engine to tighten the belt, toward it to loosen it). Repeat the procedure until the drivebelt tension is correct and tighten the bolt.

REPLACEMENT

Four-cylinder engine

7 Remove the right front wheel.

8 Remove the right side fender apron seal.

9 Remove the right side engine cover assembly.

10 Remove the upper engine control rod and bracket (see Chapter 2A, Section 17).

11 Slowly turn the drive belt tensioner clockwise and remove the drivebelt from the pulleys.

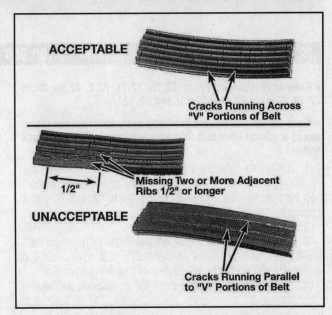

21.3 Check a multi-ribbed belt for signs like these - if the belt looks worn, replace it

12 Take the old belt to the parts store in order to make a direct comparison for length, width and design.

13 After replacing the drivebelt, make sure that it fits properly in the ribbed grooves in the pulleys. It is essential that the belt be properly centered.

V6 engine

14 To replace a belt, follow the procedures for V6 engine drivebelt adjustment, but slip the belt off the crankshaft pulley and remove it. If you are replacing the power steering pump belt, you will have to remove the air conditioning compressor belt first because of the way they are arranged on the crankshaft pulley. Because of this and because belts tend to wear out more or less together, it is a good idea to replace both belts at the same time. Mark each belt and its appropriate pulley

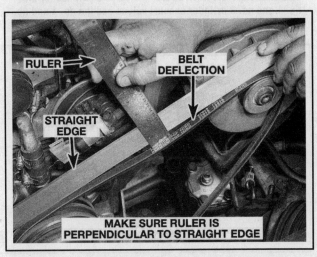

21.5 Measuring drivebelt deflection with a straightedge and ruler

groove so the replacement belts can be fitted in their proper positions.

15 Take the old belts to the parts store in order to make a direct comparison for length, width and design.

16 After replacing a drivebelt, make sure that it fits properly in the ribbed grooves in the pulleys. It is essential that the belt be properly centered.

17 Adjust the belt(s) in accordance with the procedure outlined earlier in this Section.

22 Valve clearance check and adjustment (60,000 miles [96,000 km] or 72 months)

▶ Refer to illustrations 22.7a, 22.7b, 22.7c, 22.8, 22.9a, 22.9b, 22.10, 22.11a, 22.11b, 22.11c and 22.12

➡Note: On V6 models, the following procedure requires the use of a special lifter tool. It is impossible to perform this task without it.

1 Remove the right-side front fender apron, engine cover and any other components that will interfere with valve cover removal. Disconnect the cable from the negative terminal of the battery (see Chapter 5, Section 1).

2 Remove the ignition coil(s).

3 On V6 models, drain the coolant (see Section 26), remove the radiator hose inlet, air cleaner assembly, upper suspension brace, and any other components that will interfere with valve cover removal.

4 Blow out the spark plug recesses with compressed air, if avail-

able, to remove any debris that might fall into the cylinders, then remove the spark plugs (see Section 27).

✳✳ WARNING:

Always wear eye protection when using compressed air!

5 Remove the valve cover(s) (see Chapter 2A or 2B).

6 Refer to Chapter 2A or 2B and position the number 1 piston at TDC on the compression stroke.

7 Measure the clearance of the indicated valves with a feeler gauge of the specified thickness (see illustrations). Record the clearance of each valve and note which are out of specification. This information will be used later to determine the required replacement shims or lifters.

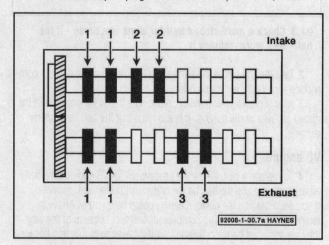

22.7a On four-cylinder engines, when the no. 1 piston is at TDC on the compression stroke, the valve clearance for the no. 1 and no. 3 cylinder exhaust valves and the no. 1 and no. 2 cylinder intake valves can be measured

22.7b Measure the clearance for each valve with a feeler gauge of the specified thickness - if the clearance is correct, you should feel a slight drag on the gauge as you pull it out

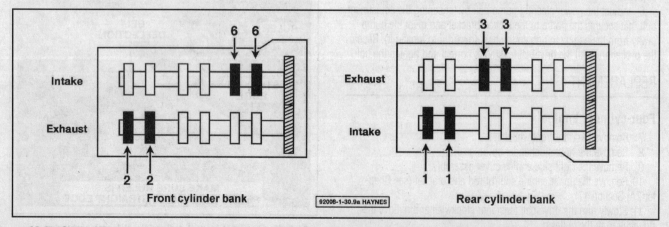

22.7c On the V6 engine, when the no. 1 piston is at TDC on the compression stroke, the clearance of the indicated valves can be measured

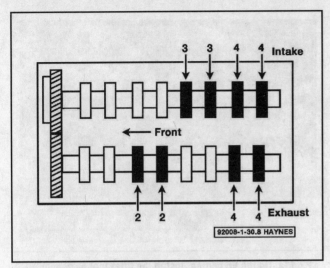

22.8 On four-cylinder engines, when the no. 4 piston is at TDC on the compression stroke, the valve clearance for the no. 2 and no. 4 exhaust valves and the no. 3 and no. 4 intake valves can be measured

8 On four-cylinder engines, turn the crankshaft one complete revolution and realign the timing marks. Measure the remaining valves (see illustration).

9 On V6 engines, turn the crankshaft 2/3-turn (240-degrees) clockwise. Measure the valve clearance on the valves shown (see illustration). Rotate the crankshaft a further 2/3-turn and measure the clearance on the remaining valves (see illustration).

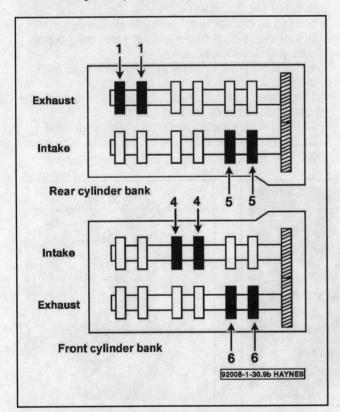

22.9b On the V6 engine, rotate the crankshaft an additional 2/3 of a revolution (240-degrees) and measure the clearance of the remaining valves

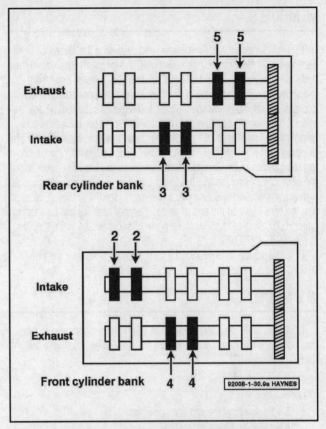

22.9a After the V6 engine has been rotated 240-degrees from TDC for the no. 1 piston, on the compression stroke, measure the clearance of the indicated valves

FOUR-CYLINDER MODELS

10 Remove the camshaft(s) for the valve(s) which you intend to adjust (refer to Chapter 2A). Remove and measure each lifter with a micrometer (see illustration). Place each lifter back into its bore in the cylinder head before moving onto the next lifter. Record the measurements for each lifter.

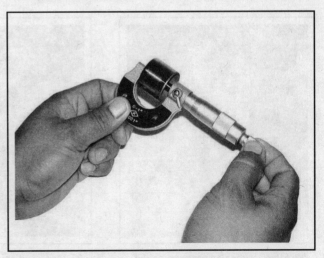

22.10 On four-cylinder models, measure the thickness of the lifter head with a micrometer

V6 MODELS

11 On V6 models, after measuring and recording the clearance of each valve, turn the crankshaft pulley until the camshaft lobe above the first valve which you intend to adjust is pointing upward, away from the shim. Position the notch in the lifter toward the spark plug. Then depress the lifter with the special lifter tools (see illustration). Place the special lifter tool in position as shown, with the longer jaw of the tool gripping the lower edge of the cast lifter boss and the upper, shorter jaw gripping the upper edge of the lifter itself. Depress the lifter by squeezing the handles of the lifter tool together, then hold the lifter down with the smaller tool and remove the larger one. Remove the adjusting shim with a small screwdriver or a pair of tweezers (see illustrations). Note that the wire hook on the end of some lifter tool handles can be used to clamp both handles together to keep the lifter depressed while the shim is removed.

12 Measure the thickness of the shim with a micrometer (see illustration).

ALL MODELS

13 To calculate the correct thickness of a replacement shim or lifter that will place the valve clearance within the specified value, use the following formula:

$$N = T + (A - V)$$

Whereas:

T = thickness of the old shim or lifter
A = valve clearance measured
N = thickness of the new shim or lifter
V = desired valve clearance (see this Chapter's Specifications)

14 Select a shim or lifter with a thickness as close as possible to the valve clearance calculated. Shims for V6 models, which are available in 17 sizes in increments of 0.0020-inch (0.050 mm), range in size from 0.0984-inch (2.500 mm) to 0.1299-inch (3.300 mm). Lifters for four-cylinder engines are available in 35 sizes, in increments of 0.0008-inch (0.020 mm), range in size from 0.1992-inch (5.060 mm) to 0.2260 inch (5.740 mm).

➡Note: **Through careful analysis of the shim or lifter sizes needed to bring the out-of-specification valve clearance within specification, it is often possible to simply move a shim or lifter**

22.11a On the V6 engine, install the lifter tool as shown and squeeze the handles together to depress the lifter, then hold the lifter down with the smaller tool so the shim can be removed . . .

that has to come out anyway to another lifter requiring a shim or lifter of that particular size, thereby reducing the number of new shims that must be purchased.

15 On V6 models, place the special lifter tool in position as shown in illustration 22.11a, with the longer jaw of the tool gripping the lower edge of the cast lifter boss and the upper, shorter jaw gripping the upper edge of the lifter itself, press down the lifter by squeezing the handles of the lifter tool together and install the new adjusting shim (note that the wire hook on the end of one lifter tool handle can be used to clamp the handles together to keep the lifter depressed while the shim is inserted). Measure the clearance with a feeler gauge to make sure that your calculations are correct.

16 Repeat this procedure until all the valves which are out of clearance have been corrected.

17 Installation of the spark plugs, valve cover, camshaft(s), accelerator cable bracket, etc. is the reverse of removal.

18 On four-cylinder models, after the camshafts are reinstalled, check the valve clearances again to be sure they are now within specification.

22.11b . . . keeping pressure on the lifter with the smaller tool and remove the shim with a small screwdriver . . .

22.11c . . . a pair of tweezers or a magnet as shown here

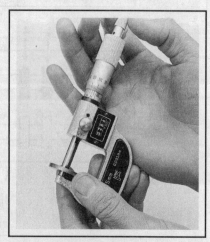

22.12 Measure the shim thickness with a micrometer

23 Automatic transaxle fluid change (60,000 miles [96,000 km] or 72 months)

▶ **Refer to illustrations 23.7, 23.8a, 23.8b, 23.9 and 23.12**

1   At the specified time intervals, the automatic transaxle and differential fluid should be drained and replaced.

➡**Note: Although the manufacturer doesn't specify it, it is a good idea to clean the transaxle fluid strainer periodically to remove accumulated dirt and metal particles.**

2   Before beginning work, purchase the specified transaxle fluid (see *Recommended fluids and lubricants* at the end of this Chapter).

3   Other tools necessary for this job include jackstands to support the vehicle in a raised position, a 10 mm hex bit or Allen wrench, a drain pan capable of holding at least eight pints, newspapers and clean rags.

4   The fluid should be drained immediately after the vehicle has been driven. Hot fluid is more effective than cold fluid at removing built up sediment.

✳✳ WARNING:

Fluid temperature can exceed 350-degrees F in a hot transaxle. Wear protective gloves.

5   After the vehicle has been driven to warm up the fluid, raise it and place it on jackstands for access to the transaxle and differential drain plugs.

✳✳ WARNING:

If the vehicle is equipped with electronically modulated air suspension, make sure that the height control switch is turned off

6   Move the necessary equipment under the vehicle, being careful not to touch any of the hot exhaust components.

7   Place the drain pan under the drain plug in the transaxle pan and remove the drain plug (see illustration). Be sure the drain pan is in position, as fluid will come out with some force. Once the fluid is drained, reinstall the drain plug securely. If you aren't going to clean the strainer, proceed to Step 14.

8   To clean the strainer, remove the front transaxle pan bolts, then loosen the rear bolts and carefully pry the pan loose with a screwdriver

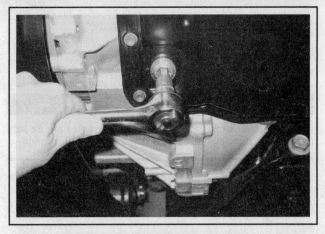

23.7 Remove the automatic transaxle drain plug

23.8a After loosening the front bolts, remove the rear pan bolts . . .

and allow the remaining fluid to drain (see illustrations). Once the fluid has drained, remove the bolts and lower the pan.

9   Remove the strainer retaining bolts, disconnect the clip (some models) and lower the strainer from the transaxle (see illustration). Be careful when lowering the strainer as it contains residual fluid.

23.8b . . . and allow the remaining fluid to drain out

23.9 Remove the strainer bolts and lower the strainer (be careful, there will be some residual fluid)

10 Wash the strainer thoroughly in clean transmission fluid.

11 Place the strainer in position, connect the clip (if equipped) and install the bolts. Tighten the bolts to the torque listed in this Chapter's Specifications.

12 Carefully clean the gasket surfaces of the fluid pan, removing all traces of old gasket material. Noting their location, remove the magnets, wash the pan in clean solvent and dry it with compressed air.

❊❊ WARNING:

Always wear eye protection when using compressed air! Be sure to clean and reinstall the magnets in the pan (see illustration).

13 Install a new gasket, place the fluid pan in position and install the bolts in their original positions. Tighten the bolts to the torque listed in this Chapter's Specifications.

14 Lower the vehicle.

15 With the engine off, add new fluid to the transaxle through the dipstick tube (see *Recommended fluids and lubricants* for the recommended fluid type and capacity). Use a funnel to prevent spills. It is best to add a little fluid at a time, continually checking the level with the dipstick (see Section 4). Allow the fluid time to drain into the pan.

16 Start the engine and shift the gear-change selector into all positions from P through L, then shift the gearchange into P and apply the parking brake.

23.12 Noting their locations, remove the magnets and wash them and the pan in solvent before installing them

17 With the engine idling, check the fluid level. Add fluid up to the Cool level on the dipstick.

18 Drive the vehicle to warm up the transaxle to normal operating temperature, then recheck the fluid level.

24 Transfer case lubricant change (4WD models) (60,000 miles [96,000 km] or 72 months)

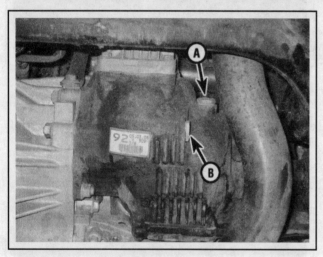

24.2 Location of the transfer casecheck/fill plug (A) and drain plug (B)

▶ Refer to illustration 24.2

1 Raise the vehicle and support it securely on jackstands.

❊❊ WARNING:

If the vehicle is equipped with electronically modulated air suspension, make sure that the height control switch is turned off.

2 Remove the check/fill plug, then remove the drain plug and drain the lubricant (see illustration).

3 Reinstall the drain plug and tighten it securely.

4 Add new lubricant until it is even with the lower edge of the filler hole. See *Recommended lubricants and fluids* for the specified lubricant type.

5 Reinstall the check/fill plug and tighten it securely.

25 Rear differential lubricant change (4WD models) (60,000 miles [96,000 km] or 72 months)

▶ Refer to illustration 25.2

1 Raise the rear of the vehicle and support it securely on jackstands.

❊❊ WARNING:

If the vehicle is equipped with electronically modulated air suspension, make sure that the height control switch is turned off.

2   Remove the check/fill plug, then remove the drain plug and drain the lubricant (see illustration).

3   Reinstall the drain plug and tighten it securely.

4   Add new lubricant until it is even with the lower edge of the filler hole (see Section 4). See *Recommended lubricants and fluids* for the specified lubricant type.

5   Reinstall the check/fill plug and tighten it securely.

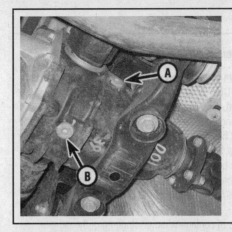

25.2 Location of the rear differential check/fill plug (A) and drain plug (B)

26   Cooling system servicing (draining, flushing and refilling) (at 100,000 miles [160,000 km] or 120 months and every 50,000 miles [80,500 km] or 60 months thereafter)

✳✳ WARNING 1:

Wait until the engine is completely cool before beginning this procedure.

WARNING 2:

Do not allow engine coolant (antifreeze) to come in contact with your skin or painted surfaces of the vehicle. Rinse off spills immediately with plenty of water. Antifreeze is highly toxic if ingested. Never leave antifreeze laying around in an open container or in puddles on the floor; children and pets are attracted by it's sweet smell and may drink it. Check with local authorities about disposing of used antifreeze. Many communities have collection centers which will see that antifreeze is disposed of safely.

1   Periodically, the cooling system should be drained, flushed and refilled to replenish the antifreeze mixture and prevent formation of rust and corrosion, which can impair the performance of the cooling system and cause engine damage. When the cooling system is serviced, all hoses and the radiator cap should be checked and replaced if necessary.

DRAINING

▶ **Refer to illustrations 26.4 and 26.5**

2   Apply the parking brake and block the wheels. If the vehicle has just been driven, wait several hours to allow the engine to cool down before beginning this procedure.

3   Once the engine is completely cool, remove the radiator cap.

4   Move a large container under the radiator drain to catch the coolant. Attach a 3/8-inch inner diameter hose to the drain fitting to direct the coolant into the container (some models are already equipped with a hose), then open the drain fitting (a pair of pliers may be required to turn it) (see illustration).

5   After the coolant stops flowing out of the radiator, move the container under the engine block drain plug(s). Loosen the plug(s) and allow the coolant in the block to drain. On four-cylinder models, the block drain plug is on the front side of the engine block. On V6 models, there's one on each side of the block (see illustration).

6   While the coolant is draining, check the condition of the radiator hoses, heater hoses and clamps (refer to Section 11 if necessary).

7   Replace any damaged clamps or hoses (see Chapter 3).

26.4 On most models you will have to remove a splash panel for access to the drain fitting located at the bottom of the radiator

26.5 The V6 engine has a coolant drain like this located on both the front and rear sides of the block

FLUSHING

▶ **Refer to illustration 26.10**

8   Once the system is completely drained, remove the thermostat from the engine (see Chapter 3). Then reinstall the thermostat housing without the thermostat. This will allow the system to be flushed.

9   Reinstall the engine block drain plug(s) and tighten the radiator drain plug. Turn your heating system controls to Hot, so that the heater core will be flushed at the same time as the rest of the cooling system.

10 Disconnect the upper radiator hose from the radiator. Place a garden hose in the upper radiator inlet, turn the water on and flush the system until the water runs clear out of the upper radiator hose (see illustration).

11 In severe cases of contamination or clogging of the radiator, remove the radiator (see Chapter 3) and have a radiator repair facility clean and repair it if necessary. Many deposits can be removed by the chemical action of a cleaner available at auto parts stores. Follow the procedure outlined in the manufacturer's instructions.

➡ **Note: When the coolant is regularly drained and the system refilled with the correct antifreeze/water mixture, there should be no need to use chemical cleaners or descalers.**

12 After flushing, drain the radiator and remove the block drain plugs once again to drain the water from the system.

REFILLING

13 Close and tighten the radiator drain. Install and tighten the block drain plug.

14 Place the heater temperature control in the maximum heat position.

15 Slowly add new coolant to the radiator until it's full. Add coolant to the reservoir up to the lower mark.

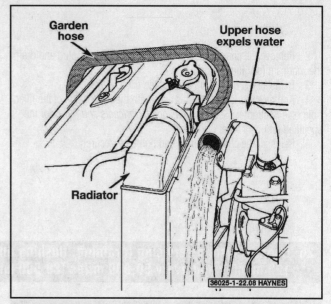

26.10 With the thermostat removed, disconnect the upper radiator hose and flush the radiator and engine block with a garden hose

16 Leave the radiator cap off and run the engine in a well-ventilated area until the thermostat opens (coolant will begin flowing through the radiator and the upper radiator hose will become hot).

17 Turn the engine off and let it cool. Add more coolant mixture to bring the level back up to the lip on the radiator filler neck.

18 Squeeze the upper radiator hose to expel air, then add more coolant mixture if necessary. Replace the radiator cap.

19 Start the engine, allow it to reach normal operating temperature and check for leaks.

27 Spark plug check and replacement (every 120,000 miles [193,000 km] or 144 months)

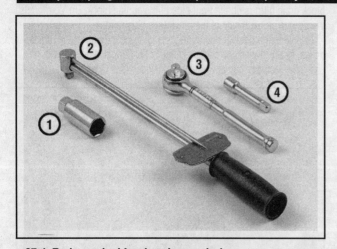

27.1 Tools required for changing spark plugs

1   *Spark plug socket* - This will have special padding inside to protect the spark plug's porcelain insulator
2   *Torque wrench* - Although not mandatory, using this tool is the best way to ensure the plugs are tightened properly
3   *Ratchet* - Standard hand tool to fit the spark plug socket
4   *Extension* - Depending on model and accessories, you may need special extensions and universal joints to reach one or more of the plugs

▶ **Refer to illustrations 27.1, 27.5, 27.7, 27.8, 27.9a and 27.9b**

➡ **Note: Do not adjust the gap on iridium spark plugs. Using a gapping tool on them could damage the iridium plating on the electrodes. These spark plugs are pre-gapped by the manufacturer.**

1   Spark plug replacement requires a spark plug socket that fits onto a ratchet. This socket is lined with a rubber grommet to protect the porcelain insulator of the spark plug and to hold the plug while you insert it into the spark plug hole (see illustration).

2   If you are replacing the plugs, purchase the new plugs and replace each plug one at a time.

➡ **Note: The manufacturer specifies that only iridium-tipped spark plugs be used on these models. When buying new spark plugs, it's essential that you obtain the correct plugs for your specific vehicle. This information can be found in the Specifications Section at the end of this Chapter, on the Vehicle Emissions Control Information (VECI) label located on the underside of the hood or in the owner's manual. If these sources specify different plugs, purchase the spark plug type specified on the VECI label because that information is provided specifically for your engine.**

3   Inspect each of the new plugs for defects. If there are any signs of

27.5 Remove the retaining bolt (A), disconnect the electrical connector (B) and detach the individual coils to reach the spark plugs

27.7 Because they are deeply recessed, the proper spark plug socket and an extension will be required when removing or installing the spark plugs

cracks in the porcelain insulator of a plug, don't use it.

4   Remove the engine cover(s) and disconnect any hoses or components that would interfere with access and move them out of the way. If you're working on a V6 model, to access the rear bank plugs, remove the upper intake manifold (see Chapter 2B).

5   Remove the bolts and detach each ignition coil assembly from the spark plugs (see illustration).

6   If compressed air is available, blow any dirt or foreign material away from the spark plug area before proceeding.

❋❋❋ WARNING:

Always wear eye protection when using compressed air!

7   Remove the spark plug (see illustration).

8   Whether you are replacing the plugs at this time or intend to reuse the old plugs, compare each old spark plug with those shown in this chart (see illustration) to determine the overall running condition of the engine.

A normally worn spark plug should have light tan or gray deposits on the firing tip.

A carbon fouled plug, identified by soft, sooty, black deposits, may indicate an improperly tuned vehicle. Check the air cleaner, ignition components and engine control system.

An oil fouled spark plug indicates an engine with worn piston rings and/or bad valve seals allowing excessive oil to enter the chamber.

This spark plug has been left in the engine too long, as evidenced by the extreme gap- Plugs with such an extreme gap can cause misfiring and stumbling accompanied by a noticeable lack of power.

A physically damaged spark plug may be evidence of severe detonation in that cylinder. Watch that cylinder carefully between services, as a continued detonation will not only damage the plug, but could also damage the engine.

A bridged or almost bridged spark plug, identified by a buildup between the electrodes caused by excessive carbon or oil build-up on the plug.

27.8 Inspect the spark plug to determine engine running conditions

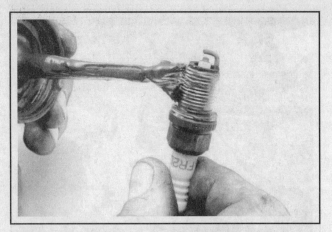

27.9a A light coat of anti-seize compound applied to the threads of the spark plugs will keep the threads in the cylinder head from being damaged the next time the plugs are removed

27.9b A section of rubber hose will aid in getting the spark plug threads started

9   Apply a small amount of anti-seize compound to the spark plug threads (see illustration). It's often difficult to insert spark plugs into their holes without cross-threading them. To avoid this possibility, fit a short piece of rubber hose over the end of the spark plug (see illustration). The flexible hose acts as a universal joint to help align the plug with the spark plug hole. Should the plug begin to cross-thread, the hose will slip on the spark plug, preventing thread damage. Tighten the plug to the torque listed in this Chapter's Specifications.

10   Attach the ignition coil assembly to the new spark plug.

Specifications

Recommended lubricants and fluids

➡Note: Listed here are manufacturer recommendations at the time this manual was written. Manufacturers occasionally upgrade their fluid and lubricant specifications, so check with your auto parts store for current recommendations.

Engine oil type	API "certified for gasoline engines"
Viscosity	SAE 5W-30
Fuel	
Four-cylinder engine	Unleaded fuel, 87 octane or higher
V6 engine	Unleaded fuel, 91 octane or higher
Coolant	Toyota Super Long Life Coolant or equivalent ethylene glycol based non-silicate/non-amine/non-nitrate/non-borate coolant with long-life Hybrid Organic Acid Technology (HOAT)
Highlander	
2005 and earlier models	Toyota ATF Type T-IV automatic transmission fluid
2006 models	Toyota ATF Type WS automatic transmission fluid
RX 300/330 models	Toyota ATF Type T-IV automatic transmission fluid
Brake fluid type	DOT 3 brake fluid
Power steering system fluid	DEXRON® II or III
Transfer case (4WD models)	API GL-5 SAE 80W-90 hypoid gear oil
Rear differential (4WD models)	API GL-5 SAE 80W-90 hypoid gear oil

Capacities

Engine oil (including filter)	
Four-cylinder engine	4.0 quarts (3.8 liters)
V6 engine	4.1 quarts (4.7 liters)
Coolant	
Highlander (2003 and earlier models)	
Four-cylinder engine	up to 6.4 quarts (6.4 liters)
V6 engine	up to 9.9 quarts (9.4 liters)
Highlander (2004 and later models)	
Four-cylinder engine	
With rear heater	up to 9.0 quarts (8.5 liters)
Without rear heater	up to 7.5 quarts (7.1 liters)
V6 engine	
With rear heater	up to 12.0 quarts (11.4 liters)
Without rear heater	up to 10.7 quarts (10.1 liters)
RX 300/330 models	up to 10.1 quarts (9.6 liters)
Automatic Transaxle (drain and refill)	up to 4.1 quarts (3.9 liters)
Transfer case (4WD models)	1 quart (0.9 liters)
Rear differential (4WD models)	1 quart (0.9 liters)

Ignition system

Spark plug	
Type 1	NGK IFR6A11 or equivalent
Type 2	Nippondenso SK20R11 or equivalent
Gap	0.043 inch (1.1 mm)

Ignition system (continued)

Engine firing order
    Four-cylinder engine   1-3-4-2
    V6 engine   1-2-3-4-5-6

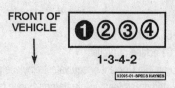

Cylinder numbering - four-cylinder engine

Cylinder numbering - V6 engine

Valve clearance (engine cold)

Four-cylinder engine
    Intake   0.008 to 0.011 inch (0.19 to 0.29 mm)
    Exhaust   0.012 to 0.016 inch (0.30 to 0.40 mm)
V6 engine
    Intake   0.006 to 0.010 inch (0.15 to 0.25 mm)
    Exhaust   0.010 to 0.014 inch (0.25 to 0.35 mm)

Brakes

Disc brake pad lining thickness (minimum)   1/16 inch (1.5 mm)

Torque specifications	Ft-lbs (unless otherwise indicated)	Nm
Automatic transaxle		
Pan bolts	67 in-lbs	7.6
Strainer bolts	96 in-lbs	11
Spark plugs		
Four-cylinder engine	168 in-lbs	19
V6 engine		
2003 and earlier models	156 in-lbs	18
2004 and later models	18.5	25
Wheel lug nuts	76	103

Section

Reference to other Chapters

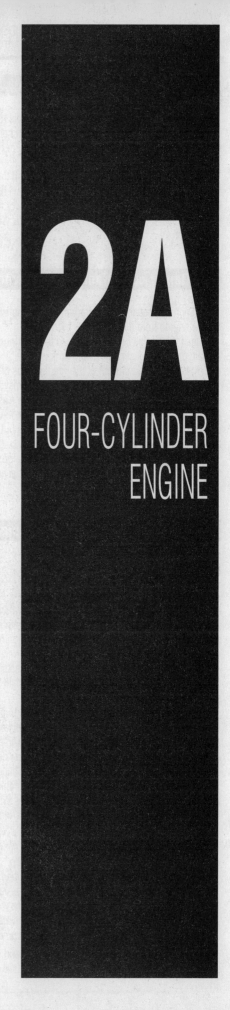

2A

FOUR-CYLINDER ENGINE

1 General information

This Part of Chapter 2 is devoted to in-vehicle repair procedures for the 2AZ-FE four-cylinder engine. Information concerning engine removal, installation and overhaul can be found in Part C of this Chapter.

The following repair procedures are based on the assumption that the engine is installed in the vehicle. If the engine has been removed from the vehicle and mounted on a stand, many of the steps outlined in this Part of Chapter 2 will not apply. The Specifications included in this Part of Chapter 2 apply only to the in-vehicle procedures contained in this Part.

This engine incorporates an aluminum cylinder block with a lower crankcase to strengthen the lower half of the block. The cylinder head utilizes dual overhead camshafts (DOHC) with four valves per cylinder. The camshafts are driven from a single timing chain off the crankshaft, and a Variable Valve Timing (VVT) system is incorporated on the intake camshaft to increase horsepower and decrease emissions.

2 Repair operations possible with the engine in the vehicle

Many major repair operations can be accomplished without removing the engine from the vehicle.

Clean the engine compartment and the exterior of the engine with some type of degreaser before any work is done. It will make the job easier and help keep dirt out of the internal areas of the engine.

Depending on the components involved, it may be helpful to remove the hood to improve access to the engine as repairs are performed (refer to Chapter 11 if necessary). Cover the fenders to prevent damage to the paint. Special pads are available, but an old bedspread or blanket will also work.

If vacuum, exhaust, oil or coolant leaks develop, indicating a need for gasket or seal replacement, the repairs can generally be made with the engine in the vehicle. The intake and exhaust manifold gaskets, oil pan gasket, crankshaft oil seals and cylinder head gasket are all accessible with the engine in place.

Exterior engine components, such as the intake and exhaust manifolds, the oil pan, the oil pump, the water pump, the starter motor, the alternator and the fuel system components can be removed for repair with the engine in place.

Since the cylinder head can be removed without pulling the engine, camshaft and valve component servicing can also be accomplished with the engine in the vehicle. Replacement of the timing chain and sprockets is also possible with the engine in the vehicle.

3 Top Dead Center (TDC) for number one piston - locating

♦ Refer to illustrations 3.5 and 3.8

1 Top Dead Center (TDC) is the highest point in the cylinder that each piston reaches as it travels up the cylinder bore. Each piston reaches TDC on the compression stroke and again on the exhaust stroke, but TDC generally refers to piston position on the compression stroke.

2 Positioning the piston(s) at TDC is an essential part of many procedures such as valve adjustment and camshaft and timing chain/sprocket removal.

3 Before beginning this procedure, be sure to place the transmission in Neutral and apply the parking brake or block the rear wheels. Also, disable the fuel injection system by relieving the fuel pressure (see Chapter 4) and the ignition system by disconnecting the primary electrical connectors at the ignition coils (see Chapter 5).

4 In order to bring any piston to TDC, the crankshaft must be turned using one of the methods outlined below. When looking at the front of the engine, normal crankshaft rotation is clockwise.

  a) The preferred method is to turn the crankshaft with a socket and ratchet attached to the bolt threaded into the front of the crankshaft. Turn the bolt in a clockwise direction.

  b) If an assistant is available to turn the ignition switch to the Start position in short bursts, you can get the piston close to TDC without a remote starter switch. Make sure your assistant is out of the vehicle, away from the ignition switch, then use a socket and ratchet as described in Paragraph (a) to complete the procedure.

5 Remove the spark plugs (see Chapter 1) and install a compression gauge in the number one spark plug hole (see illustration). It should be a gauge with a screw-in fitting and a hose at least six inches long.

6 Rotate the crankshaft using one of the methods described above while observing for pressure on the compression gauge. The moment the gauge shows pressure indicates that the number one cylinder has begun the compression stroke.

7 Once the compression stroke has begun, TDC for the compression stroke is reached by bringing the piston to the top of the cylinder.

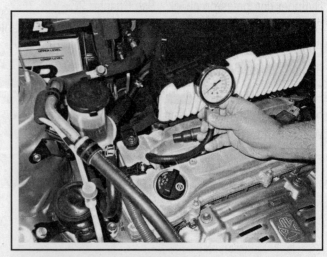

3.5 A compression gauge can be used in the number one spark plug hole to assist in finding TDC

8   Continue turning the crankshaft until the notch in the crankshaft damper is aligned with the "TDC" or the "0" mark on the timing chain cover (see illustration). At this point, the number one cylinder is at TDC on the compression stroke. If the marks are aligned but there was no compression, the piston was on the exhaust stroke; continue rotating the crankshaft 360-degrees (1-turn).

➡ **Note: If a compression gauge is not available, you can simply place a blunt object over the spark plug hole and listen for compression as the engine is rotated. Once compression at the No.1 spark plug hole is noted, the remainder of the Step is the same.**

9   After the number one piston has been positioned at TDC on the compression stroke, TDC for any of the remaining cylinders can be located by turning the crankshaft 180-degrees and following the firing order (refer to the Specifications). For example, rotating the engine 180-degrees past TDC #1 will put the engine at TDC compression for cylinder #3.

3.8 Align the notch in the damper with the "0" mark on the timing chain cover

---

4   Valve cover - removal and installation

REMOVAL

1   Disconnect the cable from the negative terminal of the battery (see Chapter 5, Section 1).
2   Remove the engine cover.
3   Disconnect the electrical connectors from the ignition coils, remove the nuts securing the wiring harness to the valve cover and position the ignition coil wiring harness aside. Then remove the ignition coil assemblies from each of the spark plugs (see Chapter 5).
4   Detach the PCV hoses from the valve cover.
5   Remove the valve cover mounting bolts and nuts (see illustration 4.7b), then detach the valve cover and gasket from the cylinder head. If the valve cover is stuck to the cylinder head, bump the end with a wood block and a hammer to jar it loose. If that doesn't work, try to slip a flexible putty knife between the cylinder head and valve cover to break the seal.

❊❊ CAUTION:

Don't pry at the valve cover-to-cylinder head joint or damage to the sealing surfaces may occur, leading to oil leaks after the valve cover is reinstalled.

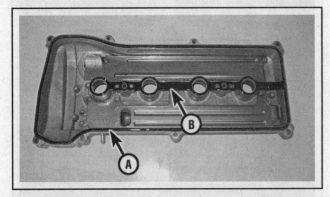

4.6 The valve cover gasket (A) and the spark plug tube seals (B) are incorporated into a single rubber O-ring-like seal - press the gasket evenly into the grooves around the underside of the valve cover and the spark plug openings

INSTALLATION

▶ **Refer to illustrations 4.6, 4.7a and 4.7b**

6   Remove the valve cover gasket from the valve cover and clean the mating surfaces with lacquer thinner or acetone. Install a new rubber gasket, pressing it evenly into the grooves around the underside of the valve cover.

➡ **Note: Make sure the spark plug tube seals are in place on the underside of the valve cover before reinstalling it (see illustration).**

The mating surfaces of the timing chain cover, the cylinder head and valve cover must be perfectly clean when the valve cover is installed. If there's residue or oil on the mating surfaces when the valve cover is installed, oil leaks may develop.

7   Apply RTV sealant at the timing chain cover-to-cylinder head joint, then install the valve cover and fasteners (see illustration).

4.7a Apply sealant at the timing chain cover-to-cylinder head joint before installing the valve cover

4.7b Valve cover mounting fastener locations and identification

8   Tighten the nuts/bolts to the torque listed in this Chapter's Specifications in three or four equal steps.

9   Reinstall the remaining parts, run the engine and check for oil leaks.

---

5   Variable Valve Timing (VVT) system - description and component replacement

▶ **Refer to illustrations 5.2a and 5.2b**

1   The VVT system varies intake camshaft timing by directing oil pressure to advance or retard the intake camshaft sprocket/actuator assembly. Changing the intake camshaft timing during certain engine conditions increases engine power output, fuel economy and reduces emissions.

2   System components include the Powertrain Control Module (PCM), the VVT oil control valve (OCV) and the intake camshaft sprocket/actuator assembly (see illustrations).

3   The PCM uses inputs from the following sensors to turn the oil control valve ON or OFF:

   a) *Vehicle Speed Sensor (VSS)*
   b) *Throttle Position Sensor (TPS)*
   c) *Mass Airflow (MAF) sensor*
   d) *Engine Coolant Temperature (ECT) sensor*

4   Once the VVT oil control valve is actuated by the PCM it directs the specified amount of oil pressure from the engine to advance or retard the intake camshaft sprocket/actuator assembly.

5   The intake camshaft sprocket/actuator assembly is equipped with an inner hub that is attached to the camshaft. The inner hub consists of a series of fixed vanes that use oil pressure as a wedge against the vanes to rotate the camshaft. The higher the oil pressure (or flow) the more the actuator assembly will rotate, thereby advancing or retarding the camshaft.

6   When oil is applied to the advance side of the vanes, the actuator can advance the camshaft up to 21 degrees in a clockwise direction. When oil is applied to the retard side of the vanes, the actuator will start to rotate the camshaft counterclockwise back to 0 degrees which is the normal position of the actuator during engine operation under no load or at idle. The PCM can also send a signal to the oil control valve to stop oil flow to both (advance and retard) passages to hold camshaft advance in its current position.

7   Under light engine loads, the VVT system will retard the camshaft timing to decrease valve overlap and stabilize engine output. Under medium engine loads, the VVT system will advance the camshaft timing to increase valve overlap, thereby increasing fuel economy and decreasing exhaust emissions. Under heavy engine loads at low RPM, the VVT system will advance the camshaft timing to help close the intake valve faster, which improves low to midrange torque. Under heavy engine

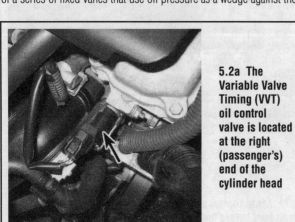

5.2a The Variable Valve Timing (VVT) oil control valve is located at the right (passenger's) end of the cylinder head

5.2b The intake camshaft actuator assembly is only visible with the valve cover removed

loads at high RPM, the VVT system will retard the camshaft timing to slow the closing of the intake valve to improve engine horsepower.

COMPONENT REPLACEMENT

➡**Note 1: A problem in the VVT oil control valve circuit will set a diagnostic trouble code and turn on the CHECK ENGINE light on the dash. Refer to Chapter 6 for accessing trouble codes.**

➡**Note 2: Most problems in the VVT system originate from the oil control valve and filter. Regular engine oil and filter changes are necessary for trouble-free operation of the OCV.**

Oil control valve (OCV) filter

8   A clogged OCV filter screen is often the cause of VVT system problems. Remove the OCV filter from the rear of the cylinder head and then inspect the filter for clogging. Clean the filter if necessary and reinstall it using a new O-ring. Make sure that the big end of the filter faces toward the head.

Oil control valve (OCV)

9   To replace the OCV, remove the hold-down bolt and pull the OCV out of the cylinder head. There is one valve at the front (timing chain) of the cylinder head. Use a new O-ring when installing the new OCV. Be sure to tighten the OCV hold-down bolt to the torque listed in this Chapter's Specifications.

Camshaft sprocket/actuator assembly

10   Remove the valve cover (see Section 4), the timing chain (see Section 6) and the intake camshaft (see Section 7).

11   Mount the camshaft in a bench vise. Secure the camshaft in the vise by clamping up the hexagonal nut part of the cam.

✳✳ **CAUTION:**

Be careful not to damage the camshaft or any of the cam lobes or journals.

12   Verify that the sprocket/actuator will not rotate from the locked position. The locked position is a neutral position in which the actuator is placed during idle and no load conditions, and anytime that the VVT system is not activated by the PCM.

13   Using brake system cleaner, remove all traces of oil from the front cam journals and the VVT oil control orifices. Apply vinyl tape over all the oil control orifices except the advance side oil port.

14   Apply 21 psi of air pressure to the advance side oil port and try to rotate the actuator assembly by hand. The actuator should rotate freely, with no obvious binding in the advance angle direction from the locked position.

➡**Note: It is critical to have an air tight seal between the air gun nozzle and the advance oil port hole, because if air leaks out at the air nozzle, or at any of the other oil control orifices, the lock pin in the actuator won't be forced out of its locating hole. If leakage occurs, apply a little more air pressure to the advance side oil port to force the lock pin from the locating hole.**

15   If the actuator does not rotate freely as described, replace the Intake camshaft sprocket/actuator assembly.

16   To replace the sprocket/actuator assembly, put the camshaft in a bench vise and remove the nut.

✳✳ **CAUTION:**

Do not remove the four bolts from the VVT actuator.

17   Remove the sprocket/actuator assembly. If it's hard to pull off the camshaft, tap it lightly with a plastic-tip hammer.

✳✳ **CAUTION:**

Do NOT try to disassemble the sprocket/actuator assembly - it's NOT rebuildable.

18   Lubricate the sprocket/actuator seating surface on the camshaft with clean engine oil. Install the sprocket/actuator onto the camshaft with the groove aligned with the straight pin. When the pin locks into the groove, further rotate the sprocket/actuator to the right side (retarded angle side). Install the sprocket/actuator assembly retaining nut, tighten the retaining nut to the torque listed in this Chapter's Specifications.

➡**Note: Check that the sprocket/actuator can move toward the retarded angle side and is locked in the extreme retarded side.**

19   Install the intake camshaft (see Section 7), the timing chain (see Section 6) and the valve cover (see Section 4).

6   Timing chain and sprockets - removal, inspection and installation

✳✳ **WARNING:**

Wait until the engine is completely cool before beginning this procedure.

➡**Note: Special tools are required for this procedure. Read through the entire procedure and acquire the necessary tools and equipment before beginning work.**

REMOVAL

♦ **Refer to illustrations 6.5, 6.8, 6.12, 6.14a, 6.14b, 6.15a, 6.15b, 6.15c, 6.15d, 6.16 and 6.17**

1   Detach the cable from the negative terminal of the battery (see Chapter 5, Section 1).

2   Remove the drivebelt (see Chapter 1) and the alternator (see Chapter 5).

6.5 Location of the engine splash shield fasteners

3   Remove the valve cover (see Section 4).

4   With the parking brake applied and the rear wheels blocked, loosen the right front wheel lug nuts, then raise the front of the vehicle and support it securely on jackstands. Remove the right front wheel and the right splash shield from the wheelwell.

❋❋ WARNING:

If the vehicle is equipped with electronically modulated air suspension, make sure that the height control switch is turned off.

5   Remove the engine splash shield (see illustration). Remove the fender splash shields (see Chapter 11).

6   Drain the cooling system (see Chapter 1). While the coolant is draining, refer to Chapter 10 and remove the power steering pump from the engine without disconnecting the fluid lines. Tie the power steering pump to the body with a piece of wire and position it out of the way.

7   Disconnect the front exhaust pipe from the exhaust manifold/catalytic converter assembly (see Chapter 4).

8   Position the number one piston at TDC on the compression stroke (see Section 3). Visually confirm the engine is at TDC on the compression stroke by verifying that the timing mark on the crankshaft pulley/vibration damper is aligned with the "0" mark on the timing chain cover and the camshaft sprocket marks are aligned and parallel with the top of the timing chain cover (see illustration 3.8).

6.8 Verify the engine is at TDC by observing the position of the camshaft sprocket marks (lower arrows) - they must be aligned with the marks on the camshaft bearing caps (upper arrows)

➡Note: There are two sets of marks on the camshaft sprockets. The marks that align at TDC are for TDC reference only; the other two marks are used to align the sprockets with the timing chain during installation.

9   Remove the crankshaft pulley/vibration damper, being careful not to rotate the engine from TDC (see Section 11). If the engine rotates off TDC during this step, reposition the engine back to TDC before proceeding. The engine should be left at TDC for the No. 1 piston during this entire procedure.

10 Support the engine from above, using an engine support fixture (available at rental yards).

11 Remove the passenger side engine mount and engine mount control rod (see Section 17). Remove the front engine mount and the transaxle mount (see Section 17).

12 Remove the drivebelt tensioner (see illustration) and the crankshaft position sensor from the timing chain cover. Also remove the bolt securing the crankshaft position sensor wiring harness to the timing chain cover.

13 Remove the oil pan (see Section 13).

14 Detach the main wiring harness junction and remove the timing chain tensioner from the rear side of the timing chain cover (see illustrations).

6.12 Drivebelt tensioner mounting fasteners

6.14a Remove the two bolts securing the main harness to the timing chain cover and position the harness aside

6.14b Timing chain tensioner mounting nuts

6.15a Timing chain cover upper fasteners

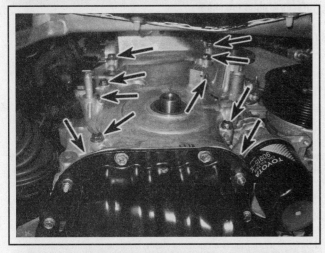

6.15b Timing chain cover lower fasteners - make a note of the fastener sizes, locations and lengths as you're removing them, as they must be installed back in their original positions

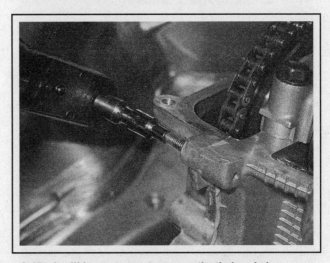

6.15c It will be necessary to remove the timing chain cover studs . . .

15 Remove the timing chain cover fasteners and pry the cover off the engine (see illustrations).

16 Slide the crankshaft position sensor reluctor ring off the crankshaft (see illustration).

6.15d . . . before prying the timing chain cover off the engine

6.16 Slide the crankshaft position sensor reluctor ring off the crankshaft - note the "F" mark on the front (it must be facing outward upon installation)

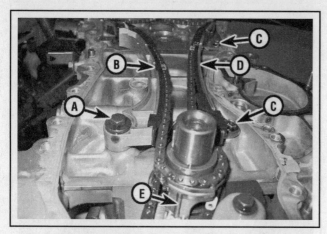

6.17 Timing chain guide mounting details

 A Pivot bolt
 B Tensioner pivot arm/chain guide
 C Stationary chain guide mounting bolts
 D Stationary chain guide
 E Lower timing chain guide

 17 Remove the timing chain tensioner pivot arm/chain guide and the lower chain guide (see illustration).

 18 Lift the timing chain off the camshaft sprockets and remove the timing chain and the crankshaft sprocket as an assembly from the engine. The crankshaft sprocket should slip off the crankshaft by hand. If not, carefully pry the sprocket off the crankshaft.

➡**Note: If you intend to reuse the timing chain, use white paint or chalk to make a mark indicating the front of the chain. If a used timing chain is reinstalled with the wear pattern in the opposite direction, noise and increased wear may occur.**

 19 Remove the stationary timing chain guide (see illustration 6.17).

 20 To remove the camshaft sprockets, loosen the bolts while holding the lug on the camshaft with a wrench, on the hex portion of the camshaft only. Note the identification marks on the camshaft sprockets before removal, then remove the bolts. Pull on the sprockets by hand until they slip off the dowels. If necessary, use a small puller, with the legs inserted in the relief holes, to pull the sprockets off.

➡**Note: These models are equipped with variable valve timing, which consists of an actuator assembly attached to the intake camshaft sprocket. When removing the intake camshaft sprocket on these models only loosen and remove the center bolt, which fastens the sprocket to the camshaft. Do not loosen the outer four bolts that secure the actuator to the sprocket.**

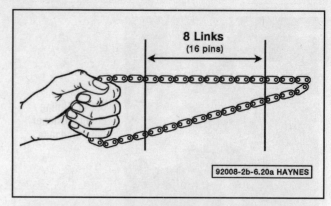

6.21a Timing chain stretch is measured by checking the length of the chain between 8 links (16 pins) at three or more places around the chain

INSPECTION

▶ **Refer to illustrations 6.21a, 6.21b and 6.22**

 21 Visually inspect all parts for wear and damage. Check the timing chain for loose pins, cracks, worn rollers and side plates. Check the sprockets for hook-shaped, chipped and broken teeth. Also check the timing chain for stretching and the diameter of the timing sprockets for wear with the chain assembled on the sprockets. Timing chain stretch is measured by checking the length of the chain between 8 links (16 pins) at 3 or more places (selected randomly) around the chain - if chain stretch exceeds the specifications between any 8 links, the chain must be replaced (see illustration). Be sure to measure across the chain rollers when checking the sprocket diameter and to measure chain stretch at three or more places around the chain. Maximum chain elongation and minimum sprocket diameter (with chain) should not exceed the amount listed in this Chapter's Specifications (see illustration). Replace the timing chain and sprockets as a set if the engine has high mileage or fails inspection.

 22 Check the chain guides for excessive wear (see illustration). Replace the chain guides if scoring or wear exceeds the amount listed in this Chapter's Specifications. Note that some scoring of the timing chain guide shoes is normal. If excessive wear is indicated, it will also be necessary to inspect the chain guide oil hole on the front of the block for clogging (see illustration 14.5). The oil pump is driven by a second chain driven from the crankshaft sprocket. If the timing chain is to be replaced, replace the oil pump drive chain at the same time (see Section 14).

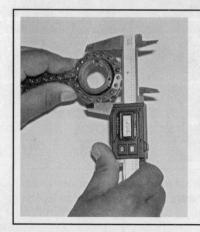

6.21b Wrap the chain around each of the timing sprockets and measure the diameter of the sprockets across the chain rollers - if the measurement is less than the minimum sprocket diameter, the chain and the timing sprockets must be replaced

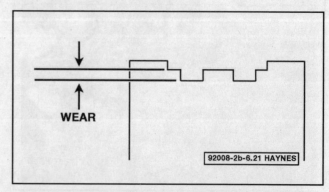

6.22 Timing chain guide wear is measured from the top of the chain contact surface to the bottom of the wear grooves

6.28 Loop the timing chain around the crankshaft sprocket (crankshaft keyway straight up) and align the No.1 colored link with the mark on the crankshaft sprocket, then install the chain and crankshaft sprocket as an assembly on the engine and install the lower chain guide

INSTALLATION

▶ **Refer to illustrations 6.28, 6.29, 6.30, 6.33a, 6.33b, 6.35, 6.36, 6.38 and 6.40**

23 Remove all traces of old sealant from the timing chain cover and the mating surfaces of the engine block and cylinder head.

24 Make sure the camshafts are positioned with the dowel pins at the top in the 12 o'clock position, then install both camshaft sprockets in their original locations by aligning the dowel pin hole on the rear of the sprockets with dowel pin on the camshaft. Apply medium strength thread locking compound to the camshaft sprocket bolt threads and make sure the washers are in place. Hold the camshaft from turning as described in Step 20 and tighten the bolts to the torque listed in this Chapter's Specifications.

25 Rotate the camshafts as necessary to align the TDC marks on camshaft sprockets (see illustration 6.8).

26 If the crankshaft has been rotated off TDC during this procedure, it will be necessary to rotate the crankshaft until the keyway is pointing straight up in the 12 o'clock position with the centerline of the cylinder bores.

27 Install the stationary timing chain guide (see illustration 6.17).

28 Loop the timing chain around the crankshaft sprocket and align the No.1 colored link (blue or orange) with the mark on the crankshaft

6.30 After the tensioner pivot arm is installed, make sure the tab on the pivot arm can't move past the stopper on the cylinder head

6.29 Loop the timing chain up over the exhaust camshaft and around the intake camshaft, while aligning the remaining two colored links with the marks on the camshaft sprockets

sprocket. Install the chain and crankshaft sprocket as an assembly on the engine, then install the lower timing chain guide (see illustration)

➡**Note: There are three colored links on the timing chain. The No.1 colored link is the link farthest away from the two colored links that are closest together.**

29 Slip the timing chain into the lip of the stationary timing chain guide and over the exhaust camshaft sprocket, then around the intake camshaft sprocket making sure to align the remaining two colored links with the marks on the camshaft sprockets (see illustration). Make sure to remove all slack from the right side of the chain when doing so.

30 Use one hand to remove the slack from the left side of the chain and install the timing chain tensioner pivot arm/chain guide. Tighten the pivot bolt to the torque listed in this Chapter's Specifications. After installation, make sure the tab on the pivot arm can't move past the stopper on the cylinder head (see illustration).

31 Reconfirm that the number one piston is still at TDC on the compression stroke and that the timing marks on the crankshaft and camshaft sprockets are aligned with the colored links on the chain.

32 Install the crankshaft position sensor reluctor ring with the "F" mark facing outward.

33 Apply a bead of RTV sealant to the timing chain cover sealing

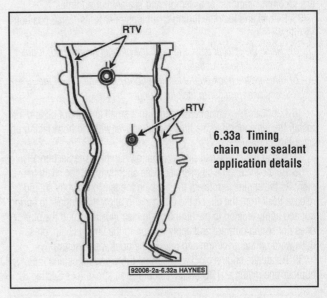

6.33a Timing chain cover sealant application details

6.33b Apply a bead of sealant on each side of the parting line between the cylinder head and the engine block

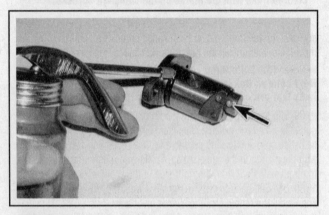

6.36 Apply a small amount of oil to the tensioner O-ring and insert the tensioner into the timing chain cover with the hook facing upward

6.35 Raise the ratchet pawl and push the plunger inward until the hook on the tensioner body can be engaged with the pin on the plunger to lock the plunger in place

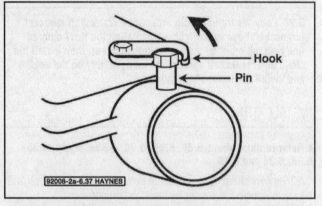

6.38 Rotate the engine counterclockwise to disengage the hook from the plunger pin on the tensioner, then rotate it clockwise and confirm that the plunger has extended outward against the pivot arm/chain guide

Visually confirm that the timing mark on the crankshaft pulley/vibration damper is aligned with the "0" mark on the timing chain cover and the camshaft sprocket marks are aligned and parallel with the top of the timing chain cover as shown in illustration 6.8.

40 Tighten the timing chain cover fasteners to the torque listed in this Chapter's Specifications (see illustration). The remainder of the installation is the reverse of removal.

surfaces, where the two central bolts go and around the perimeter of the cover's flange (see illustrations). Place the timing chain cover in position on the engine and install the bolts in their original locations.

34 Tighten the bolts evenly in several steps to the torque listed in this Chapter's Specifications. Be sure to follow the sealant manufacturer's recommendations for assembly and sealant curing times.

35 Reload and lock the timing chain tensioner to its "zero" position as follows:

  a) *Raise the ratchet pawl and push the plunger inward until it bottoms out (see illustration).*

  b) *Engage the hook on the tensioner body with the pin on the tensioner plunger to lock the plunger in place.*

36 Lubricate the tensioner O-ring with a small amount of oil and install the tensioner into the timing chain cover with the hook facing up (see illustration).

37 Install the crankshaft pulley/vibration damper (see Section 11).

38 Rotate the engine counterclockwise slightly to set the chain tension. As the engine is rotated, the hook on the tensioner body should release itself from the pin on the plunger and allow the plunger to spring out and apply tension to the timing chain (see illustration). If the plunger does not spring outward and apply tension to the timing chain, press downward on the pivot arm and release the hook with a screwdriver.

39 Rotate the engine clockwise several turns and reposition the number one piston at TDC on the compression stroke (see Section 3).

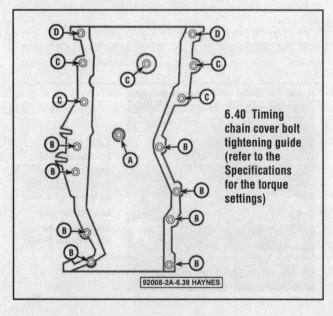

6.40 Timing chain cover bolt tightening guide (refer to the Specifications for the torque settings)

7 Camshafts and lifters - removal, inspection and installation

➡Note: The camshafts should always be thoroughly inspected before installation and camshaft endplay should always be checked prior to camshaft removal (see Step 13).

REMOVAL

▶ **Refer to illustrations 7.4, 7.8, 7.10, 7.11a and 7.11b**

1 Disconnect the cable from the negative terminal of the battery (see Chapter 5, Section 1).

2 Remove the valve cover (see Section 4).

3 Refer to Section 3 and place the engine on TDC for number 1 cylinder. Visually confirm the engine is at TDC on the compression stroke by verifying that the timing mark on the crankshaft pulley/vibration damper is aligned with the "0" mark on the timing chain cover and the camshaft sprocket TDC marks are aligned and parallel with the top of the timing chain cover (see illustrations 3.8 and 6.8).

4 With the TDC marks aligned, apply a dab of paint to the timing chain links where they meet the upper timing marks on the camshaft sprockets (see illustration).

➡Note: There are two sets of marks on the camshaft sprockets. The marks that align at TDC are for TDC reference only, the other two marks are used to align the sprockets with the timing chain during installation.

5 Remove the timing chain tensioner from the timing chain cover (see illustrations 6.14a and 6.14b) and the camshaft position sensor from the cylinder head (see Chapter 6).

6 Using a large wrench (on the hex portion of the camshaft only) to hold the camshaft from turning, loosen the exhaust camshaft sprocket bolt several turns. If the camshaft sprockets have rotated during the bolt loosening process, rotate the engine clockwise until the "TDC" marks on the cam sprockets are realigned.

➡Note: If intake camshaft sprocket (VVT actuator) removal is necessary, follow the procedure and remove the camshaft sprocket bolt later with the assembly on the bench.

7 Remove the exhaust camshaft sprocket retaining bolt. Loosen the exhaust camshaft cap bolts in two or three steps starting with the outer camshaft caps and work toward the center (opposite of the tightening sequence). Remove the cap bolts and the exhaust camshaft from

7.4 With the TDC marks aligned, apply a dab of paint to the timing chain links where they meet the upper timing marks on the camshaft sprockets (A) - the other two marks (B) are used to align the sprockets with the timing chain during installation

the cylinder head. Disengage the timing chain from the sprocket and remove the exhaust camshaft sprocket from the engine.

8 Verify the markings on the intake camshaft bearing caps. The caps should be marked from 1 to 5 with an "I" mark on the cap indicating they're for the intake camshaft (see illustration).

✳✳ CAUTION:

Keep the caps in order. They must go back in the same location they were removed from.

9 Loosen the intake camshaft bearing cap bolts in two or three steps, starting with the outer caps and work toward the center (reverse order of the tightening sequence) (see illustration 7.22).

10 Remove the intake camshaft bearing caps, then remove the intake camshaft from the cylinder head. Mark the camshaft(s) "Intake" or "Exhaust" to avoid mixing them up.

7.8 The camshaft bearing caps are numbered and have an arrow that should face the timing chain end of the engine

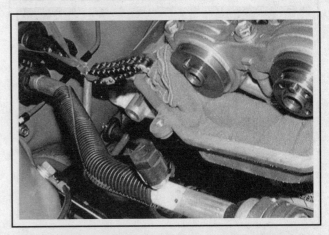

7.10 Hang the timing chain out of the way with a piece of wire and place a shop rag in the timing chain cover opening to prevent foreign objects from falling into the engine

7.11a Mark the lifters (I for intake, E for exhaust, and number their location) and remove them with a magnetic retrieval tool

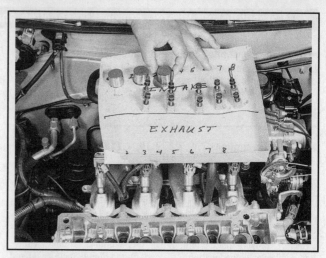

7.11b Mark up a cardboard box to store the lifters and bearing caps in order

→Note: When looking at the engine from the front of the vehicle, the forward facing cam is the exhaust camshaft and the cam nearest the firewall is the intake camshaft. It is very important that the camshafts are returned to their original locations during installation.

After removing the camshafts, hang the timing chain up with a piece of wire and attach it to an object on the firewall (see illustration). This will prevent the timing chain from falling into the engine as the remaining steps in this procedure are performed. Also place a rag into the opening of the timing chain cover to prevent any foreign objects from falling into the engine.

11 Remove the lifters from the cylinder head, keeping them in order with their respective valve and cylinder (see illustrations).

✳✳ CAUTION:

Keep the lifters in order. They must go back in the position from which they were removed.

12 Inspect the camshafts, camshaft bearings and lifters as described below. Also inspect the camshaft sprockets for wear on the teeth.

Inspect the chains for cracks or excessive wear of the rollers, and for stretching (see Section 6). If any of the components show signs of excessive wear they must be replaced.

INSPECTION

▶ Refer to illustrations 7.13, 7.14, 7.15, 7.16, 7.18 and 7.19

13 Before the camshafts are removed from the engine, check the camshaft endplay by placing a dial indicator with the stem in line with the camshaft and touching the snout (see illustration). Push the camshaft all the way to the rear and zero the dial indicator. Next, pry the camshaft to the front as far as possible and check the reading on the dial indicator. The distance it moves is the endplay. If the endplay for the intake camshaft is greater than the Specifications listed in this Chapter, check the thrust surfaces of the No.1 journal bearing for wear. If the thrust surface is worn, the bearings must be replaced. If the endplay for the exhaust camshaft is greater than the Specifications listed in this Chapter, the camshaft or the cylinder head (or both) may need to be replaced.

14 With the camshafts removed, visually check the camshaft bear-

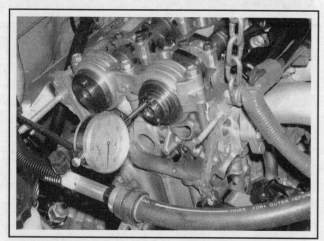

7.13 Mount a dial indicator as shown to measure camshaft endplay - pry the camshaft forward and back and read the endplay on the dial

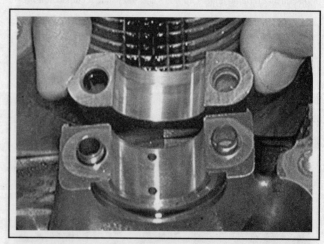

7.14 Inspect the No.2 through No.5 cam bearing surfaces in the cylinder head for pits, score marks and abnormal wear - if wear or damage is noted, the cylinder head must be replaced

7.15 Measure each journal diameter with a micrometer - if any journal measures less than the specified limit, replace the camshaft

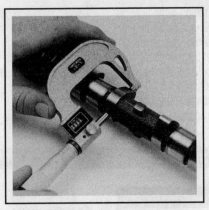

7.16 Measure the lobe heights on each camshaft - if any lobe height is less than the specified allowable minimum, replace that camshaft

7.18 Wipe off the oil and inspect each lifter for wear and scuffing

ing surfaces in the cylinder head for pitting, score marks, galling and abnormal wear (see illustration). If the bearing surfaces are damaged, the cylinder head or the No.1 journal bearings of the intake camshaft may have to be replaced. If so, it will be necessary to replace them with the same size as originally installed.

➡**Note: The size markings are located on the machined surface of the cylinder head near the no. 1 journals, and the corresponding marks are located on the back sides of the bearing shells.**

15 Measure the outside diameter of each camshaft bearing journal and record your measurements (see illustration). Compare them to the journal outside diameter specified in this Chapter, then measure the inside diameter of each corresponding camshaft bearing and record the measurements. Subtract each cam journal outside diameter from its respective cam bearing bore inside diameter to determine the oil clearance for each bearing. Compare the results to the specified journal-to-bearing clearance. If any of the measurements fall outside the standard specified wear limits in this Chapter, either the camshaft or the cylinder head, or both, must be replaced.

➡**Note: If precision measuring tools are not available, Plastigage may be used to determine the bearing journal oil clearance.**

16 Using a micrometer, measure the height of each camshaft lobe (see illustration). Compare your measurements with this Chapter's Specifications. If the height for any one lobe is less than the specified minimum, replace the camshaft.

17 Check the camshaft runout by placing the camshaft back into the cylinder head and set up a dial indicator on the center journal. Zero

the dial indicator. Turn the camshaft slowly and note the dial indicator readings. Runout should not exceed 0.0012 inch (0.03 mm). If the measured runout exceeds the specified runout, replace the camshaft.

18 Inspect each lifter for scuffing and score marks (see illustration).

19 Measure the outside diameter of each lifter (see illustration) and the corresponding lifter bore inside diameter. Subtract the lifter diameter from the lifter bore diameter to determine the oil clearance. Compare it to this Chapter's Specifications. If the oil clearance is excessive, a new cylinder head and/or new lifters will be required.

INSTALLATION

▶ **Refer to illustration 7.22**

20 If the No.1 journal bearings for the intake camshaft were removed or replaced, install them into the cylinder head and the bearing cap now. Apply moly-based engine assembly lubricant to the camshaft lobes and journals and install the camshaft into the cylinder head with the No.1 cylinder camshaft lobes pointing outward away from each other, and the dowel pins facing upward. If the old camshafts are being used, make sure they're installed in the same location from which they came.

21 Install the bearing caps and bolts and tighten them hand tight.

22 Tighten the bearing cap bolts in several equal steps, to the torque listed in this Chapter's Specifications, using the proper tightening sequence (see illustration).

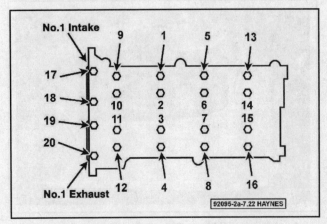

7.22 Camshaft bearing cap bolt TIGHTENING sequence

7.19 Measure the outside diameter of each lifter and the inside diameter of each lifter bore to determine the oil clearance measurement

23 Engage the camshaft sprocket teeth with the timing chain links so that the match marks made during removal align with the upper timing marks on the sprockets, then position the sprockets over the dowels on the camshaft hubs and install the camshaft sprocket bolts finger tight. At this point, the mark on the crankshaft pulley should be aligned with the "0" mark on the timing chain cover, the camshaft sprocket TDC marks should be aligned and parallel with the top of the timing chain cover and the timing chain match marks should be aligned with the upper timing sprocket marks with all of the slack in the chain positioned towards the tensioner side of the engine (see illustration 7.4).

24 Double check that the timing sprockets are returned to the proper camshaft and tighten the camshaft sprocket bolts to the torque listed in this Chapter's Specifications.

25 Install the timing chain tensioner as described in Section 6, Steps 38 and 39.

26 The remainder of installation is the reverse of removal.

8 Intake manifold - removal and installation

✳✳ WARNING:

Wait until the engine is completely cool before beginning this procedure.

REMOVAL

▸ **Refer to illustrations 8.4, 8.8 and 8.9**

1   Relieve the fuel system pressure (see Chapter 4), then disconnect the negative cable from the battery (see Chapter 5, Section 1).

8.4 Label and disconnect the vacuum hose(s) (A) and the wire harness retainer(s) (B) from the intake manifold

2   Remove the air intake duct and resonator (see Chapter 4).

3   Remove the fuel rail and injectors as an assembly (see Chapter 4). Remove the throttle body (see Chapter 4).

4   Label and detach the PCV and vacuum hoses connected to the intake manifold (see illustration). Raise the vehicle and support it securely on jackstands.

✳✳ WARNING:

If the vehicle is equipped with electronically modulated air suspension, make sure that the height control switch is turned off.

➡Note: If you're working on a 4WD model, loosen the right-side driveaxle/hub nut (see Chapter 8) and the right front wheel lug nuts before raising the vehicle.

4WD models

5   Drain the transfer case lubricant (see Chapter 1).

6   Remove the right-side driveaxle (see Chapter 8).

7   Separate the transfer case mount from the transfer case. Remove the transfer case mounting brace.

All models

8   Working below the vehicle, remove the manifold lower mounting bolts (see illustration).

9   Working from above, remove the intake manifold mounting nuts and bolts. Remove the manifold, the gasket and the manifold insulator from the engine (see illustration).

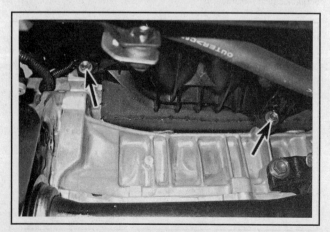

8.8 Intake manifold lower fastener locations

8.9 Intake manifold upper fastener locations

INSTALLATION

▶ **Refer to illustration 8.11**

10 Clean the mating surfaces of the intake manifold and the cylinder head mounting surface with lacquer thinner or acetone. If the gasket shows signs of leaking, check the manifold for warpage with a straight-edge and feeler gauges. If the manifold is warped it must be replaced.

11 Press a new gasket into the grooves on the intake manifold (see illustration). Install the manifold and gasket over the studs on the cylinder head.

12 Tighten the manifold-to-cylinder head nuts/bolts in three or four equal steps to the torque listed in this Chapter's Specifications. Work from the top bolts down to avoid warping the manifold.

13 Install the remaining parts in the reverse order of removal. Check the coolant level, adding as necessary (see Chapter 1).

14 On 4WD models, refill the transfer case (see Chapter 1).

15 Before starting the engine, check the accelerator cable, if equipped, for smooth operation.

8.11 Press the gasket into the groove on the intake manifold

16 Run the engine and check for coolant and vacuum leaks.

17 Road test the vehicle and check for proper operation of all accessories, including the cruise control system, if equipped.

---

9   Exhaust manifold/catalytic converter assembly - removal and installation

❋❋ **WARNING:**

The engine must be completely cool before beginning this procedure.

❋❋ **WARNING:**

If the vehicle is equipped with electronically modulated air suspension, make sure that the height control switch is turned off.

REMOVAL

▶ **Refer to illustrations 9.3, 9.5 and 9.6**

1 Disconnect the negative cable from the battery (see Chapter 5, Section 1).

2 Raise the front of the vehicle and support it securely on jackstands. Working below the vehicle, remove the engine splash shield (see illustration 6.5).

3 Apply penetrating oil to the bolts and springs retaining the exhaust pipe to the manifold. After the bolts have soaked, remove the bolts retaining the exhaust pipe to the manifold. Separate the front exhaust pipe from the manifold, being careful not to damage the oxygen sensor (see illustration).

4 Unbolt the lower exhaust manifold braces and remove them from the engine. Also disconnect the oxygen sensor connectors.

5 Working in the engine compartment, remove the upper heat shield from the manifold (see illustration).

9.3 Working below the vehicle, remove the exhaust pipe-to-manifold mounting bolts (A) and lower the front exhaust pipe. Be careful not to damage the oxygen sensor (C) - (B) indicates the mounting bolts for the exhaust manifold lower brace

9.5 Working from the engine compartment, remove the upper heat shield mounting bolts . . .

9.6 . . . and the exhaust manifold retaining nuts, then pull the manifold off the studs on the cylinder head and remove from above

6   Remove the nuts/bolts and detach the manifold and gasket (see illustration).

INSTALLATION

7   Use a scraper to remove all traces of old gasket material and carbon deposits from the manifold and cylinder head mating surfaces. If the gasket shows signs of leaking, check the manifold for warpage with a straight edge. If the manifold is warped it must be replaced.

8   Position a new gasket over the cylinder head studs, noting any directional marks or arrows on the gasket that may be present.

9   Install the manifold and thread the mounting nuts into place.

10 Working from the center out, tighten the nuts/bolts to the torque listed in this Chapter's Specifications in three or four equal steps.

11 Reinstall the remaining parts in the reverse order of removal.

12 Run the engine and check for exhaust leaks.

10 Cylinder head - removal and installation

❊❊ WARNING:

The engine must be completely cool before beginning this procedure.

REMOVAL

1   Relieve the fuel system pressure (see Chapter 4), then disconnect the cable from the negative terminal of the battery (see Chapter 5, Section 1).

2   Remove the strut brace from the engine compartment, if equipped (see Chapter 10, Section 3).

3   Drain the engine coolant (see Chapter 1).

4   Detach the coolant hoses from the cylinder head (see Chapter 3).

5   Remove the drivebelt (see Chapter 1).

6   Remove the alternator (see Chapter 5).

7   Remove the intake manifold (see Section 8) and the intake manifold insulator.

8   Remove the valve cover (see Section 4).

9   Remove the throttle body, fuel injectors and fuel rail (see Chapter 4).

10 Remove the exhaust manifold (see Section 9) and the exhaust manifold brackets.

11 Remove the timing chain and camshaft sprockets (see Section 6).

➡**Note: Instead of supporting the engine from above with an engine support fixture or hoist, it will have to be supported from below with a floor jack and block of wood, after the pressed-steel portion of the oil pan has been removed.**

12 Remove the camshafts and lifters (see Section 7).

13 Remove the variable valve timing control valve (see illustration 5.2b).

14 Label and detach the electrical connections from the cylinder head.

15 Using a 10 mm bi-hexagon bit and a breaker bar, loosen the cylinder head bolts in 1/4-turn increments until they can be removed by hand. Loosen the cylinder head bolts in the reverse order of the recommended tightening sequence (see illustration 10.27) to avoid warping

or cracking the cylinder head.

16 Lift the cylinder head off the engine block. If it's stuck, very carefully pry up at the transaxle end, beyond the gasket surface.

17 Remove any remaining external components from the cylinder head to allow for thorough cleaning and inspection.

INSTALLATION

▸ **Refer to illustration 10.27**

18 The mating surfaces of the cylinder head and block must be perfectly clean when the cylinder head is installed.

19 Use a gasket scraper to remove all traces of carbon and old gasket material, then clean the mating surfaces with lacquer thinner or acetone. If there's oil on the mating surfaces when the cylinder head is installed, the gasket may not seal correctly and leaks could develop. When working on the block, stuff the cylinders with clean shop rags to keep out debris. Use a vacuum cleaner to remove material that falls into the cylinders.

20 Check the block and cylinder head mating surfaces for nicks, deep scratches and other damage. If damage is slight, it can be removed with a file; if it's excessive, machining may be the only alternative.

21 Use a tap of the correct size to chase the threads in the cylinder head bolt holes, then clean the holes with compressed air - make sure that nothing remains in the holes.

❊❊ WARNING:

Wear eye protection when using compressed air!

22 Using a wire brush, clean the threads on each bolt to remove corrosion and restore the threads. Dirt, corrosion, sealant and damaged threads will affect torque readings. Measure each bolt, from the underside of the head to the end, and compare the lengths to the value listed in this Chapter's Specifications. Replace any bolts that have stretched beyond the maximum permissible length. If the bolts are damaged in any way, replace them with new cylinder head bolts.

23 Install the components that were removed from the cylinder head.

24 Position the new gasket over the dowel pins in the block, with the "lot number" identification UP. Then apply RTV sealant to the areas at the end of the cylinder head gasket as shown in illustration 6.33b.

25 Carefully set the cylinder head on the block without disturbing the gasket.

26 Before installing the cylinder head bolts, apply a small amount of clean engine oil to the threads and under the bolt heads.

27 Install the bolts and their washers in their original locations and tighten them finger tight. Following the recommended sequence, tighten the bolts to the torque listed in this Chapter's Specifications (see illustration). Step 2 of the tightening sequence requires each bolt to be tightened an additional 90-degrees. If you don't have an angle-torque attachment for your torque wrench, simply apply a paint mark at one edge of each cylinder head bolt and tighten the bolt until that mark is 90-degrees (1/4-turn) from where you started.

28 The remaining installation steps are the reverse of removal.

29 Check and adjust the valve clearances as necessary (see Chapter 1).

30 Change the engine oil and filter (see Chapter 1).

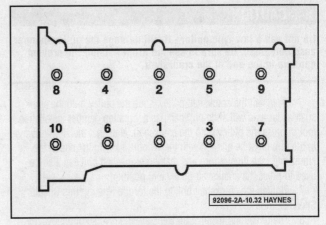

10.27 Cylinder head bolt TIGHTENING sequence

31 On 4WD models, refill the transfer case (see Chapter 1).

32 Refill the cooling system (see Chapter 1), run the engine and check for leaks.

11 Crankshaft pulley/vibration damper - removal and installation

▶ **Refer to illustrations 11.4, 11.5 and 11.6**

1 Disconnect the cable from the negative terminal of the battery (see Chapter 5, Section 1).

2 Remove the drivebelt (see Chapter 1). Unbolt the bracket from the engine movement control rod (see Section 17).

3 With the parking brake applied and the shifter in Park (automatic) or in gear (manual), loosen the lug nuts from the right front wheel, then raise the front of the vehicle and support it securely on jackstands. Remove the engine splash shield (see illustration 6.5), the right front wheel and the right splash shield from the wheelwell.

✳✳ WARNING:

If the vehicle is equipped with electronically modulated air suspension, make sure that the height control switch is turned off.

4 Remove the bolt from the front of the crankshaft. A breaker bar will probably be necessary, since the bolt is very tight (see illustration).

5 Using a puller that bolts to the crankshaft hub, remove the crankshaft pulley from the crankshaft (see illustration).

➡**Note: Depending on the type of puller you have it may be necessary to support the engine from above, remove the right side engine mount and lower the engine to gain sufficient clearance to use the puller.**

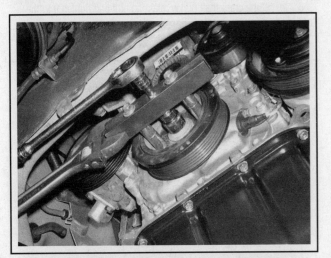

11.4 A puller base and several spacers can be mounted to the center hub of the pulley to keep the crankshaft from turning as the pulley retaining bolt is loosened - install the socket over the crankshaft bolt head before installing the puller, then insert the extension through the center hole of the puller

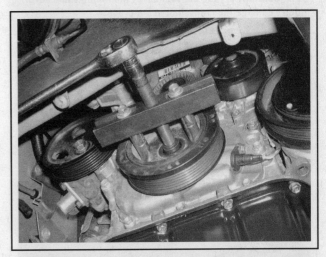

11.5 Reinstall the puller back onto the crankshaft pulley with the center bolt attached and remove the crankshaft pulley

❈❈ CAUTION:

Do not use a jaw-type puller - it will damage the pulley/damper assembly. Also, be sure to use the proper adapter to prevent damage to the end of the crankshaft.

6   To install the crankshaft pulley, slide the pulley onto the crankshaft as far as it will slide on, then use a vibration damper installation tool to press the pulley onto the crankshaft. Note that the slot (keyway) in the hub must be aligned with the Woodruff key in the end of the crankshaft (see illustration) and that the crankshaft bolt can also be used to press the crankshaft pulley into position.

7   Tighten the crankshaft bolt to the torque listed in this Chapter's Specifications.

8   The remaining installation steps are the reverse of removal.

11.6 Align the keyway in the crankshaft pulley hub with the Woodruff key in the crankshaft

12 Crankshaft front oil seal - replacement

▶ Refer to illustrations 12.2, 12.3 and 12.4

1   Remove the crankshaft pulley (see Section 11).
2   Note how the seal is installed - the new one must be installed to the same depth and facing the same way. Carefully pry the oil seal out of the cover with a seal puller or a large screwdriver (see illustration). Be very careful not to distort the cover or scratch the crankshaft! Wrap electrician's tape around the tip of the screwdriver to avoid damage to the crankshaft.

3   Apply clean engine oil or multi-purpose grease to the outer edge of the new seal, then install it in the cover with the lip (spring side) facing IN. Drive the seal into place with a seal driver or a large socket and a hammer (see illustration). Make sure the seal enters the bore squarely and stop when the front face is at the proper depth.

4   Check the surface on the pulley hub that the oil seal rides on. If the surface has been grooved from long-time contact with the seal, a press-on sleeve may be available to renew the sealing surface (see illustration). This sleeve is pressed into place with a hammer and a block of wood and is commonly available at auto parts stores for various applications.

5   Lubricate the pulley hub with clean engine oil and reinstall the crankshaft pulley (see Section 11).

6   Install the crankshaft pulley retaining bolt and tighten it to the torque listed in this Chapter's Specifications.

7   The remainder of installation is the reverse of the removal.

12.2 Carefully pry the old seal out of the timing chain cover - don't damage the crankshaft in the process

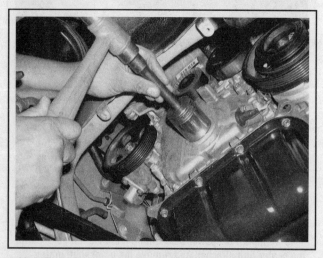

12.3 Drive the new seal into place with a seal driver or a large socket and hammer

12.4 If the sealing surface of the pulley hub has a wear groove from contact with the seal, repair sleeves are available at most auto parts stores

13 Oil pan - removal and installation

REMOVAL

▶ **Refer to illustrations 13.6a and 13.6b**

1   Disconnect the cable from the negative terminal of the battery (see Chapter 5, Section 1).
2   Set the parking brake and block the rear wheels.
3   Raise the front of the vehicle and support it securely on jackstands.

❊❊ WARNING:

If the vehicle is equipped with electronically modulated air suspension, make sure that the height control switch is turned off.

4   Remove the engine splash shield (see illustration 6.5).
5   Drain the engine oil and remove the oil filter (see Chapter 1). Remove the oil dipstick.
6   Remove the bolts and detach the oil pan. If it's stuck, pry it loose very carefully with a small screwdriver or putty knife (see illustrations). Don't damage the mating surfaces of the pan and block or oil leaks could develop.

INSTALLATION

7   Use a scraper to remove all traces of old sealant from the block and oil pan. Clean the mating surfaces with lacquer thinner or acetone.
8   Make sure the threaded bolt holes in the block are clean.
9   Check the oil pan flange for distortion, particularly around the bolt holes. Remove any nicks or burrs as necessary.
10 Inspect the oil pump pick-up tube assembly for cracks and a blocked strainer. If the pick-up was removed, clean it thoroughly and install it now, using a new gasket. Tighten the nuts/bolts to the torque listed in this Chapter's Specifications.
11 Apply a 3/16-inch wide bead of RTV sealant to the mating surface of the oil pan, following the groove but going to the inside where the bolt holes are located.
12 Carefully position the oil pan on the engine block and install the oil pan-to-engine block bolts loosely.
13 Working from the center out, tighten the oil pan-to-engine block

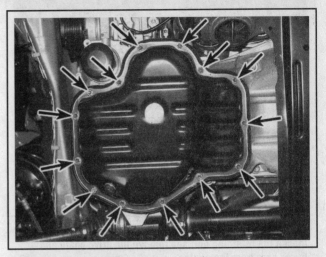

13.6a Oil pan mounting bolts

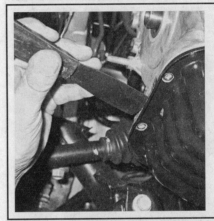

13.6b Pry the oil pan loose with a screwdriver or putty knife - be careful not to damage the mating surfaces of the pan and block or oil leaks may develop

bolts to the torque listed in this Chapter's Specifications in three or four steps.
14 The remainder of installation is the reverse of removal.

➡**Note: Be sure follow the sealant manufacturer's recommendations for assembly and sealant curing times.**

15 Run the engine and check for oil pressure and leaks.

14 Oil pump - removal and installation

REMOVAL

▶ **Refer to illustrations 14.2a, 14.2b, 14.3 and 14.5**

1   Refer to Section 6, Steps 1 through 18 and remove the timing chain and the crankshaft sprocket.
2   Remove the oil pump drive chain and sprockets (see illustrations).

14.2a Rotate the engine 90-degrees (1/4-turn) counterclockwise (from TDC) and set the crankshaft key to the left horizontal position (A), then remove the oil pump drive chain tensioner and bolt (B)

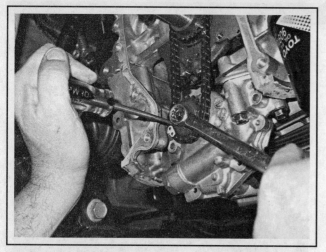

14.2b Using a screwdriver to lock the lower gear in place, loosen the retaining bolt and remove the drive chain and lower gear from the engine

14.3 Oil pump mounting bolts

3   Remove the three bolts and detach the oil pump body from the engine (see illustration). You may have to pry carefully between the front of the block and the pump body with a screwdriver to remove it.

4   Use a scraper to remove all traces of sealant and old gasket material from the pump body and engine block, then clean the mating surfaces with lacquer thinner or acetone.

5   Inspect the oil pump and chain for wear and damage. If the oil pump shows signs of wear or you're in doubt about its condition, it is simply best to replace it. Check the oil pump chain for wear or scoring. Replace if necessary. Wrap the chain around each of the oil pump drive sprockets and measure the diameter of the sprockets across the chain rollers (see illustration 6.21b) - if the measurement is less than the minimum sprocket diameter, the chain and the sprockets must be replaced. Oil pump chain guide wear is measured from the top of the chain contact surface to the bottom of the wear grooves (see illustration 6.22). Also check the oil jet for blockage (see illustration).

14.5 Check that the oil jet is free of debris - a blockage here will lead to an excessively worn timing chain, oil pump drive chain and guides

INSTALLATION

▶ **Refer to illustration 14.9**

6   Lubricate the pump cavity by pouring a small amount of clean engine oil into the inlet side of the oil pump and turning the drive gear shaft.

7   Position the oil pump and a new gasket against the block and install the mounting bolts.

8   Tighten the bolts to the torque listed in this Chapter's Specifications in several steps. Follow a criss-cross pattern to avoid warping the body.

9   With the crankshaft key still set in the left horizontal position and the flat on the oil pump drive shaft facing upward, install the drive chain and sprockets so that the colored links on the chain align with the alignment marks on the drive gears (see illustration).

10 Tighten the lower drive sprocket retaining bolt to the torque listed in this Chapter's Specifications.

11 Reinstall the remaining parts in the reverse order of removal.

12 Add oil to the proper level, start the engine and check for oil pressure and leaks.

14.9 Install the oil pump drive chain with the colored links aligned with the marks on the sprockets

15 Driveplate - removal and installation

REMOVAL

1 Disconnect the cable from the negative terminal of the battery (see Chapter 5, Section 1).
2 Remove the transaxle (see Chapter 7).
3 Using a center-punch or paint, apply alignment marks on the crankshaft flange and driveplate to ensure correct alignment on installation.
4 Remove the bolts retaining the driveplate to the crankshaft. Use a driveplate holding tool (available at auto parts stores) or wedge a screwdriver or prybar through one of the holes in the driveplate to keep it from turning while you loosen the bolts.
5 Remove the driveplate, taking note of spacers used and on which side of the driveplate they were installed.

INSTALLATION

6 Inspect the driveplate. Look for any fractures in the driveplate. Inspect the driveplate carefully for any other type of damage.
7 Position the driveplate on the crankshaft flange, aligning the marks made during removal. Align the bolt holes; note that some models may have a staggered bolt pattern to ensure correct installation.
8 Apply non-hardening thread locking compound to the threads of the bolts. Install the bolts and tighten them in a crossing pattern to the torque listed in this Chapter's Specifications. Work up to the final torque in several steps.
9 Install the transaxle (see Chapter 7).

16 Rear main oil seal - replacement

1 Remove the transaxle (see Chapter 7).
2 Remove the driveplate (see Section 15).
3 Pry the oil seal from the rear of the engine with a seal removal tool or a screwdriver. Be careful not to nick or scratch the crankshaft or the seal bore. Thoroughly clean the seal bore in the block with a shop towel. Remove all traces of oil and dirt.
4 Lubricate the outside diameter of the seal and install the seal over the end of the crankshaft. Make sure the lip of the seal points toward the engine. Preferably, a seal installation tool (available at most auto parts stores) should be used to press the new seal back into place. If the proper seal installation tool is unavailable, use a large socket and carefully drive the new seal squarely into the seal bore and flush with the edge of the engine block.
5 Install the driveplate (see Section 15).
6 Install the transaxle (see Chapter 7).

17 Powertrain mounts - check and replacement

1 Powertrain mounts seldom require attention, but broken or deteriorated mounts should be replaced immediately or the added strain placed on driveline components may cause damage or wear.

CHECK

2 During the check, the engine (or transaxle) must be raised slightly to remove the weight from the mounts.
3 Raise the vehicle and support it securely on jackstands, then remove the engine splash shield (see Section 6) and position a jack under the engine oil pan. Place a large block of wood between the jack and the oil pan, then carefully raise the engine just enough to take the weight off the mounts. Do not position the wood block under the oil drain plug.

※ WARNING 1:

DO NOT place any part of your body under the engine when it is supported only by a jack!

※ WARNING 2:

If the vehicle is equipped with electronically modulated air suspension, make sure that the height control switch is turned off.

4 Check the mounts to see if the rubber is cracked, hardened or separated from the bushing in the center of the mount.
5 Check for relative movement between the mount plates and the engine or frame, (use a large screwdriver or pry bar to attempt to move the mounts).
6 If movement is noted, lower the engine and tighten the mount fasteners.

REPLACEMENT

7 Disconnect the cable from the negative terminal of the battery (see Chapter 5, Section 1), then raise the vehicle and support it securely on jackstands, if not already done. Support the engine as described in Step 3.

※ WARNING:

If the vehicle is equipped with electronically modulated air suspension, make sure that the height control switch is turned off.

→Note: If several mounts need replacement, only replace one at a time and tighten them as you go. Do not remove all the mounts at once.

17.8 Right-side powertrain mount nuts (A) and bolts (B) (under the bolt hole covers)

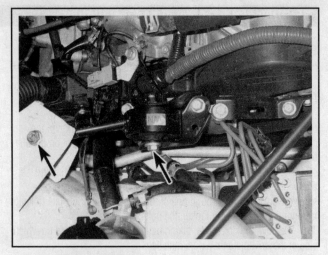

17.20 Location of the engine movement control rod mounting bolts

Passenger side (right-hand) engine mount

♦ Refer to illustration 17.8

8   Working below the vehicle, remove the nuts securing the mount to the upper and lower brackets (see illustration).

9   Remove the bolts securing the mount to the frame, then raise the engine enough to allow removal of the mount.

➡Note: It will be necessary to loosen the mounting nuts on the other powertrain mounts at the subframe to allow the engine to be raised enough for clearance for the passenger side engine mount removal.

10 Installation is the reverse of the removal. Use thread-locking compound on the bolts and be sure to tighten them securely.

Driver's side (left-hand) transaxle mount

11 Working below the vehicle, remove the nut securing the mount to the upper and lower brackets.

12 Remove the bolts securing the mount to the frame, then raise the transaxle enough to allow removal of the mount.

➡Note: It will be necessary to loosen the mounting nuts on the other powertrain mounts at the subframe to allow the engine to be raised enough for clearance for the driver's side transaxle mount removal.

13 Installation is the reverse of the removal. Use thread-locking compound on the bolts and be sure to tighten them securely.

Front engine mount

14 Working below the vehicle, remove the nut securing the engine bracket to the mount.

15 Remove the bolts securing the mount to the frame, then raise the engine enough to allow removal of the mount.

➡Note: It will be necessary to loosen the mounting nuts on the other powertrain mounts at the subframe to allow the engine to be raised enough for clearance for the front engine mount removal.

16 Installation is the reverse of the removal. Use thread-locking compound on the bolts and be sure to tighten them securely.

Rear engine mount

17 Working below the vehicle, remove the nut securing the engine bracket to the mount.

18 Remove the bolts securing the mount to the frame, then raise the engine enough to allow removal of the mount.

19 Installation is the reverse of the removal. Use thread-locking compound on the bolts and be sure to tighten them securely.

Engine movement control rod

♦ Refer to illustration 17.20

20 Working in the engine compartment, remove the bolts securing the engine movement control rod and its bracket (see illustration).

21 Installation is the reverse of the removal. Use thread-locking compound on the bolts and be sure to tighten them securely.

Specifications

General

Engine designation	2AZ-FE
Displacement	144 cubic inches (2.36 liters)
Cylinder numbers (drivebelt end-to-transaxle end)	1-2-3-4
Firing order	1-3-4-2

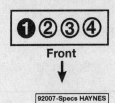

Front

92007-Specs HAYNES

Cylinder numbering

Cylinder head

Warpage limits	
Cylinder head	0.0031 inch (0.08 mm)
Intake manifold	0.0079 inch (0.20 mm)
Exhaust manifolds	0.0276 inch (0.70 mm)
Cylinder head bolt length (maximum)	6.350 to 6.390 inches (161.3 to 162.3 mm)

Timing chain

Timing chain sprocket wear limit	
Camshaft sprocket(s) (w/chain)	3.831 inches (97.3 mm)
Crankshaft sprocket (w/chain)	2.031 inches (51.6 mm)
Timing chain stretch limit	
8 links (16 pins)	4.543 inches (115.4 mm)
Timing chain guide wear limit	0.039 inch (1.0 mm)

Camshaft and lifters

Journal diameter	
No. 1 journal	1.4162 to 1.4167 inches (35.971 to 35.985 mm)
All others	0.9039 to 0.9045 inch (22.959 to 22.975 mm)
Bearing oil clearance	
No.1 journal	
Intake	
Mark 1 size	0.0028 to 0.0015 inch (0.070 to 0.038 mm)
Mark 2 and Mark 3 sizes	0.0031 to 0.0015 inch (0.080 to 0.038 mm)
Exhaust	0.0016 to 0.0031 inch (0.040 to 0.080 mm)
All others	0.0010 to 0.0024 inch (0.025 to 0.062 mm)
Runout limit	0.0012 inch (0.03 mm)
Thrust clearance (endplay)	
Intake	0.0016 to 0.0037 inch (0.040 to 0.095 mm)
Exhaust	0.0032 to 0.0053 inch (0.080 to 0.135 mm)
Lobe height	
Intake camshaft	
Standard	1.8305 to 1.8345 inches (46.495 to 46.595 mm)
Service limit (minimum)	1.8262 inches (46.385 mm)
Exhaust camshaft	
Standard	1.8104 to 1.8143 inches (45.983 to 46.083 mm)
Service limit (minimum)	1.8060 inches (45.873 mm)

Camshaft and lifters (continued)

Valve lifter
    Diameter   1.2191 to 1.2195 inches (30.966 to 30.976 mm)
    Bore diameter   1.2208 to 1.2215 inches (31.009 to 31.025 mm)
Lifter oil clearance
    Standard   0.0013 to 0.0023 inch (0.033 to 0.059 mm)
    Service limit   0.0028 inch (0.070 mm)

Oil pump

Drive chain sprocket wear limit (diameter w/chain)   1.898 inches (48.2 mm)

Torque specifications

	Ft-lbs (unless otherwise indicated)	Nm
Camshaft bearing cap bolts		
Journal No.1 (intake and exhaust)	22	30
All others	80 in-lbs	9
Camshaft sprocket bolts	40	54
Crankshaft pulley/vibration damper bolt		
2004 and earlier models	125	170
2005 and later models	133	180
Cylinder head bolts (in sequence - see illustration 10.32)		
Step 1	58	79
Step 2	Tighten an additional 90-degrees	
Drivebelt tensioner	44	60
Exhaust manifold nuts/bolts	27	37
Exhaust manifold brace bolts	32	44
Exhaust manifold heat shield bolts	108 in-lbs	12
Exhaust pipe to manifold-converter nuts	36	48
Driveplate bolts	72	98
Intake manifold nuts/bolts	22	30
Lower crankcase-to-engine block bolts	24	33
Oil control valve mounting bolt	80 in-lbs	9
Oil pump bolts		
10 mm bolts	71 in-lbs	8
12 mm bolts	14	20
Oil pump drive chain tensioner	108 in-lbs	12
Oil pump sprocket bolt	20	30
Oil pan bolts	80 in-lbs	9
Timing chain guide bolts (stationary)	80 in-lbs	9
Timing chain tensioner pivot arm bolt	168 in-lbs	19
Timing chain cover bolts (see illustration 6.40)		
Bolt A	80 in-lbs	9
Bolts B	15	21
Bolts C	32	43
Nuts D	80 in-lbs	9
Timing chain tensioner nuts	80 in-lbs	9
Valve cover fasteners		
Bolts/nuts A	96 in-lbs	11
Bolts B	120 in-lbs	14

Section

Reference to other Chapters

2B

V6 ENGINE

1 General information

The models covered by this manual are available with the 3.0L 1MZ-FE V6 engine or the 3MZ-FE, which is virtually the same engine as the 1MZ-FE but with a larger displacement of 3.3L. The design is a DOHC (dual overhead cam) with four valves per cylinder (24 in all), an aluminum engine block, distributorless ignition, and two-piece oil pan.

This Part of Chapter 2 is devoted to in-vehicle repair procedures for the V6 engine. Information concerning engine removal and installation and engine overhaul can be found in Part C of this Chapter.

The following repair procedures are based on the assumption that the engine is installed in the vehicle. If the engine has been removed from the vehicle and mounted on a stand, many of the steps outlined in this Part of Chapter 2 will not apply.

2 Repair operations possible with the engine in the vehicle

Many major repair operations can be accomplished without removing the engine from the vehicle.

Clean the engine compartment and the exterior of the engine with some type of degreaser before any work is done. It will make the job easier and help keep dirt out of the internal areas of the engine.

Depending on the components involved, it may be helpful to remove the hood to improve access to the engine as repairs are performed (refer to Chapter 11 if necessary). Cover the fenders to prevent damage to the paint. Special pads are available, but an old bedspread or blanket will also work.

If vacuum, exhaust, oil or coolant leaks develop, indicating a need for gasket or seal replacement, the repairs can generally be made with the engine in the vehicle. The intake and exhaust manifold gaskets, oil pan gasket, crankshaft oil seals and cylinder head gaskets are all accessible with the engine in place.

Exterior engine components, such as the intake and exhaust manifolds, the oil pan, the oil pump, the water pump, the starter motor, the alternator, and the fuel system components can be removed for repair with the engine in place.

Since the cylinder heads can be removed without pulling the engine, valve component servicing can also be accomplished with the engine in the vehicle. Replacement of the camshafts, timing belt and sprockets is also possible with the engine in the vehicle.

3 Top Dead Center (TDC) for number one piston - locating

▶ **Refer to illustrations 3.5 and 3.6**

1   Top Dead Center (TDC) is the highest point in the cylinder that each piston reaches as it travels up the cylinder bore. Each piston reaches TDC on the compression stroke and again on the exhaust stroke, but TDC generally refers to piston position on the compression stroke.

2   Positioning the piston(s) at TDC is an essential part of many procedures such as valve timing and camshaft and timing belt/sprocket removal.

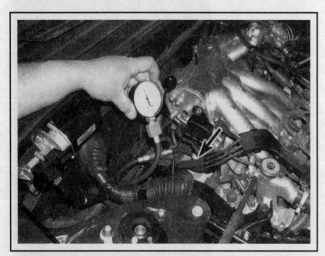

3.5 A compression gauge can be used in the number one plug hole to assist in finding TDC

3   Before beginning this procedure, be sure to place the transaxle in Neutral and apply the parking brake or block the rear wheels. Disable the fuel system by relieving the fuel system pressure (see Chapter 4).

4   In order to bring any piston to TDC, the crankshaft must be turned using one of the methods outlined below. When looking at the front of the engine, normal crankshaft rotation is clockwise.

   a) *The preferred method is to turn the crankshaft clockwise with a socket and ratchet attached to the bolt threaded into the front of the crankshaft.*

   b) *A remote starter switch, which may save some time, can also be used. Follow the instructions included with the switch. Once the piston is close to TDC, use a socket and ratchet as described in the previous paragraph.*

   c) *If an assistant is available to turn the ignition switch to the Start position in short bursts, you can get the piston close to TDC without a remote starter switch. Make sure your assistant is out of the vehicle, away from the ignition switch, then use a socket and ratchet as described in Paragraph (a) to complete the procedure.*

5   Remove the spark plug and install a compression pressure gauge in the number one spark plug hole. It should be a gauge with a screw-in fitting and a hose at least six inches long (see illustration).

✳✳ **CAUTION:**

It is possible to check the compression on cylinder number 1 on the V6 engine with the upper intake manifold and throttle body installed on the engine. The spark plugs can remain in the cylinder heads (except for number 1) if all of the ignition coils and the fuel pump have been disabled.

6   Rotate the crankshaft using one of the methods described above while observing the compression gauge. When the compression stroke of the number one cylinder is reached, compression pressure will begin to show on the gauge; continue to rotate the crankshaft and align the notch on the crankshaft pulley with the 0 mark on the timing plate (see illustration). If you go past the marks, release the gauge pressure and rotate the crankshaft around two more revolutions.

7   After the number one piston has been positioned at TDC on the compression stroke, TDC for the remaining cylinders can be located by turning the crankshaft 120-degrees (1/3-turn) at a time and following the firing order (refer to this Chapter's Specifications).

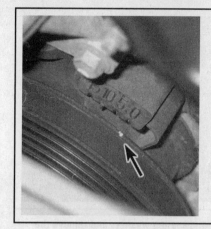

3.6 Turn the crankshaft until the notch in the pulley aligns with the zero on the timing plate

4   Valve covers - removal and installation

REMOVAL

♦ **Refer to illustrations 4.2a, 4.2b, 4.5a, 4.5b and 4.6**

1   Disconnect the cable from the negative terminal of the battery (see Chapter 5, Section 1).

2   Remove the engine cover(s) (see illustrations).

3   Remove the ignition coils (see Chapter 5).

4   Remove the upper intake manifold to access the rear valve cover (see Section 5).

5   Detach the engine wiring harness from the right side of the engine, the number 3 timing belt cover, the rear of the engine and left side of the engine compartment (see illustrations).

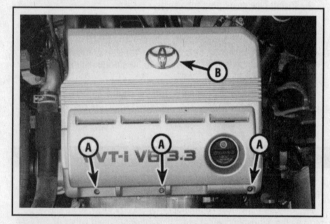

4.2a Remove the oil filler cap, then the three fasteners (A) and the clip (B) to remove the engine cover

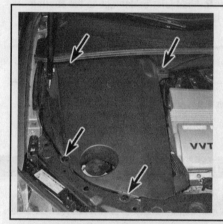

4.2b Location of the fasteners on the engine side cover

4.5a Remove nuts and disconnect the left-hand engine harness . . .

4.5b . . . disconnect the wire clips at the timing belt cover and the five bolts retaining the right-hand harness, then move the harness away from the rear valve cover

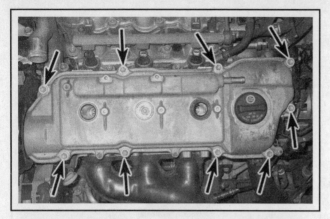

4.6 Remove the bolts and sealing washers and remove the valve cover - 3.0L V6 engine shown, 3.3L V6 engine similar

6   Remove the retaining bolts and sealing washers, then detach the cover(s) (see illustration). If the cover is stuck to the head, bump the end with a wood block and a hammer to jar it loose. If that doesn't work, try to slip a flexible putty knife between the head and cover to break the seal.

✳✳ CAUTION:

Don't pry at the cover-to-head joint or damage to the sealing surfaces may occur, leading to oil leaks after the cover is reinstalled.

INSTALLATION

▸ **Refer to illustration 4.9**

7   The mating surfaces of the cylinder head and cover must be clean

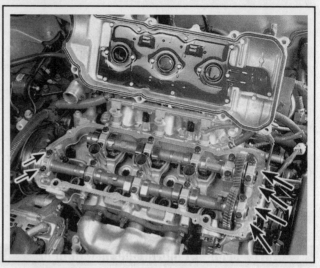

4.9 Apply RTV sealant to the areas indicated and install the cover with a new gasket

when the cover is installed. Use a gasket scraper to remove all traces of sealant and old gasket material, then clean the mating surfaces with lacquer thinner or acetone. If there's residue or oil on the mating surfaces when the cover is installed, oil leaks may develop.

8   Install new spark plug tube seals.

9   Apply RTV sealant to the gasket/seal joints at the front and rear camshaft-to-head mounts and install the valve cover with a new gasket (see illustration).

10 Tighten the bolts a little at a time to the torque listed in this Chapter's Specifications.

11 Reinstall the remaining parts, run the engine and check for oil leaks.

5   Intake manifold - removal and installation

✳✳ WARNING:

Wait until the engine is completely cool before beginning this procedure.

REMOVAL

1   Relieve the fuel system pressure (see Chapter 4), then disconnect the cable from the negative terminal of the battery (see Chapter 5, Section 1).

2   Remove the engine cover(s) (see illustrations 4.2a and 4.2b).

Upper intake manifold

▸ **Refer to illustration 5.6**

➡Note: **The following procedure describes upper intake manifold removal for access to the rear valve cover, the rear cylinder head and the sensors. However, it is possible to separate the upper intake manifold from the throttle body leaving the air filter housing and the throttle body intact for spark plug removal, compression check and fuel rail servicing.**

3   On 2004 and later Highlander models and all Lexus models,

remove the top cowl/ventilation cover (see Chapter 11).

4   On 2004 and later Highlander models and all Lexus models, remove the windshield wiper arms and the windshield wiper motor (see Chapter 12).

5   On 2004 and later Highlander models and all Lexus models, remove the lower cowl cover (see Chapter 11).

6   Disconnect the PCV hose from the valve cover (see illustration).

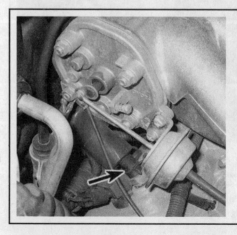

5.6 Location of the PCV hose at the rear valve cover

5.7 Remove the nuts, lift the vacuum switching assembly from the upper intake manifold and position the assembly off to the side

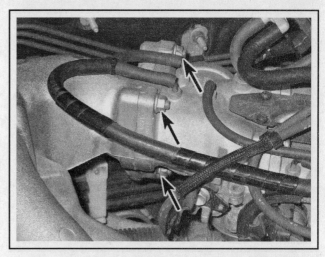

5.8 Location of the throttle body mounting nuts (2001 model shown, later models similar)

For spark plug removal

▸ **Refer to illustrations 5.7, 5.8, 5.10 and 5.12**

7 Disconnect the ground strap, the ACIS system components (see Chapter 6), the electrical connectors and the vacuum lines (see illustration) from the upper intake manifold. Label each connector using tape and a marker to ensure correct reassembly.

8 Remove the throttle body mounting nuts (see illustration).

9 Unbolt the upper intake manifold brace(s) at the back of the upper intake manifold assembly.

10 Remove the bolts and nuts mounting the upper intake manifold to the lower intake manifold (see illustration).

11 Remove the two Torx mounting studs (A) from the cylinder head (see illustration 5.10).

12 Move the upper intake manifold to the side and separate it from the throttle body and lower intake manifold (see illustration).

For all other procedures

13 Disconnect the ground strap, the ACIS system components (see

Chapter 6), the electrical connectors, vacuum lines (see illustration 5.7) and coolant bypass hoses from the upper intake manifold/throttle body assembly.

➡**Note: Clamp-off the coolant hoses before detaching them, or plug them as soon as they are detached. Be prepared for coolant spillage.**

The throttle body can remain attached to the upper intake manifold unless it is being removed for cleaning or gasket service. Label each connector using tape and a marker to insure correct reassembly.

14 Remove the air filter housing and air intake ducts (see Chapter 4).

15 Disconnect the accelerator cable (see Chapter 4) or the electronic throttle control system electrical connector (see Chapter 6) from the throttle body.

16 Remove the strut bar between the two shock towers (see Chapter 10, Section 3).

17 Unbolt the upper intake manifold brace(s) and the throttle body brace at the back of the upper intake manifold.

18 Remove the bolts and nuts mounting the upper intake manifold

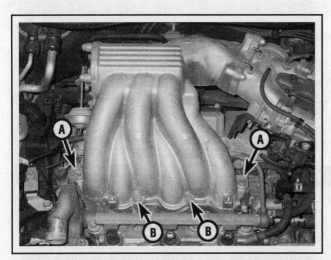

5.10 Location of the upper intake manifold bolts (B) and the mounting nuts (A) - 3.0L V6 engine shown, 3.3L V6 engines similar

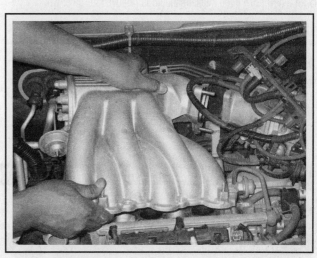

5.12 Slide the upper intake manifold to the side to detach it from the throttle body

5.23 This water transfer hose should be replaced whenever the intake manifold is off for other repairs

5.26 Bolt tightening sequence for the lower intake manifold

to the lower intake manifold and lift the upper intake manifold/throttle body assembly from the engine compartment (see illustration 5.10).

Lower intake manifold

▶ **Refer to illustration 5.23**

19 Drain the coolant into a clean container (see Chapter 1).

20 Disconnect the electrical connectors from the fuel injectors (see Chapter 4). Also detach the fuel line from the fuel rail (see Chapter 4).

➡Note: **The intake manifold can be removed with the injectors and fuel rails in place or removed, depending on the work to be done.**

21 Disconnect the heater hoses from the lower intake manifold.

22 Remove the mounting bolts and two nuts following the reverse of the tightening sequence, then detach the lower intake manifold from the engine (see illustration 5.26). If the manifold is stuck, don't pry between the gasket mating surfaces or damage may result.

23 There is a water transfer hose (see illustration) that is exposed only when the intake manifold is removed. Because of the difficulty in getting at this hose for replacement, we recommend that it be replaced with a new hose if the intake manifold is removed for other work.

INSTALLATION

▶ **Refer to illustrations 5.26 and 5.27**

24 Use a scraper to remove all traces of old gasket material and sealant from the lower intake manifold and cylinder heads, then clean the mating surfaces with lacquer thinner or acetone.

5.27 After the surface of the lower intake manifold has been prepared properly, install a new gasket

25 Install new gaskets, then position the manifold on the engine. Make sure the gaskets haven't shifted, then install the nuts/bolts.

26 Tighten the nuts/bolts, in three or four equal steps, to the torque listed in this Chapter's Specifications. Tighten the bolts in the correct sequence (see illustration).

27 Install the remaining parts in the reverse order of removal. Use a new gasket between the lower intake manifold and the upper intake manifold (see illustration).

28 Refill the cooling system. Run the engine and check for fuel, vacuum and coolant leaks.

---

6   Exhaust manifold/catalytic converter assemblies - removal and installation

▶ **Refer to illustrations 6.3a, 6.3b, 6.5 and 6.6**

✳✳ WARNING:

The engine must be completely cool before beginning this procedure.

➡Note: **2004 and later models are equipped with exhaust manifold/catalytic converter assemblies while 2003 and earlier models are equipped with separate catalytic converters.**

1 Disconnect the cable from the negative terminal of the battery (see Chapter 5, Section 1).

2 Spray penetrating oil on the exhaust manifold fasteners and allow it to soak in.

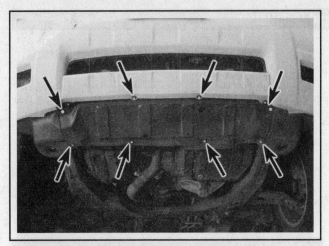

6.3a Locations of the engine splash shield fasteners - Highlander

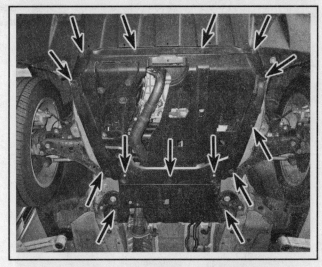

6.3b Locations of the main and rear splash shield fasteners - RX 300/330

3 If you're removing the rear (firewall side) exhaust manifold, raise the front of the vehicle and support it securely on jackstands, then remove the engine splash shields (see illustration).

☀☀ WARNING:

If the vehicle is equipped with electronically modulated air suspension, make sure that the height control switch is turned off.

4 Remove the heated oxygen sensors from the manifold(s) (see Chapter 6).

5 Remove the bolts and the heat shield over the exhaust manifold, on models so equipped (see illustration).

6 Remove the exhaust manifold brace, if equipped (see illustration).

7 Remove the nuts retaining the exhaust pipe(s) to the exhaust manifold(s).

8 Unbolt the exhaust manifold(s) from the cylinder head(s), working from the ends toward the middle. Slip the manifold(s) off the

mounting studs.

9 Carefully inspect the manifold(s) and fasteners for cracks and damage.

10 Use a scraper to remove all traces of old gasket material and carbon deposits from the manifold and cylinder head mating surfaces. If the gasket was leaking, check the manifold for warpage on the cylinder head mounting surface by placing a straightedge over the surface and trying to insert a feeler gauge. If the clearance exceeds the limit listed in this Chapter's Specifications, have the manifold resurfaced at an automotive machine shop.

11 Position a new gasket over the cylinder head studs.

12 Install the manifold(s) and thread the mounting nuts into place.

13 Working from the center out, tighten the nuts to the torque listed in this Chapter's Specifications in three or four equal steps.

14 Reinstall the remaining parts in the reverse order of removal. Use new gaskets when connecting the exhaust pipes.

15 Run the engine and check for exhaust leaks.

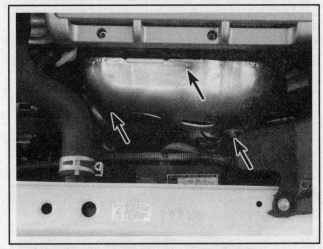

6.5 Remove the exhaust heat shield bolts (not all models are equipped with a heat shield)

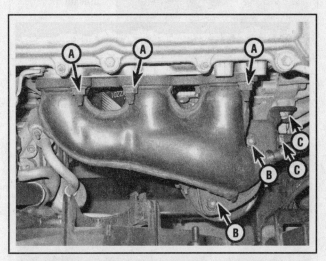

6.6 Location of the front exhaust manifold mounting nuts (A), the exhaust flange nuts (B) and the brace bolts (C) - lower exhaust manifold mounting nuts hidden from view

7 Timing belt and sprockets - removal, inspection and installation

REMOVAL

▶ **Refer to illustrations 7.13, 7.14a, 7.14b, 7.16, 7.19, 7.21, 7.23 and 7.25**

1 Disconnect the cable from the negative terminal of the battery (see Chapter 5, Section 1).

7.13 Remove the number 2 (upper) timing belt cover

7.14a Remove the mounting bolts and lift the engine mount spacer and side bracket from the engine mount bracket

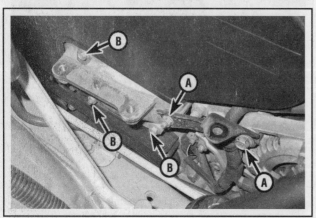

7.14b First remove the alternator bracket bolts (A) and the bracket, then remove the engine mount bracket bolts (B) and separate the bracket from the engine block

2 On 2004 and later Highlander models and all Lexus models, remove the top cowl/ventilation cover (see Chapter 11).

3 On 2004 and later Highlander models and all Lexus models, remove the windshield wiper arms and the windshield wiper motor (see Chapter 12).

4 On 2004 and later Highlander models and all Lexus models, remove the lower cowl cover (see Chapter 11).

5 Loosen the lug nuts on the right front wheel, but don't remove them yet.

6 Raise the front of the vehicle and support it securely on jackstands. Apply the parking brake and block the rear wheels. Remove the right front wheel.

❊❊ WARNING:

If the vehicle is equipped with electronically modulated air suspension, make sure that the height control switch is turned off.

7 Remove the right front inner fender apron (see Chapter 11).

8 Remove the engine splash shields from below the engine compartment (see illustration 6.3).

9 Position the number one cylinder at TDC (see Section 3).

10 Remove the drivebelts (see Chapter 1).

11 Support the engine with a jack from below and remove the engine movement control rod and its bracket on the engine (see illustration 16.1f). Place a wood block on the jack head and do not place the jack directly under the oil pan drain plug.

12 Remove the power steering pump and position it off to the side without disconnecting the fluid lines. Remove the power steering pump bracket.

13 Remove the upper (number 2) timing belt cover and gasket (see illustration).

➡**Note: Unclip the wiring harness above the cover and push it back enough to remove the belt cover.**

14 Remove the engine mounting brackets (see illustrations).

15 Remove the alternator bracket (see illustration 7.14b).

16 Check to see if there are installation marks on the timing belt - if you intend to re-use the belt and the marks have been obscured, make new ones (see illustration).

7.16 If you intend to re-use the belt and the original installation marks are obscured or missing, make new ones

7.19 After loosening the crankshaft pulley with a puller, it should come off by hand

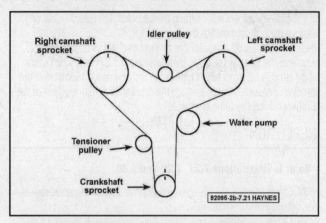

7.21 Timing belt alignment marks on the crankshaft sprocket and camshaft sprockets

7.23 Remove the two bolts and detach the timing belt tensioner

17 Remove the crankshaft pulley bolt. Wedge a large screwdriver into the driveplate ring gear teeth or against a converter bolt to keep the engine from turning, or use a chain wrench to hold the pulley stationary. Use a breaker bar and socket to loosen the pulley bolt.

18 When the crankshaft pulley bolt is loosened, the TDC position of the crankshaft may be disturbed. Check and align again, if necessary.

➡Note: The crankshaft timing belt sprocket has a TDC alignment mark that lines up with a mark on the oil pump housing, making it easy to check the TDC alignment even after the crankshaft pulley and lower timing belt cover are removed.

19 Once the pulley bolt is removed, remove the crankshaft pulley from the crankshaft. The pulley should slide off the crankshaft by hand (see illustration), but if it's stuck, use a bolt-type puller to remove it.

❊❊ CAUTION:

Do not use a jaw-type puller - it will damage the pulley/damper assembly. Also, be sure to use the proper adapter to prevent damage to the end of the crankshaft.

20 Remove the lower (number 1) timing belt cover and gasket.
21 Make sure the camshaft marks are properly aligned with the marks on the rear timing belt cover (see illustration).
22 Rotate the crankshaft counterclockwise approximately 60-degrees BTDC.

❊❊ CAUTION:

The engine pistons must be moved from the TDC number 1 position where they will not accidentally contact the valves when the timing belt tension has been released and the timing belt removed.

➡Note: The crankshaft pulley bolt must be installed and tightened enough to be able to rotate the engine in a counterclockwise direction.

23 Remove the timing belt tensioner (see illustration).

❊❊ CAUTION:

Loosen the bolts a little at a time, alternating from side to side until the tension has been released from the timing belt.

24 Remove the timing belt in the correct order (see illustration 7.21):
a) Step 1: Slide the timing belt off the tensioner pulley
b) Step 2: Lift the timing belt over the right side camshaft sprocket
c) Step 3: Slide the timing belt from under the idler pulley
d) Step 4: Lift the timing belt over the left side camshaft sprocket
e) Step 5: Slide the timing belt off the water pump sprocket
f) Step 6: Release the timing belt from the crankshaft sprocket

25 The camshaft sprockets can be removed at this point. Remove the valve cover(s) (see Section 5) and hold the camshaft with a wrench

7.25 If the camshaft sprockets are to be removed, hold the hex portion of the camshaft with a wrench while removing the sprocket bolt

on the cast-in hex while loosening the sprocket bolt (see illustration). Remove the bolt and detach the sprocket.

26 The crankshaft sprocket can be removed at this point, after removing the sprocket retainer (see illustration 7.37). If it won't come off by hand, a steering wheel type puller may be needed to remove the sprocket. Be careful not to damage the crankshaft sensor portion of the sprocket during the removal process.

INSPECTION

▶ Refer to illustrations 7.27, 7.29 and 7.30

27 Check the belt for the presence of oil or dirt, and inspect for visible defects (see illustration).

28 Check the belt tensioner for visible oil leakage. If there's only a faint trace of oil on the pushrod side, the tensioner seal is in satisfactory condition.

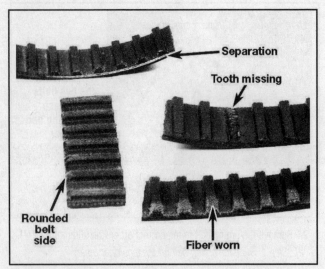

7.27 Check the timing belt for cracked or missing teeth - if the belt is cracked or worn, also check the pulleys for nicks or burrs - wear on one side of the belt indicates pulley misalignment problems

29 Hold the tensioner in both hands and push it forcefully against an immovable object (see illustration). If the pushrod moves, replace the tensioner.

30 On 1999 through 2003 models, measure the protrusion of the pushrod from the housing end (see illustration). Compare your measurement to this Chapter's Specifications. If the protrusion is not as specified, replace the tensioner.

31 Check that the idler pulleys turn smoothly.

INSTALLATION

▶ Refer to illustrations 7.39 and 7.41

32 Remove all dirt, oil and grease from the timing belt area at the front of the engine.

33 Install the camshaft sprocket(s) (if they were removed) on the camshaft(s) with the flange side facing OUT. Align the pin hole in the sprocket with the pin in the end of the camshaft. Do not interchange the camshaft sprockets. Install the intake camshaft sprocket onto the intake camshaft and the exhaust camshaft sprocket onto the exhaust camshaft.

34 Install the camshaft sprocket-retaining bolt(s) and tighten it to the torque listed in this Chapter's Specifications.

35 Install the tensioner pulley. Apply thread-locking compound to the first two or three threads of the bolt, then position the pulley and washer and install the bolt. Tighten the bolt to the torque listed in this Chapter's Specifications.

36 Install the upper idler pulley. Tighten the bolt to the torque listed in this Chapter's Specifications. Make sure the pulley turns smoothly.

37 Align the camshaft sprocket alignment marks (see illustration 7.21).

38 Install the crankshaft sprocket with the flange side up against the engine. Be careful not to damage the crankshaft sensor portion of the crankshaft sprocket.

39 Rotate the crankshaft back to TDC number 1 alignment. Check the alignment of the TDC marks on the sprocket and the oil pump housing, and reinstall the retainer and bolt (see illustration).

40 Install the timing belt, starting at the crankshaft sprocket, in the correct order:

a) Step 1: Engage the timing belt with the crankshaft sprocket
b) Step 2: Slide the timing belt over the water pump sprocket

7.29 Check the tensioner for signs of leakage and test for leakdown by forcing it against an immovable object

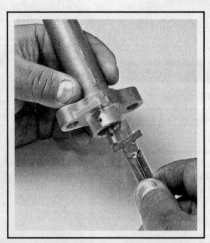

7.30 Measure the tensioner pushrod protrusion and compare it to the Specifications

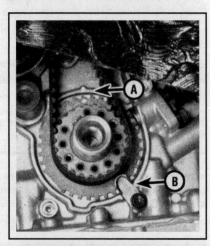

7.39 Align the marks on the crankshaft timing sprocket with the marks on the oil pump case (A), and install the sprocket retainer (B) and bolt

c) Step 3: Lift the timing belt over the left side camshaft sprocket
d) Step 4: Slide the timing belt under the idler pulley
e) Step 5: Lift the timing belt over the right side camshaft sprocket
f) Step 6: Slide the timing belt over the tensioner pulley

41 Using a press or vise, slowly compress the timing belt tensioner pushrod (see illustration). Insert a metal pin, drill bit or Allen wrench through the holes in the pushrod and housing. Remove the tensioner from the press or vise.

42 Install the timing belt tensioner and tighten the bolts to the torque listed in this Chapter's Specifications. Remove the retaining pin.

43 Using a socket and breaker bar on the crankshaft pulley bolt, turn the crankshaft slowly (clockwise) through two complete revolutions (720-degrees). Recheck the timing marks (see illustration 7.21).

✺✺ CAUTION:

If the timing marks are not aligned exactly as shown, repeat the timing belt installation procedure. DO NOT start the engine until you're absolutely certain that the timing belt is installed correctly. Serious and costly engine damage could occur if the belt is installed incorrectly.

44 Slip the belt guide over the end of the crankshaft with the cupped side facing out.

45 Install the lower (number 1) timing belt cover and gasket.

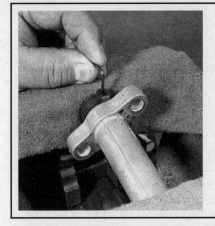

7.41 Restrain the tensioner pushrod by compressing the unit in a vise and inserting a pin approximately 0.060-inch (1.5 mm) in diameter - make sure the rubber boot is in place

46 Slip the crankshaft (drivebelt) pulley onto the crankshaft, aligning the pulley keyway with the crankshaft key. Install the bolt and tighten it to the torque listed in this Chapter's Specifications. Prevent the crankshaft from turning by using the method described in Step 17.

47 Reinstall the engine movement control rod's mounting bracket (see Section 16).

48 Install the upper (number 2) timing belt cover and gasket.

49 Install the engine movement control rod and braces and tighten the bolts securely (see Section 16).

50 Reinstall the remaining parts in the reverse order of removal.

8   Variable Valve Timing (VVT) system - description, check and component replacement

DESCRIPTION

▶ **Refer to illustration 8.4**

➡**Note: The following procedures apply to the oil control valve, oil filter or intake camshaft sprocket/actuator on either cylinder head.**

1   The VVT system varies intake camshaft timing by directing oil pressure to advance or retard the intake camshaft sprocket/actuator assembly. Changing the intake camshaft timing during certain engine conditions increases engine torque and fuel economy and reduces emissions.

2   System components include the Powertrain Control Module (PCM), an oil control valve (OCV) in each cylinder head, an oil filter for each OCV and an intake camshaft sprocket/actuator assembly.

3   The PCM uses inputs from the vehicle speed sensor (VSS), the throttle position sensor (TPS), the mass air flow (MAF) sensor and the engine coolant temperature (ECT) sensor to turn the oil control valve ON or OFF.

4   When the OCV is energized by the PCM, it directs a specified amount of oil pressure from the engine to advance or retard the intake camshaft sprocket/actuator assembly (see illustration).

5   The intake camshaft sprocket/actuator assembly is equipped with an inner hub that is attached to the camshaft. The inner hub consists of a series of fixed vanes that use oil pressure as a wedge against the vanes to rotate the camshaft. The higher the oil pressure, the greater the amount of actuator rotation and, therefore, the amount of camshaft advance or retard. How does greater oil pressure advance or retard the cam? By directing the oil to the advance or retard side of the fixed vanes.

6   When oil is applied to the advance side of the vanes, the actuator

can advance the camshaft up to 21 degrees in a clockwise direction. When oil is applied to the retard side of the vanes, the actuator will start to rotate the camshaft counterclockwise back to 0 degrees, which is the normal position of the actuator during engine operation under no load or at idle. The PCM can also send a signal to the oil control valve to stop oil flow to both (advance and retard) passages to hold camshaft advance in its current position.

7   Under light engine loads, the VVT system retards the camshaft timing to decrease valve overlap and stabilize engine output. Under medium engine loads, the VVT system advances the camshaft timing to increase valve overlap, thereby increasing fuel economy and decreasing exhaust emissions. Under heavy engine loads at low RPM, the VVT system advances the camshaft timing to help close the intake valve

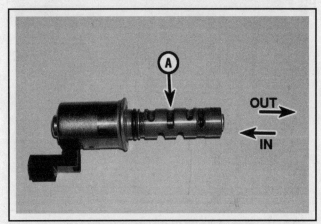

8.4 When voltage is applied, the OCV plunger (A), which you can see through the slots in the housing, should move out; when voltage is cut, the plunger should move in

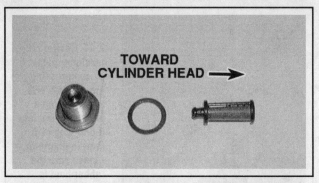

8.8 If you suspect a problem with the VVT system, inspect the OCV filter for obstructions; when you install the filter, use a new O-ring and make sure that the big end of the filter faces toward the head

faster, which improves low to midrange torque. Under heavy engine loads at high RPM, the VVT system retards the camshaft timing to slow the closing of the intake valve to improve engine horsepower.

COMPONENT REPLACEMENT

➡Note 1: A problem in the VVT oil control valve circuit will set a diagnostic trouble code and turn on the CHECK ENGINE light on the dash. Refer to Chapter 6 for accessing trouble codes.

➡Note 2: Most problems in the VVT system originate from the oil control valve(s) and filter(s). Regular engine oil and filter changes are necessary for trouble-free operation of the OCVs.

➡Note 3: Some checks and inspections of the VVT system require removal of the valve cover and the intake camshaft.

Oil control valve (OCV) filter

▶ Refer to illustration 8.8

8   A clogged OCV filter screen is often the cause of VVT system problems. Remove the OCV filter from the rear of the cylinder head and then inspect the filter for clogging. Clean the filter if necessary and reinstall it using a new O-ring. Make sure that the big end of the filter faces toward the head (see illustration). Be sure to tighten the filter plug to the torque listed in this Chapter's Specifications.

Oil control valve (OCV)

▶ Refer to illustrations 8.9a and 8.9b

9   To replace the OCV, remove the hold-down bolt and pull the OCV out of the cylinder head (see illustrations). There is one valve at the rear (transaxle end) of each cylinder head. Use a new O-ring when installing the new OCV. Be sure to tighten the OCV hold-down bolt to the torque listed in this Chapter's Specifications.

Camshaft sprocket/actuator assembly

10   Remove the valve cover (see Section 4), the timing belt (see Section 7) and the intake camshaft (see Section 10).

➡Note: Do NOT remove the exhaust camshaft or sprocket from the engine.

11   Mount the camshaft in a bench vise. Secure the camshaft in the vise by clamping up the hexagonal nut part of the cam.

✳✳ CAUTION:

Be careful not to damage the camshaft or any of the cam lobes or journals.

12   Verify that the sprocket/actuator will not rotate from the locked position. The locked position is a neutral position in which the actuator is placed during idle and no load conditions, and anytime that the VVT system is not activated by the PCM.

13   Using brake system cleaner, remove all traces of oil from the front cam journals and the VVT oil control orifices. Apply vinyl tape over all the oil control orifices except the advance side oil port.

14   Apply 14 psi of air pressure to the advance side oil port and try to rotate the actuator assembly by hand. The actuator should rotate freely, with no obvious binding, for about 30-degrees in the advance angle direction from the locked position.

➡Note: It is critical to have an air tight seal between the air gun nozzle and the advance oil port hole, because if air leaks out at the air nozzle, or at any of the other oil control orifices, the lock pin in the actuator won't be forced out of its locating hole. If leakage occurs, apply a little more air pressure to the advance side oil port to force the lock pin from the locating hole.

15   If the actuator does not rotate freely as described, replace the intake camshaft sprocket/actuator assembly.

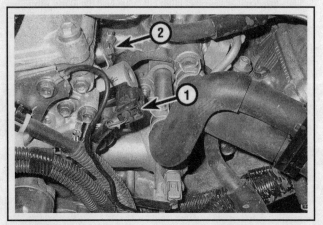

8.9a Location of the Oil Control Valve for the rear cylinder bank

1   *Electrical connector*   2   *Mounting bolt*

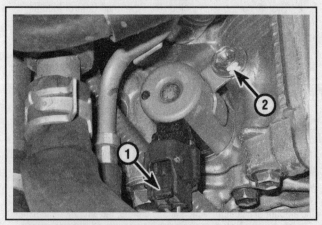

8.9b Location of the Oil Control Valve for the front cylinder bank

1   *Electrical connector*   2   *Mounting bolt*

16 To replace the sprocket/actuator assembly, put the camshaft in a bench vise and remove the 46mm nut by turning it clockwise (the nut is reverse-threaded).

17 Remove the sprocket/actuator assembly. If it's hard to pull off the camshaft, tap it lightly with a plastic-tip hammer.

⁂ CAUTION:

Do NOT try to disassemble the sprocket/actuator assembly - it's NOT rebuildable.

18 Lubricate the sprocket/actuator seating surface on the camshaft with clean engine oil. Before installing the sprocket/actuator assembly, make sure that the lock pin (in the camshaft) and lock pin groove (inside the sprocket/actuator) are aligned. Install the sprocket/actuator assembly. Using a NEW sprocket/actuator assembly retaining nut, tighten the retaining nut to the torque listed in this Chapter's Specifications (don't forget, it's reverse-threaded).

19 Install the intake camshaft (see Section 10), the timing belt (see Section 7) and the valve cover (see Section 4).

9   Oil seals - replacement

CRANKSHAFT OIL SEAL

1 Remove the timing belt and crankshaft sprocket (see Section 7).

2 Note how far the seal is recessed in the bore, then cut away the seal lip with a razor knife.

3 Carefully pry the seal out of the engine with a screwdriver or seal removal tool. If you use a screwdriver, wrap tape around the tip - don't scratch the housing bore or damage the crankshaft (if the crankshaft is damaged, the new seal will end up leaking).

4 Clean the bore in the engine and coat the outer edge of the new seal with engine oil or multi-purpose grease. Apply the same grease to the seal lip.

5 Using a seal driver or a socket with an outside diameter slightly smaller than the outside diameter of the seal, carefully drive the new seal into place with a hammer. Make sure it's installed squarely and driven in to the same depth as the original. Check the seal after installation to make sure the spring didn't pop out of place.

6 Reinstall the crankshaft timing sprocket and timing belt (see Section 8).

7 Run the engine and check for oil leaks at the front seal.

CAMSHAFT OIL SEALS

8 Remove the timing belt and camshaft sprocket(s) (see Section 7).

9 Remove the bolts and detach the rear timing belt cover.

10 Note how far the seal is seated in the bore, then carefully pry it out with a screwdriver. Wrap the screwdriver tip with tape - don't scratch the bore or damage the camshaft (if the camshaft is damaged, the new seal will end up leaking).

11 Clean the bore and coat the outer edge of the new seal with engine oil or multi-purpose grease. Apply multi-purpose grease to the seal lip.

12 Using a seal driver or a socket with an outside diameter slightly smaller than the outside diameter of the seal, carefully drive the new seal into place with a hammer. Make sure it's installed squarely and driven in to the same depth as the original.

13 Reinstall the rear timing belt cover and tighten the bolts.

14 Reinstall the camshaft sprocket(s) and timing belt (see Section 7).

15 Run the engine and check for oil leaks at the camshaft seal.

10 Camshafts and lifters - removal, inspection and installation

➡Note: Before beginning this procedure, obtain two 6 x 1.0 mm bolts 16 to 20 mm long. They will be referred to as service bolts in the text.

REMOVAL

▶ Refer to illustrations 10.3, 10.10, 10.12, 10.13 and 10.14

1 Position the engine at TDC (see Section 3), then remove the valve covers (see Section 4), the timing belt and the camshaft sprocket (see Section 7).

2 The following steps apply to the removal of each of the four camshafts. On each head, the exhaust camshaft subgear is secured first, the intake cam removed, then the exhaust camshaft is removed.

3 Make sure the cam timing marks on the drive and driven gears are in alignment (see illustration).

➡Note: On the right (rear) cylinder head, align the two dots on the intake camshaft gear with the two dots on the exhaust camshaft gear. On the left (front) cylinder head, align the one dot on the intake camshaft gear with the one dot on the exhaust camshaft gear.

10.3 Timing marks on the backside of the camshaft gears (typical markings)

10.10 Mark up a cardboard box to store the lifters/shims and camshaft bearing caps - use a separate box for each set to avoid mix-ups and mark the FRONT, INTAKE and EXHAUST orientation

4   Secure the exhaust camshaft sub-gear to the driven gear with a service bolt installed in the threaded hole (see illustration 10.12).

✱✱ CAUTION:

Since the camshaft thrust clearance is minimal, the camshafts must be held level as they are being removed. If they aren't, the portion of the cylinder head next to the cam gears may crack or be damaged by the gear leverage. Before lifting a camshaft out of the head, make certain that the torsional spring force of the sub-gear has been eliminated by the service bolt.

5   Loosen the intake camshaft bearing cap bolts in 1/4-turn increments until they can be removed by hand. Start with the outer caps and work inward.

6   Remove the bearing caps and gently lift out the intake camshaft. Be sure to keep it level.

7   Loosen the exhaust camshaft bearing cap bolts in 1/4-turn increments until they can be removed by hand. Start with the outer caps and work inward.

8   Remove the exhaust camshaft bearing caps and oil seal and gently lift out the exhaust camshaft. Be sure to keep it level.

10.12 With the hex portion of the camshaft held in a vise, use a two-pin spanner to remove the tension from the sub-gear and remove the service bolt, then release the sub-gear

9   Repeat the steps for the other cylinder head.
10   Store the bearing caps in the correct order.

➡Note: If necessary, the valve lifters and shims can now be removed with a magnetic tool. Be sure to store them separately so they can be reinstalled in their original locations (see illustration).

11   To disassemble an exhaust camshaft gear, mount the cam in a vise with the jaws gripping the large hex on the shaft.

12   Install a second service bolt in the unthreaded hole in the camshaft sub-gear. Using a screwdriver positioned against the service bolt just installed, rotate the sub-gear clockwise and remove the first service bolt. The second bolt isn't needed if you have a two-pin spanner (see illustration).

13   Remove the sub-gear snap-ring (see illustration).

14   The wave washer, sub-gear and camshaft gear spring can now be removed from the exhaust camshaft (see illustration). Be sure to keep the parts from the rear cylinder head cams separate from the front cylinder head parts. The front of the intake camshaft has the VVT assembly, which is secured to the camshaft with a large nut. To remove the VVT assembly, hold the hex portion of the camshaft in a vise and use a breaker bar and 46mm socket.

10.13 Remove the snap-ring with a pair of snap-ring pliers

10.14 Remove the wave washer (1), the camshaft sub-gear (2) and the gear spring (3) - exhaust camshaft shown

→Note: The nut is a left-hand thread. Do NOT remove the nut or the VVT assembly unless either the camshaft or VVT assembly is to be replaced.

INSPECTION

15 Refer to Chapter 2, Part A for camshaft, lifter and related component inspection procedures. Be sure to use the Specifications in this Part of Chapter 2 for the V6 engines.

INSTALLATION

16 Reassemble the exhaust camshaft gear(s) by installing the camshaft gear spring, sub-gear, wave washer and snap-ring.

17 Mount the camshaft in a padded vise.

18 Insert a service bolt into the unthreaded hole in the camshaft sub-gear. Using a screwdriver, align the holes of the camshaft driven gear and sub-gear by turning the camshaft sub-gear clockwise. Install a second service bolt in the threaded hole, tightening it to clamp the gears together. Remove the service bolt from the unthreaded hole. Repeat the procedure for the other exhaust camshaft.

19 Apply moly-base grease or engine assembly lube to the lifters, then install them in their original locations in the cylinder heads. Make sure the valve adjustment shims are in place in the lifters, and that all lifters are installed in their original bores.

20 Apply camshaft installation lubricant to the exhaust camshaft lobes, bearing journals and gear thrust faces.

21 Set the exhaust camshaft in place in the cylinder head with the timing mark on the subgear facing the center of the head.

22 Apply a thin coat of RTV sealant to the outer edges of the front bearing cap-to-cylinder head mating surfaces.

23 Install the bearing caps in numerical order with the arrows pointing toward the timing belt end of the engine.

24 Tighten the bearing cap bolts in 1/4-turn increments to the torque listed in this Chapter's Specifications. Start with the center cap

and work your way out to the ends.

25 Refer to Section 9 and install a new camshaft oil seal.

26 Apply camshaft installation lubricant to the intake camshaft lobes, bearing journals and gear thrust faces. If the VVT assembly was removed from the intake camshaft, install it with its groove aligned with the pin on the camshaft. Oil the threads and install a new nut and torque to this Chapter's Specifications.

27 Set the intake camshaft in place in the cylinder head, with the timing mark aligned with the exhaust camshaft timing mark (see illustration 10.3).

→Note: On the right (rear) cylinder head, align the two dots on the intake camshaft gear with the two dots on the exhaust camshaft gear. On the left (front) cylinder head, align the one dot on the intake camshaft gear with the one dot on the exhaust camshaft gear.

28 Install the bearing caps in numerical order with the arrows pointing toward the front (timing belt end) of the engine.

29 Tighten the bearing cap bolts in 1/4-turn increments to the torque listed in this Chapter's Specifications. Start with the center cap and work your way out to the ends.

30 Remove the service bolt from the exhaust camshaft sub-gear.

31 Reinstall the timing belt rear cover, the camshaft sprockets and the timing belt (see Section 8). Before installation, inspect the gasket on the timing belt rear cover. If there is minor damage, repair it with RTV sealant. If there are large sections missing, scrape off the old gasket and install a new one.

32 Reinstall the remaining components in the reverse order of removal.

33 Before reinstalling the valve covers, apply RTV sealant in the areas indicated (see illustration 4.9). Clean the rubber half-circle plugs for the back of the heads and reinstall with new RTV sealant.

34 The remainder of the installation is the reverse of the disassembly sequence.

35 Run the engine, then check for leaks and proper operation.

11 Cylinder heads - removal and installation

❋❋ WARNING:

Wait until the engine is completely cool before beginning this procedure.

REMOVAL

♦ Refer to illustrations 11.8 and 11.11

1 Disconnect the cable from the negative terminal of the battery (see Chapter 5, Section 1).

2 Drain the cooling system, including the engine block (see Chapter 1).

3 Remove the upper and lower intake manifolds (see Section 5).

4 Remove the exhaust manifold(s) (see Section 6).

5 Remove the alternator (see Chapter 5). Refer to Chapter 10 and unbolt and set aside the power steering pump.

6 Remove the timing belt, camshaft sprockets and upper idler pulley (see Section 7).

7 Remove the bolts holding the rear timing belt cover and remove it from the engine block and cylinder heads.

11.8 Remove the bolts and the coolant transfer casting (A) - the transfer hose (B) should be replaced whenever the intake manifold is removed

8 Disconnect the coolant sensor connectors and remove the water transfer casting (see illustration).

9 Remove the camshaft(s) from the head(s) you intend to remove

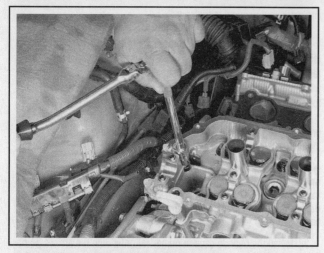

11.11 Remove the recessed bolt with an 8 mm hex bit

11.20 Be sure the new head gaskets are positioned right side up (check all holes and coolant passages for correct alignment) and over the block dowels

(see Section 10). Disconnect the electrical connectors from the camshaft position sensor and the VVT oil control valve (see Section 8).

10 Remove the bolts at the rear of the cylinder heads and move the engine wiring harnesses away from the heads.

11 Using an 8 mm hex bit or Allen wrench, remove the recessed head bolts (one in each head) (see illustration).

12 Using a 12-point socket, loosen the rest of the cylinder head bolts in 1/4-turn increments until they can be removed by hand, along with their hardened washers. Follow the reverse order of the recommended tightening sequence (see illustration 11.23).

13 Lift the cylinder head off the engine block. If the head is stuck, place a wood block against it and strike the wood with a hammer.

✳✳ CAUTION:

Don't pry between the head and block. The gasket surfaces may be damaged and leaks could result.

14 Repeat the procedure for the other head, if necessary.

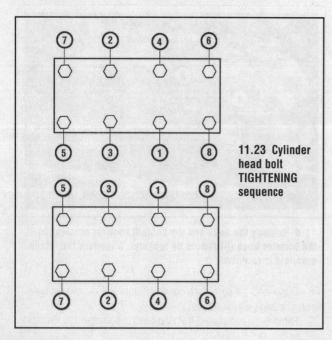

11.23 Cylinder head bolt TIGHTENING sequence

INSTALLATION

▶ **Refer to illustrations 11.20 and 11.23**

15 The mating surfaces of the cylinder heads and block must be perfectly clean when the heads are installed.

16 Use a gasket scraper to remove all traces of carbon and old gasket material, then clean the mating surfaces with lacquer thinner or acetone. If there's oil on the mating surfaces when the head is installed, the gasket may not seal correctly and leaks could develop. When working on the block, stuff the cylinders with clean shop rags to keep out debris. Use a vacuum cleaner to remove material that falls into the cylinders.

17 Check the block and head mating surfaces for nicks, deep scratches and other damage. If damage is slight, it can be removed with a file; if it's excessive, machining may be the only alternative.

18 Use a tap of the correct size to chase the threads in the cylinder head bolt holes, then clean the holes with compressed air - make sure that nothing remains in the holes.

✳✳ WARNING:

Wear eye protection when using compressed air!

19 Mount each bolt in a vise and run a die down the threads to remove corrosion and restore the threads. Dirt, corrosion, sealant and damaged threads will affect torque readings. Measure the diameter of the shoulder area of each head bolt and compare your findings with the value listed in this Chapter's Specifications. Replace any bolts that have stretched too thin.

20 Position the new gaskets over the dowel pins in the block (see illustration).

21 Carefully set the head on the block without disturbing the gasket.

22 Before installing the head bolts, apply a small amount of clean engine oil to the threads.

23 Install the bolts in their original locations and tighten them finger tight. Following the recommended sequence, tighten the bolts to the torque listed in this Chapter's Specifications (see illustration). Don't tighten the recessed bolt at this time.

24 Mark the front of each bolt head with paint. You can also mark the socket you are using.

25 Following the same sequence, tighten each 12-point bolt an additional 1/4-turn (90-degrees) (see illustration 11.23).

26 Tighten the recessed bolt to the torque listed in this Chapter's Specifications.

27 Repeat the entire procedure to install the other cylinder head.

28 The remaining installation steps are the reverse of removal.

29 Refill the cooling system, change the oil and filter (see Chapter 1), run the engine and check for leaks.

12 Oil pan - removal and installation

REMOVAL

▶ **Refer to illustrations 12.7, 12.8 and 12.9**

1   Remove the hood (see Chapter 11).

2   Disconnect the cable from the negative terminal of the battery (see Chapter 5, Section 1).

3   Raise the vehicle and support it securely on jackstands.

✳✳ WARNING:

If the vehicle is equipped with electronically modulated air suspension, make sure that the height control switch is turned off.

4   Remove the engine splash shields (see illustration 6.3).

5   Drain the engine oil and remove the oil filter. The oil pan is a two-part assembly, with an aluminum casting attached to the block and transaxle, and a lower stamped-steel pan section at the bottom.

6   Disconnect the front exhaust pipe from the catalytic converter and the support bracket, and the nuts holding the pipe to the front and rear exhaust manifolds/catalytic converter assemblies.

7   Remove the driveplate cover (see illustration).

8   Remove the two bolts holding the aluminum oil pan section to the transaxle (see illustration).

9   Remove the ten bolts and two nuts securing the steel oil pan section, and detach the steel pan (see illustration). If it's stuck, pry it loose very carefully with a small screwdriver or putty knife. Don't damage the mating surfaces of the pan or oil leaks could develop.

10 Remove the oil pump strainer/pickup (see illustration 13.4).

11 Remove the bolts securing the aluminum oil pan section to the block.

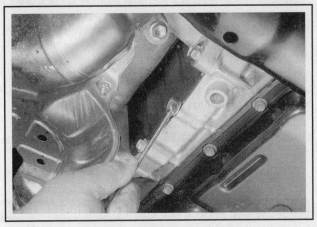

12.7 Remove two bolts and the driveplate cover

➡**Note: Some bolts are within the area formerly covered by the steel pan.**

12 Remove the oil pan baffle plate, if equipped.

INSTALLATION

▶ **Refer to illustration 12.18**

13 Use a scraper to remove all traces of old sealant from the block and oil pan. Clean the mating surfaces with lacquer thinner or acetone.

14 Make sure the threaded bolt holes in the block are clean.

15 Check the flange of the steel pan section for distortion, particularly around the bolt holes. If necessary, place the pan on a wood block and use a hammer to flatten and restore the gasket surface.

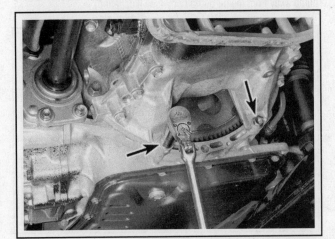

12.8 Remove the two bolts securing the aluminum upper pan to the transaxle

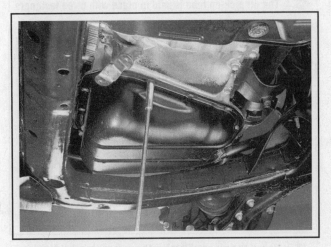

12.9 Remove the steel section of the oil pan

16 If the baffle had been removed, reinstall it now.

17 Clean the mating surfaces of the engine block and aluminum pan section, being careful not to gouge the soft metal, which could lead to leaks.

18 Apply a 4 mm wide bead of RTV sealant to the aluminum pan section (see illustration).

19 Install the aluminum pan section within five minutes and uniformly tighten the bolts to the torque listed in this Chapter's Specifications in several passes. Work from the center out towards the ends of the pan.

20 Inspect the oil pump pick-up/strainer assembly for cracks and a blocked strainer. If the pick-up was removed, clean it with solvent or thinner and install it now, using a new gasket (see Section 13). Tighten the fasteners to the torque listed in this Chapter's Specifications.

21 Apply a 3 to 4 mm wide bead of RTV sealant to the flange of the steel oil pan section.

➡ Note: The steel pan section must be installed within five minutes once the sealant has been applied.

22 Carefully position the steel pan on the aluminum section and install the bolts. Working from the center out, tighten them to the torque listed in this Chapter's Specifications in three or four steps.

23 The remainder of installation is the reverse of removal. Allow the

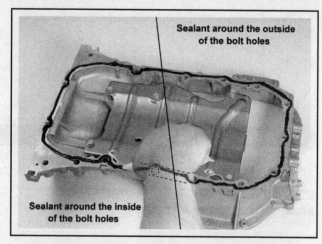

12.18 Apply the bead of RTV sealant around the perimeter of the aluminum pan section as shown

sealant to set for at least two hours before adding new oil and a new oil filter.

24 Run the engine and check for oil pressure and leaks.

13 Oil pump - removal, inspection and installation

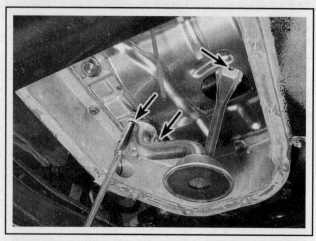

13.4 The oil pick-up tube is held in place with three fasteners

REMOVAL

▶ Refer to illustrations 13.4, 13.7, 13.8, 13.9, 13.11 and 13.12

1 Remove the oil pan (see Section 12).

2 Remove the timing belt (see Section 7).

3 Remove the crankshaft sprocket (see Section 7) and the crankshaft position sensor (see Chapter 6). Place a jack under the engine. Remove the powertrain mounts (see Section 16).

4 Remove the oil pick-up tube (see illustration).

5 Remove the alternator and its bracket (see Chapter 5).

6 Unbolt the air conditioning compressor and set it aside without disconnecting the refrigerant lines.

7 Remove the air conditioning compressor bracket (see illustration).

8 Remove the power steering pump adjusting bracket and pry the pump away from the oil pump body (see illustration).

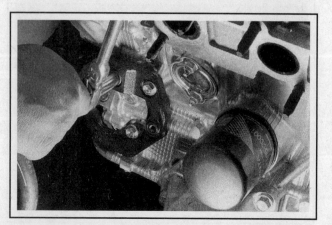

13.7 Unbolt the air conditioning compressor bracket from the block

13.8 Remove the power steering pump adjuster bar and pry the pump away from the oil pump body

13.9 Using a 10 mm hex bit or Allen wrench, remove the timing belt tensioner pulley

9   Remove the timing belt tensioner pulley (see illustration).
10 Remove the bolts and detach the oil pump from the engine. You may have to pry carefully between the front main bearing cap and the pump body with a screwdriver.
11 Remove the O-ring. Remove the oil pressure relief valve snap-ring, retainer, spring and valve (see illustration).

※※ WARNING:

The spring is tightly compressed - be careful and wear eye protection.

---

12 Use a large Phillips screwdriver to remove the screws retaining the body cover to the rear of the oil pump (see illustration).
13 Lift the cover off and remove the pump rotors.
14 Use a scraper to remove all traces of sealant and old gasket material from the pump body and engine block, then clean the mating surfaces with lacquer thinner or acetone.

INSPECTION

▶ **Refer to illustrations 13.17a, 13.17b and 13.17c**

15 Clean all components with solvent, then inspect them for wear and damage.
16 Check the oil pressure relief valve sliding surface and valve spring. If either the spring or the valve is damaged, they must be

13.11 Remove the oil pressure relief plug, spring and valve

13.12 Use a large Phillips screwdriver or bit to remove the screws retaining the pump cover

replaced as a set.
17 Check the clearance of the following components with a feeler gauge and compare the measurements to this Chapter's Specifications (see illustrations):

  a) *Driven rotor-to-oil pump body clearance*
  b) *Rotor side clearance*
  c) *Rotor tip clearance*

13.17a Measure the driven rotor-to-body clearance with a feeler gauge

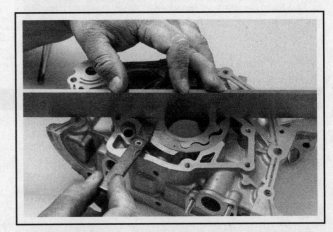

13.17b Measure the rotor side clearance with a precision straightedge and feeler gauge

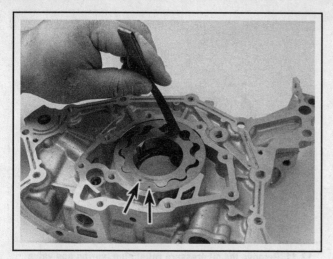

13.17c Measure the rotor tip clearance with a feeler gauge - note the rotor marks are facing out (when the pump body cover is installed, the marks will be against the cover)

13.26 Be sure to install a new O-ring on the block, and align the drive rotor and the crankshaft as the oil pump is installed

INSTALLATION

▶ **Refer to illustration 13.26**

18 Pry the old crankshaft seal out with a screwdriver.

19 Apply multi-purpose grease or engine oil to the outer edge of the new seal and carefully drive it into place with a seal driver and a hammer. Also apply multi-purpose grease to the seal lip.

20 Place the drive and driven rotors into the pump body with the marks facing out (see illustration 13.17c).

21 Pack the pump cavity with petroleum jelly and install the cover. Tighten the screws securely following a criss-cross pattern.

22 Lubricate the oil pressure relief valve with engine oil and install the valve components in the pump body.

23 Use acetone or lacquer thinner and a clean rag to remove all traces of oil from the gasket surfaces.

24 Apply a 2 to 3 mm wide bead of anaerobic sealant to the oil pump. Avoid using an excessive amount of sealant, especially around oil passages and bolt holes.

25 Position a new O-ring on the block.

26 Engage the flats on the oil pump drive rotor with the flats on the crankshaft and slide the pump into place (see illustration).

27 Install the oil pump mounting bolts in their original locations and tighten them to the torque listed in this Chapter's Specifications in a criss-cross pattern.

28 Using a new gasket, install the oil pick-up tube and tighten the fasteners to the torque listed in this Chapter's Specifications.

29 Reinstall the remaining parts in the reverse order of removal.

30 Add oil, start the engine and check for oil leaks.

31 Recheck the engine oil level.

14 Driveplate - removal and installation

Refer to Chapter 2, Part A for this procedure, but be sure to use the torque specifications in this Part of Chapter 2 for the V6 engine.

On V6 engines, there are two spacers used in driveplate mounting, one between the crankshaft flange and the driveplate, the other between the rear face of the driveplate and the driveplate mounting bolts. When removing the driveplate, mark the spacers and reinstall them in the same positions and facing the same way as originally installed.

15 Rear main oil seal - replacement

Refer to Chapter 2, Part A for this procedure, but note that the V6 engine doesn't have a gasket between the seal retainer and the engine block. Instead, apply a 2 to 3 mm wide bead of anaerobic sealant to the retainer flange before attaching the retainer to the block. Also, be sure to use the torque specifications in this Part of Chapter 2 for the V6 engine.

16 Powertrain mounts - check and replacement

▶ Refer to illustrations 16.1a, 16.1b, 16.1c, 16.1d, 16.1e and 16.1f

Refer to Chapter 2, Part A, but note that the V6 engine mounts are slightly different in ways that don't significantly affect the check and replacement procedures (see illustrations).

16.1a First remove the access plugs at the subframe, then remove the nuts from the lower engine mount

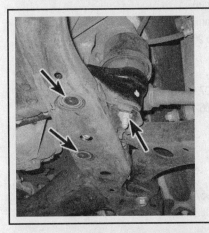

16.1c Location of the driver's side access plugs and mounting nuts for the left side transaxle mount

16.1e Location of the rear engine/driveaxle mount through-bolt

16.1b Carefully raise the engine until the engine mount studs clear the subframe, remove the nut from the top of the mount and angle the engine mount from the vehicle - it will be necessary to loosen the other mounts at the subframe in order to raise the engine the extra clearance

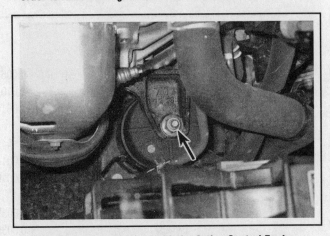

16.1d On models equipped with the Active Control Engine Mount System, disconnect the electrical connectors at the front engine mount before removing the mounting nuts

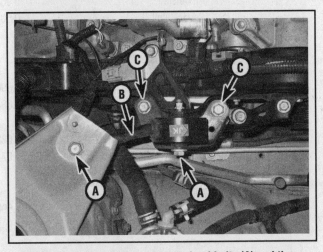

16.1f Remove the movement-control rod bolts (A) and the engine movement control rod (B), then remove the mounting bolts (C) and the engine mounting brace

Specifications

General

Engine designations	1MZ-FE and 3MZ-FE
Displacement	
1MZ-FE	183 cubic inches (3.0 liters)
3MZ-FE	202 cubic inches (3.3 liters)
Cylinder numbers (timing belt end-to-transaxle end)	
Right (firewall) side	1-3-5
Left (radiator) side	2-4-6
Firing order	1-2-3-4-5-6

❶ ③ ⑤

② ④ ⑥

Front ↓

92095-2b-specs HAYNES

Cylinder numbering

Cylinder head

Warpage limits	
Cylinder head	0.0039 inch (0.10 mm)
Intake manifold	0.0031 inch (0.08 mm)
Exhaust manifolds	0.0196 inch (0.50 mm)
Cylinder head bolt diameter (minimum)	0.3445 inch (8.75 mm)

Camshaft and related components

Valve clearance (engine cold)	
Intake	0.006 to 0.010 inch (0.15 to 0.25 mm)
Exhaust	0.010 to 0.014 inch (0.25 to 0.35 mm)
Bearing journal diameter	1.0614 to 1.0620 inches (26.959 to 26.975 mm)
Bearing oil clearance	
1999 through 2003 RX 300	
Intake	0.0014 to 0.0028 inch (0.035 to 0.072 mm)
Exhaust	0.0010 to 0.0024 inch (0.025 to 0.062 mm)
Service limit	
Intake	0.0039 inch (0.10 mm)
Exhaust	0.0035 inch (0.09 mm)
Highlander and 2004 and later RX 330	
Intake #4 and #5	0.0010 to 0.0022 inch (0.025 to 0.057 mm)
All others	0.0010 to 0.0024 inch (0.025 to 0.062 mm)
Service limit	0.0039 inch (0.10 mm)
Lobe height	
1999 through 2003	
Intake	
Standard	1.6902 to 1.6942 inches (42.932 to 43.032 mm)
Service Limit	1.6842 inches (42.78 mm)
Exhaust	
Standard	1.6836 to 1.6876 inches (42.764 to 42.864 mm)
Service limit	1.6776 inches (42.61 mm)
2004 and later	
Intake	
Standard	1.6981 to 1.7020 inches (43.132 to 43.232 mm)
Service Limit	1.6921 inches (42.98 mm)
Exhaust	
Standard	1.6933 to 1.6972 inches (43.010 to 43.110 mm)
Service limit	1.6874 inches (42.86 mm)

Thrust clearance (endplay)
    Standard   0.0016 to 0.0035 inch (0.040 to 0.090 mm)
    Service limit   0.0047 inch (0.12 mm)
Runout limit (total indicator reading)   0.0024 inch (0.06 mm)
Camshaft gear backlash
    Standard   0.0008 to 0.0079 inch (0.020 to 0.200 mm)
    Service limit   0.0118 inch (0.30 mm)
Timing belt tensioner protrusion
    (1999 through 2003 models)   0.394 to 0.425 inch (10.0 to 10.8 mm)

Oil pump

Driven rotor-to-pump body clearance
  1999 through 2003 RX 300
    Standard   0.0098 to 0.0128 inch (0.250 to 0.325 mm)
    Service limit   0.0118 inch (0.30 mm)
  2001 through 2003 Highlander and 2004 and later RX 330
    Standard   0.0050 to 0.0064 inch (0.125 to 0.152 mm)
    Service limit   0.0118 inch (0.30 mm)
  2004 and later Highlander
    Standard   0.0098 to 0.0128 inch (0.250 to 0.325 mm)
    Service limit   0.0197 inch (0.50 mm)
Rotor tip clearance
  1999 through 2003 Highlander and all RX 300, 330
    Standard   0.0043 to 0.0094 inch (0.11 to 0.24 mm)
    Service limit   0.0138 inch (0.35 mm)
  2004 and later Highlander
    Standard   0.0024 to 0.0071 inch (0.060 to 0.180 mm)
    Service limit   0.0118 inch (0.30 mm)
Rotor side clearance
    Standard   0.0012 to 0.0035 inch (0.03 to 0.09 mm)
    Service limit   0.0059 inch (0.15 mm)

Torque specifications

	Ft-lbs (unless otherwise indicated)	Nm
Crankshaft pulley bolt		
1999 through 2003	159	215
2004 and later	162	220
Camshaft bearing cap bolts	144 in-lbs	16
Cylinder head bolts (12-point)		
Step 1	40	54
Step 2	Tighten an additional 90-degrees (1/4-turn)	
Cylinder head bolt (recessed)	168 in-lbs	19
Driveplate bolts*	61	83
Exhaust camshaft sprocket bolt	94	125
Engine mounting bracket bolts/nuts	21	28
Exhaust manifold nuts	36	49
Exhaust manifold heat shield bolts	80 in-lbs	9
Idler pulley bolts		
No. 1 (to block)	25	34
No. 2 (to bracket)	32	43

Torque specifications	Ft-lbs (unless otherwise indicated)	Nm
Intake camshaft sprocket bolt		
1999 through 2003 RX 300	65	88
Highlander and 2004 and later RX 330	94	125
Intake manifold		
Lower intake manifold bolts/nuts	132 in-lbs	15
Upper intake manifold bolts/nuts		
1999 through 2003	32	43
2004 and later	21	28
Lower engine crankcase to engine block bolts		
10 mm bolt head	71 in-lbs	8
12 mm bolt head	168 in-lbs	20
Oil pan bolts		
Aluminum section		
Highlander		
10 mm bolt head	71 in-lbs	8
12 mm bolt head	168 in-lbs	20
RX 300, 330		
10 mm bolt head	71 in-lbs	8
12 mm bolt head	168 in-lbs	20
14 mm bolt head	27	37
Steel section	71 in-lbs	8
Oil pump mounting bolts		
1999 through 2003		
10 mm bolt head	71 in-lbs	8
12 mm bolt head	168 in-lbs	20
2004 and later		
10 mm bolt head	71 in-lbs	8
12 mm bolt head	168 in-lbs	20
14 mm bolt head	32	43
Oil pump cover screws	84 in-lbs	10
Oil pick-up tube mounting bolts	71 in-lbs	8
Timing belt cover (front) bolts		
RX 300, 330	75 in-lbs	9
Highlander	80 in-lbs	9
Timing belt cover (rear) bolts	80 in-lbs	9
Timing belt tensioner bolts	20	27
Rear crankshaft oil seal retainer mounting bolts	71 in-lbs	8
Valve cover bolts	71 in-lbs	8
Variable Valve Timing (VVT) system		
Oil control valve filter plug	32	43
Oil control valve hold-down bolt	71 in-lbs	8
Intake camshaft actuator retaining nut**	111	150

* Apply thread locking compound to the threads prior to installation

** Reverse-threaded.

Section

Reference to other Chapters

2C

GENERAL ENGINE OVERHAUL PROCEDURES

1   General information - engine overhaul

▶ **Refer to illustrations 1.2, 1.3, 1.4, 1.5, 1.6 and 1.7**

Included in this portion of Chapter 2 are general information and diagnostic testing procedures for determining the overall mechanical condition of your engine.

The information ranges from advice concerning preparation for an overhaul and the purchase of replacement parts and/or components to detailed, step-by-step procedures covering removal and installation.

The following Sections have been written to help you determine whether your engine needs to be overhauled and how to remove and install it once you've determined it needs to be rebuilt. For information concerning in-vehicle engine repair, see Chapter 2A or 2B.

The Specifications included in this Part are general in nature and include only those necessary for testing the oil pressure, checking the engine compression, and bottom-end torque specifications. Refer to Chapter 2A or 2B for additional engine Specifications.

It's not always easy to determine when, or if, an engine should be completely overhauled, because a number of factors must be considered.

High mileage is not necessarily an indication that an overhaul is needed, while low mileage doesn't preclude the need for an overhaul. Frequency of servicing is probably the most important consideration.

An engine that's had regular and frequent oil and filter changes, as well as other required maintenance, will most likely give many thousands of miles of reliable service. Conversely, a neglected engine may require an overhaul very early in its service life.

Excessive oil consumption is an indication that piston rings, valve seals and/or valve guides are in need of attention. Make sure that oil leaks aren't responsible before deciding that the rings and/or guides are bad. Perform a cylinder compression check to determine the extent of the work required (see Section 3). Also check the vacuum readings under various conditions (see Section 4).

Check the oil pressure with a gauge installed in place of the oil pressure sending unit and compare it to this Chapter's Specifications (see Section 2). If it's extremely low, the bearings and/or oil pump are probably worn out.

Loss of power, rough running, knocking or metallic engine noises, excessive valve train noise and high fuel consumption rates may also point to the need for an overhaul, especially if they're all present at the same time. If a complete tune-up doesn't remedy the situation, major mechanical work is the only solution.

An engine overhaul involves restoring the internal parts to the specifications of a new engine. During an overhaul, the piston rings are replaced and the cylinder walls are reconditioned (rebored and/or

1.2 An engine block being bored - an engine rebuilder will use special machinery to recondition the cylinder bores

1.3 If the cylinders are bored, the machine shop will normally hone the engine on a machine like this

honed) (see illustrations 1.2 and 1.3). If a rebore is done by an automotive machine shop, new oversize pistons will also be installed. The main bearings, connecting rod bearings and camshaft bearings are generally replaced with new ones and, if necessary, the crankshaft may be reground to restore the journals (see illustration 1.4). Generally, the valves are serviced as well, since they're usually in less-than-perfect condition at this point. While the engine is being overhauled, other components, such as the distributor, starter and alternator, can be rebuilt as well. The end result should be a like-new engine that will give many trouble-free miles.

➡ **Note: Critical cooling system components such as the hoses, drivebelts, thermostat and water pump should be replaced with new parts when an engine is overhauled. The radiator should be checked carefully to ensure that it isn't clogged or leaking (see Chapter 3). If you purchase a rebuilt engine or short block, some rebuilders will not warranty their engines unless the radiator has been professionally flushed. Also, we don't recommend overhauling the oil pump - always install a new one when an engine is rebuilt.**

Overhauling the internal components on today's engines is a difficult and time-consuming task which requires a significant amount of specialty tools and is best left to a professional engine rebuilder (see illustrations 1.5, 1.6 and 1.7). A competent engine rebuilder will handle the inspection of your old parts and offer advice concerning the reconditioning or replacement of the original engine. Never purchase parts or have machine work done on other components until the block has been thoroughly inspected by a professional machine shop. As a general rule, time is the primary cost of an overhaul, especially since the vehicle may be tied up for a minimum of two weeks or more. Be aware that some engine builders only have the capability to rebuild the engine you bring them while other rebuilders have a large inventory of rebuilt exchange engines in stock. Also be aware that many machine shops could take as much as two weeks time to completely rebuild your engine depending on shop workload. Sometimes it makes more sense to simply exchange your engine for another engine that's already rebuilt to save time.

1.4 A crankshaft having a main bearing journal ground

1.5 A machinist checks for a bent connecting rod, using specialized equipment

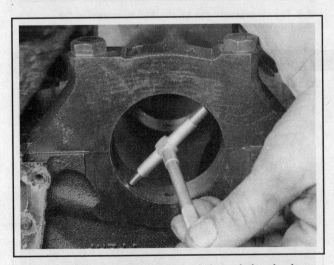
1.6 A bore gauge being used to check the main bearing bore

1.7 Uneven piston wear like this indicates a bent connecting rod

2 Oil pressure check

▶ **Refer to illustrations 2.2a, 2.2b and 2.3**

1   Low engine oil pressure can be a sign of an engine in need of rebuilding. A "low oil pressure" indicator (often called an "idiot light")

2.2a On four-cylinder models, the oil-pressure sending unit is located at the left end of the cylinder head

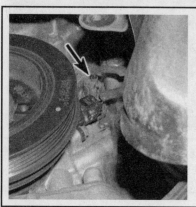

2.2b On V6 models, the oil pressure sending unit is located at the front of the engine, above the crankshaft position sensor

is not a test of the oiling system. Such indicators only come on when the oil pressure is dangerously low. Even a factory oil pressure gauge in the instrument panel is only a relative indication, although much better for driver information than a warning light. A better test is with a mechanical (not electrical) oil pressure gauge.

2   Locate the oil pressure indicator sending unit on the engine block (see illustrations).

3   Unscrew and remove the oil pressure sending unit and then screw in the hose for your oil pressure gauge (see illustration). If necessary, install an adapter fitting. Use Teflon tape or thread sealant on the threads of the adapter and/or the fitting on the end of your gauge's hose.

4   Connect an accurate tachometer to the engine, according to the tachometer manufacturer's instructions.

5   Check the oil pressure with the engine running (normal operating temperature) at the specified engine speed, and compare it to this Chapter's Specifications. If it's extremely low, the bearings and/or oil pump are probably worn out.

2.3 The oil pressure can be checked by removing the sending unit and installing a pressure gauge in its place

3 Cylinder compression check

▶ **Refer to illustration 3.6**

1   A compression check will tell you what mechanical condition the upper end of your engine (pistons, rings, valves, head gaskets) is in. Specifically, it can tell you if the compression is down due to leakage caused by worn piston rings, defective valves and seats or a blown head gasket.

➡**Note: The engine must be at normal operating temperature and the battery must be fully charged for this check.**

2   Begin by cleaning the area around the spark plugs before you remove them (compressed air should be used, if available). The idea is to prevent dirt from getting into the cylinders as the compression check is being done.

3   Unplug and remove the ignition coils (see Chapter 5), then remove all of the spark plugs from the engine (see Chapter 1).

4   Block the throttle wide open.

5   Disable the fuel pump circuit by relieving the fuel pressure (see Chapter 4).

6   Install a compression gauge in the spark plug hole (see illustration).

3.6 Use a compression gauge with a threaded fitting for the spark plug hole, not the type that requires hand pressure to maintain the seal - be sure to open the throttle valve as far as possible during the test

7   Crank the engine over at least seven compression strokes and watch the gauge. The compression should build up quickly in a healthy engine. Low compression on the first stroke, followed by gradually increasing pressure on successive strokes, indicates worn piston rings. A low compression reading on the first stroke, which doesn't build up during successive strokes, indicates leaking valves or a blown head gasket (a cracked head could also be the cause). Deposits on the undersides of the valve heads can also cause low compression. Record the highest gauge reading obtained.

8   Repeat the procedure for the remaining cylinders and compare the results to this Chapter's Specifications.

9   Add some engine oil (about three squirts from a plunger-type oil can) to each cylinder, through the spark plug hole, and repeat the test.

10   If the compression increases after the oil is added, the piston rings are definitely worn. If the compression doesn't increase significantly, the leakage is occurring at the valves or head gasket. Leakage past the valves may be caused by burned valve seats and/or faces or warped, cracked or bent valves.

11   If two adjacent cylinders have equally low compression, there's a strong possibility that the head gasket between them is blown. The appearance of coolant in the combustion chambers or the crankcase would verify this condition.

12   If one cylinder is slightly lower than the others, and the engine has a slightly rough idle, a worn lobe on the camshaft could be the cause.

13   If the compression is unusually high, the combustion chambers are probably coated with carbon deposits. If that's the case, the cylinder head(s) should be removed and decarbonized.

14   If compression is way down or varies greatly between cylinders, it would be a good idea to have a leak-down test performed by an automotive repair shop. This test will pinpoint exactly where the leakage is occurring and how severe it is.

4   Vacuum gauge diagnostic checks

▶ **Refer to illustrations 4.4 and 4.6**

A vacuum gauge provides inexpensive but valuable information about what is going on in the engine. You can check for worn rings or cylinder walls, leaking head or intake manifold gaskets, incorrect carburetor adjustments, restricted exhaust, stuck or burned valves, weak valve springs, improper ignition or valve timing and ignition problems.

Unfortunately, vacuum gauge readings are easy to misinterpret, so they should be used in conjunction with other tests to confirm the diagnosis.

Both the absolute readings and the rate of needle movement are important for accurate interpretation. Most gauges measure vacuum in inches of mercury (in-Hg). The following references to vacuum assume the diagnosis is being performed at sea level. As elevation increases (or atmospheric pressure decreases), the reading will decrease. For every 1,000 foot increase in elevation above approximately 2000 feet, the gauge readings will decrease about one inch of mercury.

Connect the vacuum gauge directly to the intake manifold vacuum, not to ported (throttle body) vacuum (see illustration). Be sure no hoses are left disconnected during the test or false readings will result.

Before you begin the test, allow the engine to warm up completely. Block the wheels and set the parking brake. With the transmission in Park, start the engine and allow it to run at normal idle speed.

✳✳ WARNING:

Keep your hands and the vacuum gauge clear of the fans and drivebelts.

Read the vacuum gauge; an average, healthy engine should normally produce about 17 to 22 in-Hg with a fairly steady needle (see illustration). Refer to the following vacuum gauge readings and what they indicate about the engine's condition:

1   A low steady reading usually indicates a leaking gasket between the intake manifold and cylinder head(s) or throttle body, a leaky vacuum hose, late ignition timing or incorrect camshaft timing. Check ignition timing with a timing light and eliminate all other possible causes,

4.4 A simple vacuum gauge can be handy in diagnosing engine condition and performance

utilizing the tests provided in this Chapter before you remove the timing chain cover to check the timing marks.

2   If the reading is three to eight inches below normal and it fluctuates at that low reading, suspect an intake manifold gasket leak at an intake port or a faulty fuel injector.

3   If the needle has regular drops of about two-to-four inches at a steady rate, the valves are probably leaking. Perform a compression check or leak-down test to confirm this.

4   An irregular drop or down-flick of the needle can be caused by a sticking valve or an ignition misfire. Perform a compression check or leak-down test and read the spark plugs.

5   A rapid vibration of about four in-Hg vibration at idle combined with exhaust smoke indicates worn valve guides. Perform a leak-down test to confirm this. If the rapid vibration occurs with an increase in engine speed, check for a leaking intake manifold gasket or head gasket, weak valve springs, burned valves or ignition misfire.

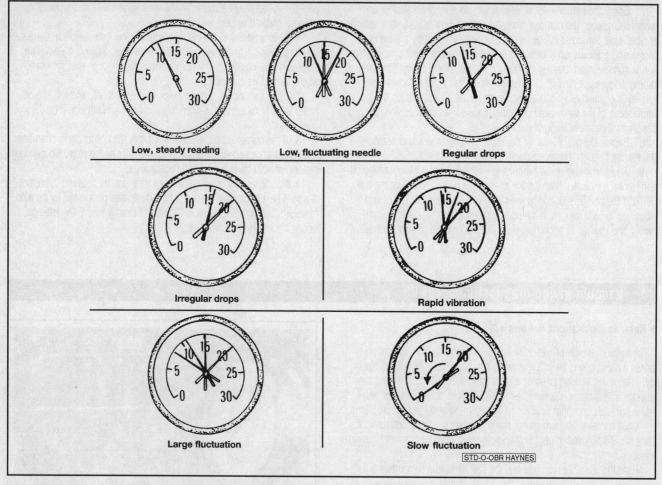

4.6 Typical vacuum gauge readings

6   A slight fluctuation, say one inch up and down, may mean ignition problems. Check all the usual tune-up items and, if necessary, run the engine on an ignition analyzer.

7   If there is a large fluctuation, perform a compression or leakdown test to look for a weak or dead cylinder or a blown head gasket.

8   If the needle moves slowly through a wide range, check for a clogged PCV system, incorrect idle fuel mixture, throttle body or intake manifold gasket leaks.

9   Check for a slow return after revving the engine by quickly snapping the throttle open until the engine reaches about 2,500 rpm and let it shut. Normally the reading should drop to near zero, rise above normal idle reading (about 5 in-Hg over) and then return to the previous idle reading. If the vacuum returns slowly and doesn't peak when the throttle is snapped shut, the rings may be worn. If there is a long delay, look for a restricted exhaust system (often the muffler or catalytic converter). An easy way to check this is to temporarily disconnect the exhaust ahead of the suspected part and redo the test.

5   Engine rebuilding alternatives

The do-it-yourselfer is faced with a number of options when purchasing a rebuilt engine. The major considerations are cost, warranty, parts availability and the time required for the rebuilder to complete the project. The decision to replace the engine block, piston/connecting rod assemblies and crankshaft depends on the final inspection results of your engine. Only then can you make a cost effective decision whether to have your engine overhauled or simply purchase an exchange engine for your vehicle.

Some of the rebuilding alternatives include:

Individual parts - If the inspection procedures reveal that the engine block and most engine components are in reusable condition, purchasing individual parts and having a rebuilder rebuild your engine may be the most economical alternative. The block, crankshaft and piston/connecting rod assemblies should all be inspected carefully by a machine shop first.

Short block - A short block consists of an engine block with a crankshaft and piston/connecting rod assemblies already installed. All new bearings are incorporated and all clearances will be correct. The existing camshafts, valve train components, cylinder head and external parts can be bolted to the short block with little or no machine shop work necessary.

Long block - A long block consists of a short block plus an oil pump, oil pan, cylinder head, valve cover, camshaft and valve train components, timing sprockets and chain or gears and timing cover. All components are installed with new bearings, seals and gaskets incorporated throughout. The installation of manifolds and external parts is all that's necessary.

Low mileage used engines - Some companies now offer low mileage used engines which is a very cost effective way to get your vehicle up and running again. These engines often come from vehicles that have been in totaled in accidents or come from other countries that have a higher vehicle turn over rate. A low mileage used engine also usually has a similar warranty like the newly remanufactured engines.

Give careful thought to which alternative is best for you and discuss the situation with local automotive machine shops, auto parts dealers and experienced rebuilders before ordering or purchasing replacement parts.

6 Engine removal - methods and precautions

▶ **Refer to illustrations 6.1, 6.2, 6.3 and 6.4**

If you've decided that an engine must be removed for overhaul or major repair work, several preliminary steps should be taken. Read all removal and installation procedures carefully prior to committing this job. These engines are removed by lowering to the floor and then raising the vehicle sufficiently to slide it out; this will require a vehicle hoist.

Locating a suitable place to work is extremely important. Adequate work space, along with storage space for the vehicle, will be needed. If a shop or garage isn't available, at the very least a flat, level, clean work surface made of concrete or asphalt is required. Cleaning the engine compartment and engine before beginning the removal procedure will help keep tools clean and organized (see illustrations 6.1 and 6.2).

An engine hoist or A-frame will also be necessary. Make sure the equipment is rated in excess of the combined weight of the engine and transmission. Safety is of primary importance, considering the potential hazards involved in lifting the engine out of the vehicle.

If you're a novice at engine removal, get at least one helper. One person cannot easily do all the things you need to do to lift a big heavy engine out of the engine compartment. Also helpful is to seek advice and assistance from someone who's experienced in engine removal.

Plan the operation ahead of time. Arrange for or obtain all of the tools and equipment you'll need prior to beginning the job (see illustrations 6.3 and 6.4). some of the equipment necessary to perform engine

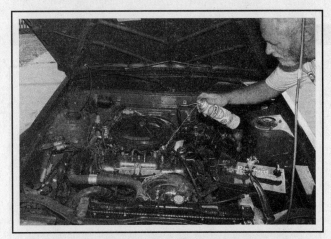

6.1 After tightly wrapping water-vulnerable components, use a spray cleaner on everything, with particular concentration on the greasiest areas, usually around the valve cover and lower edges of the block. If one section dries out, apply more cleaner

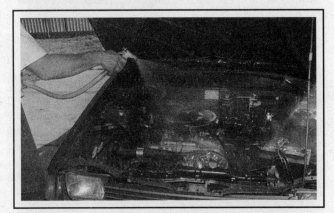

6.2 Depending on how dirty the engine is, let the cleaner soak in according to the directions and then hose off the grime and cleaner. Get the rinse water down into every area you can get at; then dry important components with a hair dryer or paper towels

6.3 Get an engine hoist that's strong enough to easily lift your engine in and out of the engine compartment; an adapter, like the one shown here, can be used to change the angle of the engine as it's being removed or installed

6.4 Get an engine stand sturdy enough to firmly support the engine while you're working on it. Stay away from three-wheeled models - they have a tendency to tip over more easily, so get a four-wheeled unit

removal and installation safely and with relative ease are (in addition to an engine hoist) a heavy duty floor jack, complete sets of wrenches and sockets as described in the front of this manual, wooden blocks, plenty of rags and cleaning solvent for mopping up spilled oil, coolant and gasoline. If the hoist must be rented, make sure that you arrange for it in advance and have everything disconnected and/or removed before bringing the hoist home. This will save you money and time.

Plan for the vehicle to be out of use for quite a while. A machine shop can do the work that is beyond the scope of the home mechanic. Machine shops often have a busy schedule, so before removing the engine, consult the shop for an estimate of how long it will take to rebuild or repair the components that may need work.

7 Engine - removal and installation

➡Note 1: Engine removal on these vehicles is a difficult job, especially for the do-it-yourself mechanic working at home. Because of the vehicle's design, the manufacturer states that the engine and transaxle have to be removed as a unit from the bottom of the vehicle, not the top. With a floor jack and jackstands, the vehicle can't be raised high enough or supported safely enough for the engine/transaxle assembly to slide out from underneath. The manufacturer recommends that removal of the engine/transaxle assembly only be performed with the use of a frame-contact type vehicle hoist.

➡Note 2: Keep in mind that during this procedure you'll have to adjust the height of the vehicle with the vehicle hoist to perform certain operations.

REMOVAL

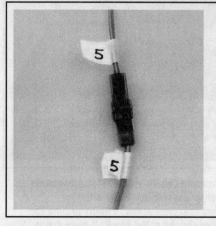

7.15 Label both ends of each wire or hose before disconnecting it

▶ **Refer to illustrations 7.15, 7.30a, 7.30b and 7.38**

1 Park the vehicle on a frame-contact type vehicle hoist, then engage the arms of the hoist with the jacking points of the vehicle. Raise the hoist arms until they contact the vehicle, but not so much that the wheels come off the ground.

2 Relieve the fuel system pressure (see Chapter 4).

3 On 2004 and later Highlander V6 models and all Lexus models, remove the top cowl/ventilation cover (see Chapter 11).

4 On 2004 and later Highlander V6 models and all Lexus models, remove the windshield wiper arms and the windshield wiper motor (see Chapter 12).

5 On 2004 and later Highlander V6 models and all Lexus models, remove the lower cowl cover (see Chapter 11).

6 Remove the battery and battery tray (see Chapter 5).

7 Remove the air filter housing and air intake duct (see Chapter 4).

8 Disconnect the accelerator cable from the throttle body (see Chapter 4), if equipped.

9 Disconnect the cruise control cable from the throttle body, if equipped.

10 Disconnect the shift cable from the transaxle (see Chapter 7).

11 Remove the alternator (see Chapter 5).

12 Remove the air conditioning compressor without disconnecting the refrigerant lines (see Chapter 3). Tie the compressor out of the way (but not to the subframe).

13 Disconnect the intermediate shaft from the steering gear (see Chapter 10).

14 Disconnect the fuel feed hose from the fuel rail (see Chapter 4). Plug the line and fitting.

15 Clearly label and disconnect all vacuum lines, emissions hoses, electrical connectors and ground straps connecting the engine and transaxle to the vehicle. Masking tape and/or a touch up paint applica-tor work well for marking items (see illustration). Take instant photos or sketch the locations of components and brackets, if necessary.

16 Using a suction gun, remove as much fluid from the power steer-ing fluid reservoir as possible. Detach the power steering hose from the return line.

17 Working on the powertrain mounts, remove the engine movement control rod (see Chapter 2A or 2B). Also remove its mounting bracket from the cylinder head.

18 Disconnect the electrical harness from the main fuse/relay box in the engine compartment (see Chapter 5, illustration 4.4b). Disconnect any other electrical connectors attached to the harness leading to the PCM.

7.30a Remove the transaxle bellhousing access cover bolts . . .

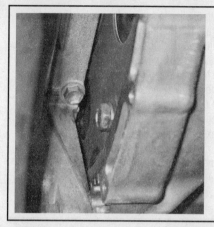

7.30b . . . then remove the torque converter bolts

19 Loosen the front wheel lug nuts and the driveaxle/hub nuts (see Chapter 8).

20 Raise the vehicle on the hoist and remove the front wheels.

21 Remove the under-vehicle splash shield(s) (see Chapter 2A or 2B). Also remove the inner fender liners and fender apron seals (see Chapter 11).

22 Disconnect the stabilizer bar links from the stabilizer bar and the tie-rod ends from the steering knuckles (see Chapter 10).

23 Remove the driveaxles (see Chapter 8).

24 On 4WD models, remove the driveshaft (see Chapter 8).

25 On 4WD models, drain the transfer case fluid (see Chapter 1).

26 Drain the engine coolant (see Chapter 1).

27 Drain the engine oil (see Chapter 1).

28 On four-cylinder models, disconnect the oil cooler lines from the external oil cooler (see Chapter 3).

29 Drain the transaxle fluid (see Chapter 1).

30 Remove the driveplate-to-torque converter bolts (see illustrations). Also disconnect the transaxle fluid cooler hoses from the transmission fluid cooler (see Chapter 7).

31 Disconnect the fluid lines from the power steering pump. Also detach the pressure and return lines from the power steering gear.

32 Unbolt the exhaust pipe(s) from the exhaust manifold/catalytic converter assemblies (see Chapter 2A or 2B), then remove the front portion of the exhaust system (see Chapter 4).

33 Lower the vehicle and disconnect the radiator hoses and heater hoses from the engine. Remove the radiator (see Chapter 3).

34 Support the engine/transaxle assembly from above with a hoist. Attach the hoist chain to the lifting brackets. If no lifting brackets or hooks are present, lifting hooks may be available from your local auto parts store or dealer parts department. If not, you will have to fasten the chains to some substantial parts of the engine - ones that are strong enough to take the weight, but in locations that will provide good balance. If you're attaching a chain to a stud on the engine, or are using a bolt passing through the chain and into a threaded hole, place a washer between the nut or bolt head and the chain and tighten the nut or bolt securely.

✳✳ WARNING:

Do not place any part of your body under the engine/transaxle when it's supported only by a hoist or other lifting device.

35 Take up the slack until there is slight tension on the hoist. Position the chain on the hoist so it balances the engine and the transaxle level with the vehicle.

➡**Note 1: Depending on the design of the engine hoist, it may be helpful to position the hoist from the side of the vehicle, so that when the engine/transaxle assembly is lowered, it will fit between the legs of the hoist.**

➡**Note 2: The sling or chain must be long enough to allow the engine hoist to lower the engine/transaxle assembly to the ground, without letting the hoist arm contact the vehicle.**

36 Support the subframe with a pair of floor jacks. Recheck to be sure that there aren't any hoses or wiring between the subframe and the vehicle. Unbolt the powertrain mounts from the subframe (see Chapter 2A or 2B), remove the subframe mounting bolts (see Chapter 10) and lower it to the floor, then remove it out from underneath the vehicle.

✳✳ WARNING:

This can be tricky, depending on the design of the engine hoist, but it is imperative that the subframe be supported securely before its mounting bolts are removed.

37 Recheck to be sure nothing is still connecting the engine or transaxle to the vehicle. Disconnect and label anything still remaining.

38 Lower the engine/transaxle assembly (see illustration). Once the engine/transaxle assembly is on the floor, disconnect the engine hoist and raise the vehicle until it clears the engine/transaxle assembly.

7.38 After the subframe has been removed, lower the engine/transaxle unit to the floor

39 Reconnect the chain or sling to support the engine and transaxle.

40 Raise the engine/transaxle assembly, then support the engine with blocks of wood or another floor jack, while leaving the sling or chain attached. Support the transaxle with another floor jack, preferably one with a transaxle jack head adapter. At this point the transfer case can be unbolted and removed (4WD models - see Chapter 7). Be very careful to ensure that the components are supported securely so they won't topple off their supports during disconnection.

41 Remove the transaxle-to-engine bolts and separate the transaxle from the engine.

42 Reconnect the lifting chain to the engine, then raise the engine and attach it to an engine stand.

INSTALLATION

43 Installation is the reverse of removal, noting the following points:

a) Check the engine/transaxle mounts. If they're worn or damaged, replace them.
b) Attach the transaxle to the engine following the procedure described in Chapter 7.
c) When installing the subframe, tighten the subframe mounting bolts to the torque listed in the Chapter 10 Specifications.
d) Tighten the driveaxle/hub nuts to the torque listed in the Chapter 8 Specifications. Tighten all steering and suspension fasteners to the torque listed in the Chapter 10 Specifications. Tighten the wheel lug nuts to the torque listed in the Chapter 1 Specifications.
e) Refill the engine coolant, oil, power steering and transaxle fluids (see Chapter 1).
f) Reconnect the battery (see Chapter 5, Section 1).
g) Run the engine and check for proper operation and leaks. Shut off the engine and recheck fluid levels.

8 Engine overhaul - disassembly sequence

1 It's much easier to disassemble the engine if it's mounted on a portable engine stand. A stand can often be rented quite cheaply from an equipment rental yard. Before the engine is mounted on a stand, the driveplate should be removed from the engine.

2 If a stand isn't available, it's possible to remove the external engine components with it blocked up on the floor. Be extra careful not to tip or drop the engine when working without a stand.

3 If you're going to obtain a rebuilt engine, all external components must come off first, to be transferred to the replacement engine. These components include:

Driveplate
Ignition system components
Emissions-related components
Engine mounts and mount brackets
Fuel injection components
Intake/exhaust manifolds
Oil filter
Thermostat and housing assembly
Water pump

➡Note: When removing the external components from the engine, pay close attention to details that may be helpful or important during installation. Note the installed position of gaskets, seals, spacers, pins, brackets, washers, bolts and other small items.

4 If you're going to obtain a short block (assembled engine block, crankshaft, pistons and connecting rods), then you should remove the timing belt, cylinder head, oil pan, oil pump pick-up tube, oil pump and water pump from your engine so that you can turn in your old short block to the rebuilder as a core. See Engine rebuilding alternatives for additional information regarding the different possibilities to be considered.

9 Balance shafts (four-cylinder models) - removal and installation

➡Note: This procedure assumes that the engine has been removed from the vehicle and the driveplate, timing chain and oil pan have also been removed (see Chapter 2A).

REMOVAL

1 The balance shafts are installed in a carrier mounted to the lower crankcase assembly. The shafts are connected through two gears which rotate them in opposite directions. These gears are driven by a chain from the crankshaft which counterbalances reciprocating masses within the engine.

2 Remove the timing chain and tensioner from the engine block (see Chapter 2A).

3 Remove the oil pump (see Chapter 2A).

4 Remove the balance shaft carrier assembly mounting bolts. Follow the reverse of the tightening sequence (see illustration 9.10).

5 Lift the balance shaft carrier assembly from the engine and remove the balance shafts.

➡Note: Note the location of the balance shaft timing marks on the front gear and rear gear on each balance shaft assembly (see illustration 9.7). These marks must align when the balance shafts are reinstalled onto the lower crankcase assembly.

CLEANING AND INSPECTION

6 Clean all components with solvent and dry thoroughly. Inspect all components for damage and wear. Pay close attention to the chain, sprocket and gear teeth and the bearing surfaces of the carrier and balance shafts. Replace defective parts as necessary.

INSTALLATION

▶ Refer to illustrations 9.7, 9.8, 9.9 and 9.10

7 Prepare each balance shaft for assembly by aligning the marks on the front gear with the marks on the rear gear (see illustration).

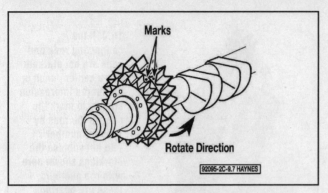

9.7 Location of the alignment marks on the front gear and the rear gear

8   Lubricate the balance shafts with clean engine oil, install them onto the lower crankcase assembly and align the balance shaft marks together (see illustration).

9   Check the alignment of the crankshaft number 1 journal. The

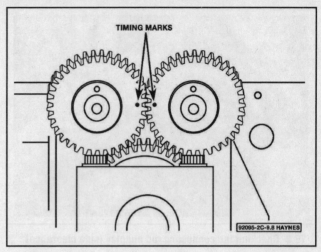

9.8 Install the balance shafts with the marks on the gears in alignment

crankshaft key should be positioned 180-degrees from the balance shafts (see illustration). Double-check the marks on the balance shafts.

10 Install the balance shaft carrier assembly and tighten the bolts to the torque listed in this Chapter's Specifications. Follow the correct bolt tightening sequence (see illustration).

11 Install the oil pump (see Chapter 2A).

12 The remainder of installation is the reverse of removal.

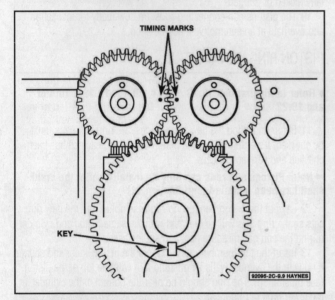

9.9 Check the balance shaft alignment marks with the crankshaft key pointing away from the balance shafts at 180-degrees

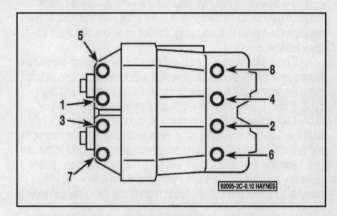

9.10 Balance shaft carrier bolt tightening sequence

10 Pistons and connecting rods - removal and installation

REMOVAL

▶ Refer to illustrations 10.1, 10.3 and 10.4

➡Note: Prior to removing the piston/connecting rod assemblies, remove the cylinder head, oil pan, and the upper (aluminum) oil pan (see Chapter 2A or 2B). Also, on four-cylinder engines, remove the balance shafts (see Section 9).

1   Use your fingernail to feel if a ridge has formed at the upper limit of ring travel (about 1/4-inch down from the top of each cylinder). If carbon deposits or cylinder wear have produced ridges, they must be completely removed with a special tool (see illustration). Follow the manufacturer's instructions provided with the tool. Failure to remove the ridges before attempting to remove the piston/connecting rod assemblies may result in piston breakage.

10.1 Before you try to remove the pistons, use a ridge reamer to remove the raised material (ridge) from the top of the cylinders

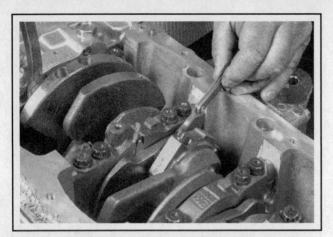

10.3 Checking the connecting rod endplay (side clearance)

10.4 If the connecting rods and caps are not marked, use a center punch or numbered impression stamps to mark the caps to the rods by cylinder number - do not confuse the markings shown here with rod numbers; these are bearing size identifications

2   After the cylinder ridges have been removed, turn the engine so the crankshaft is facing up.

3   Before the pistons and connecting rods are removed, check the connecting rod endplay with feeler gauges. Slide them between the first connecting rod and the crankshaft throw until the play is removed (see illustration). Repeat this procedure for each connecting rod. The endplay is equal to the thickness of the feeler gauge(s). Check with an automotive machine shop for the endplay service limit. If the play exceeds the service limit, new connecting rods will be required. If new rods (or a new crankshaft) are installed, the endplay may fall under the minimum allowable clearance. If it does, the rods will have to be machined to restore it. If necessary, consult an automotive machine shop for advice.

4   Check the connecting rods and caps for identification marks (see illustration). If they aren't plainly marked, use a small center-punch to make the appropriate number of indentations on each rod and cap (1, 2, 3, etc., depending on the cylinder they're associated with).

5   Loosen each of the connecting rod cap bolts 1/2-turn at a time until they can be removed by hand. Remove the number one connecting rod cap and bearing insert. Don't drop the bearing insert out of the cap.

6   Remove the bearing insert and push the connecting rod/piston assembly out through the top of the engine. Use a wooden or plastic hammer handle to push on the upper bearing surface in the connecting rod.

7   If resistance is felt, double-check to make sure that all of the ridge was removed from the cylinder.

8   Repeat the procedure for the remaining cylinders.

9   After removal, reassemble the connecting rod caps and bearing

inserts in their respective connecting rods and install the cap bolts finger tight. Leaving the old bearing inserts in place until reassembly will help prevent the connecting rod bearing surfaces from being accidentally nicked or gouged.

10 The pistons and connecting rods are now ready for inspection and overhaul at an automotive machine shop.

PISTON RING INSTALLATION

▶ **Refer to illustrations 10.13, 10.14, 10.15, 10.19a, 10.19b and 10.22**

11 Before installing the new piston rings, the ring end gaps must be checked. It's assumed that the piston ring side clearance has been checked and verified correct.

➡**Note: Pistons and rods can only be installed after the crankshaft has been installed (see Section 11).**

12 Lay out the piston/connecting rod assemblies and the new ring sets so the ring sets will be matched with the same piston and cylinder during the end gap measurement and engine assembly.

13 Insert the top (number one) ring into the first cylinder and square it up with the cylinder walls by pushing it in with the top of the piston (see illustration). The ring should be near the bottom of the cylinder, at the lower limit of ring travel.

14 To measure the end gap, slip feeler gauges between the ends of

10.13 Install the piston ring into the cylinder then push it down into position using a piston so the ring will be square in the cylinder

10.14 With the ring square in the cylinder, measure the ring end gap with a feeler gauge

10.15 If the ring end gap is too small, clamp a file in a vise as shown and file the piston ring ends - be sure to file the ends squarely and finish by removing all raised material or burrs with a fine stone

the ring until a gauge equal to the gap width is found (see illustration). The feeler gauge should slide between the ring ends with a slight amount of drag. Check with an automotive machine shop for the correct end gap for your engine. If the gap is larger or smaller than specified, double-check to make sure you have the correct rings before proceeding.

15 If the gap is too small, it must be enlarged or the ring ends may come in contact with each other during engine operation, which can cause serious damage to the engine. The end gap can be increased by filing the ring ends very carefully with a fine file. Mount the file in a vise equipped with soft jaws, slip the ring over the file with the ends contacting the file face and slowly move the ring to remove material from the ends. When performing this operation, file only by pushing the ring from the outside end of the file towards the vise (see illustration).

16 Excess end gap isn't critical unless it's greater than approximately 0.040-inch. Again, double-check to make sure you have the correct ring type.

17 Repeat the procedure for each ring that will be installed in the first cylinder and for each ring in the remaining cylinders. Remember to keep rings, pistons and cylinders matched up.

18 Once the ring end gaps have been checked/corrected, the rings can be installed on the pistons.

➡**Note: On the V6 engines the pistons for the left cylinder bank are marked either with an L or a single circular mark at the front of the piston. The pistons for the right cylinder bank are marked either with an R or two circular marks at the front of the piston.**

10.19b DO NOT use a piston ring installation tool when installing the oil control side rails

10.19a Installing the spacer/expander in the oil ring groove

19 The oil control ring (lowest one on the piston) is usually installed first. It's composed of three separate components. Slip the spacer/expander into the groove (see illustration). If an anti-rotation tang is used, make sure it's inserted into the drilled hole in the ring groove. Next, install the upper side rail in the same manner (see illustration). Don't use a piston ring installation tool on the oil ring side rails, as they may be damaged. Instead, place one end of the side rail into the groove between the spacer/expander and the ring land, hold it firmly in place and slide a finger around the piston while pushing the rail into the groove. Finally, install the lower side rail.

20 After the three oil ring components have been installed, check to make sure that both the upper and lower side rails can be rotated smoothly inside the ring grooves.

21 The number two (middle) ring is installed next. It's usually stamped with a mark that must face up, toward the top of the piston. Do not mix up the top and middle rings, as they have different cross-sections.

➡**Note: Always follow the instructions printed on the ring package or box - different manufacturers may require different approaches.**

22 Use a piston ring installation tool and make sure the identification mark is facing the top of the piston, then slip the ring into the middle groove on the piston (see illustration). Don't expand the ring any more than necessary to slide it over the piston.

23 Install the number one (top) ring in the same manner. Make sure

10.22 Use a piston ring installation tool to install the number 2 and the number 1 (top) rings - be sure the directional mark on the piston ring(s) is facing toward the top of the piston

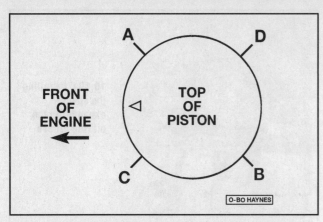

10.30a On four-cylinder engines, position the piston ring end gaps as shown here before installing the piston/connecting rod assemblies into the engine

A Top compression ring gap
and oil ring spacer gap
B Second compression ring
C Upper oil ring gap
D Lower oil ring gap

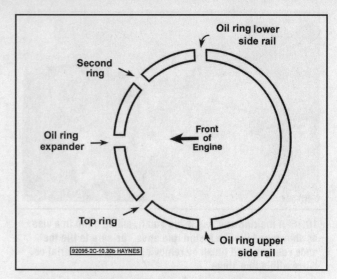

10.30b On V6 engines, position the piston ring end gaps as shown here before installing the piston/connecting rod assemblies into the engine

the mark is facing up. Be careful not to confuse the number one and number two rings.

24 Repeat the procedure for the remaining pistons and rings.

INSTALLATION

25 Before installing the piston/connecting rod assemblies, the cylinder walls must be perfectly clean, the top edge of each cylinder bore must be chamfered, and the crankshaft must be in place.

26 Remove the cap from the end of the number one connecting rod (refer to the marks made during removal).

27 Remove the original bearing inserts and wipe the bearing surfaces of the connecting rod and cap with a clean, lint-free cloth. They must be kept spotlessly clean.

Connecting rod bearing oil clearance check

▶ Refer to illustrations 10.30a, 10.30b, 10.35, 10.37 and 10.41

28 Clean the back side of the new upper bearing insert, then lay it in place in the connecting rod. Make sure the tab on the bearing fits into the recess in the rod. Don't hammer the bearing insert into place and be very careful not to nick or gouge the bearing face. Don't lubricate the bearing at this time.

29 Clean the back side of the other bearing insert and install it in the rod cap. Again, make sure the tab on the bearing fits into the recess in the cap, and don't apply any lubricant. It's critically important that the mating surfaces of the bearing and connecting rod are perfectly clean and oil free when they're assembled.

30 Position the piston ring gaps around the piston as shown (see illustrations).

31 Lubricate the piston and rings with clean engine oil and attach a piston ring compressor to the piston. Leave the skirt protruding about 1/4-inch to guide the piston into the cylinder. The rings must be compressed until they're flush with the piston.

32 Rotate the crankshaft until the number one connecting rod journal is at BDC (bottom dead center) and apply a liberal coat of engine oil to the cylinder walls.

33 With the mark (one or two dots, see the **Note** in Step 18) on top of the piston facing the front (timing belt/chain end) of the engine, gently insert the piston/connecting rod assembly into the number one

cylinder bore and rest the bottom edge of the ring compressor on the engine block.

➡Note: The connecting rod also has a mark on it that must face the correct direction. On all models, the marks on the connecting rods face the front (timing belt/chain) of the engine.

34 Tap the top edge of the ring compressor to make sure it's contacting the block around its entire circumference.

35 Gently tap on the top of the piston with the end of a wooden or plastic hammer handle (see illustration) while guiding the end of the connecting rod into place on the crankshaft journal. The piston rings may try to pop out of the ring compressor just before entering the cylinder bore, so keep some downward pressure on the ring compressor. Work slowly, and if any resistance is felt as the piston enters the cylinder, stop immediately. Find out what's hanging up and fix it before proceeding. Do not, for any reason, force the piston into the cylinder - you might break a ring and/or the piston.

36 Once the piston/connecting rod assembly is installed, the connecting rod bearing oil clearance must be checked before the rod cap is permanently installed.

10.35 Use a plastic or wooden hammer handle to push the piston into the cylinder

10.37 Place Plastigage on each connecting rod bearing journal parallel to the crankshaft centerline

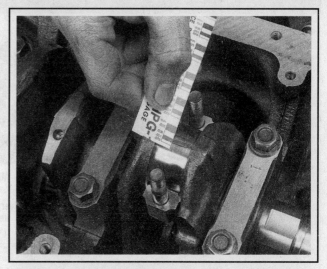

10.41 Use the scale on the Plastigage package to determine the bearing oil clearance - be sure to measure the widest part of the Plastigage and use the correct scale; it comes with both standard and metric scales

37 Cut a piece of the appropriate size Plastigage slightly shorter than the width of the connecting rod bearing and lay it in place on the number one connecting rod journal, parallel with the journal axis (see illustration).

38 Clean the connecting rod cap bearing face and install the rod cap. Make sure the mating mark on the cap is on the same side as the mark on the connecting rod.

39 Install the rod bolts and tighten them to the torque listed in this Chapter's Specifications in two steps.

➥**Note: Use a thin-wall socket to avoid erroneous torque readings that can result if the socket is wedged between the rod cap and the bolt. If the socket tends to wedge itself between the fastener and the cap, lift up on it slightly until it no longer contacts the cap. DO NOT rotate the crankshaft at any time during this operation.**

40 Remove the fasteners and detach the rod cap, being very careful not to disturb the Plastigage.

41 Compare the width of the crushed Plastigage to the scale printed on the Plastigage envelope to obtain the oil clearance (see illustration). The connecting rod oil clearance is usually about 0.001 to 0.002 inch. Consult an automotive machine shop for the clearance specified for the rod bearings on your engine.

42 If the clearance is not as specified, the bearing inserts may be the wrong size (which means different ones will be required). Before deciding that different inserts are needed, make sure that no dirt or oil was between the bearing inserts and the connecting rod or cap when the clearance was measured. Also, recheck the journal diameter. If the Plastigage was wider at one end than the other, the journal may be tapered. If the clearance still exceeds the limit specified, the bearing will have to be replaced with an undersize bearing.

※※ **CAUTION:**

When installing a new crankshaft always use a standard size bearing.

Final installation

43 Carefully scrape all traces of the Plastigage material off the rod journal and/or bearing face. Be very careful not to scratch the bearing - use your fingernail or the edge of a plastic card.

44 Make sure the bearing faces are perfectly clean, then apply a uniform layer of clean moly-base grease or engine assembly lube to both of them. You'll have to push the piston into the cylinder to expose the face of the bearing insert in the connecting rod.

45 Slide the connecting rod back into place on the journal, install the rod cap and bolts, tightening them to the torque listed in this Chapter's Specifications in two steps.

46 Repeat the entire procedure for the remaining pistons/connecting rods.

47 The important points to remember are:

a) *Keep the back sides of the bearing inserts and the insides of the connecting rods and caps perfectly clean when assembling them.*

b) *Make sure you have the correct piston/rod assembly for each cylinder.*

c) *The mark on the piston must face the front (timing belt/chain end) of the engine.*

d) *Lubricate the cylinder walls liberally with clean oil.*

e) *Lubricate the bearing faces when installing the rod caps after the oil clearance has been checked.*

48 After all the piston/connecting rod assemblies have been correctly installed, rotate the crankshaft a number of times by hand to check for any obvious binding.

49 As a final step, check the connecting rod endplay again. If it was correct before disassembly and the original crankshaft and rods were reinstalled, it should still be correct. If new rods or a new crankshaft were installed, the endplay may be inadequate. If so, the rods will have to be removed and taken to an automotive machine shop for resizing.

ENGINE BEARING ANALYSIS

Debris

Babbitt bearing embedded with debris from machinings

Microscopic detail of debris

Microscopic detail of gouges

Overplated copper alloy bearing gouged by cast iron debris

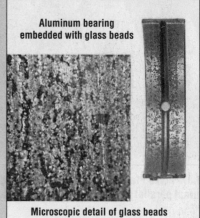

Aluminum bearing embedded with glass beads

Microscopic detail of glass beads

Damaged lining caused by dirt left on the bearing back

Misassembly

Result of a lower half assembled as an upper - blocking the oil flow

Excessive oil clearance is indicated by a short contact arc

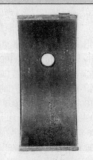

Polished and oil-stained backs are a result of a poor fit in the housing bore

Result of a wrong, reversed, or shifted cap

Overloading

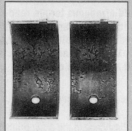

Damage from excessive idling which resulted in an oil film unable to support the load imposed

Damaged upper connecting rod bearings caused by engine lugging; the lower main bearings (not shown) were similarly affected

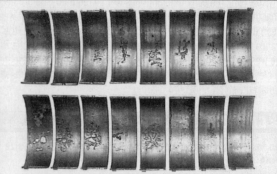

The damage shown in these upper and lower connecting rod bearings was caused by engine operation at a higher-than-rated speed under load

Misalignment

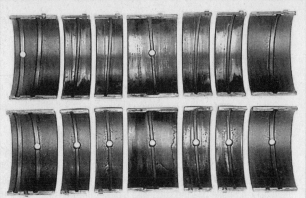

A warped crankshaft caused this pattern of severe wear in the center, diminishing toward the ends

A poorly finished crankshaft caused the equally spaced scoring shown

A bent connecting rod led to the damage in the "V" pattern

A tapered housing bore caused the damage along one edge of this pair

Lubrication

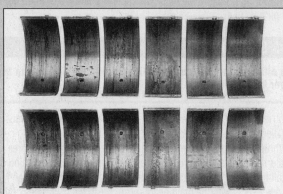

Result of dry start: The bearings on the left, farthest from the oil pump, show more damage

Result of a low oil supply or oil starvation

Severe wear as a result of inadequate oil clearance

Corrosion

Microscopic detail of corrosion

Corrosion is an acid attack on the bearing lining generally caused by inadequate maintenance, extremely hot or cold operation, or interior oils or fuels

Microscopic detail of cavitation

Example of cavitation - a surface erosion caused by pressure changes in the oil film

Damage from excessive thrust or insufficient axial clearance

Bearing affected by oil dilution caused by excessive blow-by or a rich mixture

11 Crankshaft - removal and installation

REMOVAL

▶ **Refer to illustrations 11.1 and 11.3**

➡**Note 1: The crankshaft can be removed only after the engine has been removed from the vehicle. It's assumed that the driveplate, crankshaft pulley, timing belt/chain, oil pan, oil pump body, oil filter and piston/connecting rod assemblies have already been removed. On V6 engines, the rear main oil seal retainer must be unbolted and separated from the block before proceeding with crankshaft removal.**

➡**Note 2: V6 engines have main caps that are retained by four bolts each, plus side bolts (one on each side) that pass through the block and thread into the installed main caps.**

1 Before the crankshaft is removed, measure the endplay. Mount a dial indicator with the indicator in line with the crankshaft and touching the end of the crankshaft (see illustration).

2 Pry the crankshaft all the way to the rear and zero the dial indicator. Next, pry the crankshaft to the front as far as possible and check the reading on the dial indicator. The distance traveled is the endplay. A typical crankshaft endplay will fall between 0.003 to 0.010-inch. If it's greater than that, check the crankshaft thrust surfaces for wear after it's removed. If no wear is evident, new main bearings should correct the endplay.

3 If a dial indicator isn't available, feeler gauges can be used. Gently pry the crankshaft all the way to the front of the engine. Slip feeler gauges between the crankshaft and the front face of the thrust bearing or washer to determine the clearance (see illustration).

4 Loosen the main bearing cap bolts 1/4-turn at a time each, until they can be removed by hand (see illustration 11.19a). On V6 engines, remove the side bolts first (in sequence) (see illustration 11.19c), then the inner bolts (see illustration 11.19b).

5 Gently tap the main bearing caps with a soft-face hammer. Pull the main bearing cap straight up and off the cylinder block. Try not to drop the bearing inserts if they come out with the cap.

6 Carefully lift the crankshaft out of the engine. It may be a good idea to have an assistant available, since the crankshaft is quite heavy and awkward to handle. With the bearing inserts in place inside the engine block and main bearing caps, reinstall the main bearing caps onto the engine block and tighten the bolts finger tight. Make sure you

install the main bearing cap(s) with the arrow facing the front end of the engine.

INSTALLATION

7 Crankshaft installation is the first step in engine reassembly. It's assumed at this point that the engine block and crankshaft have been cleaned, inspected and repaired or reconditioned.

8 Position the engine block with the bottom facing up.

9 Remove the mounting bolts and lift off the main bearing cap(s).

10 If they're still in place, remove the original bearing inserts from the block and from the main bearing caps. Wipe the bearing surfaces of the block and main bearing caps with a clean, lint-free cloth. They must be kept spotlessly clean. This is critical for determining the correct bearing oil clearance.

MAIN BEARING OIL CLEARANCE CHECK

▶ **Refer to illustrations 11.17, 11.19a, 11.19b, 11.19c and 11.21**

11 Without mixing them up, clean the back sides of the new upper main bearing inserts (with grooves and oil holes) and lay one in each main bearing saddle in the block. Each upper bearing has an oil groove and oil hole in it.

❋❋ CAUTION:

The oil holes in the block must line up with the oil holes in the upper bearing inserts.

If you're working on a V6 engine, the thrust bearings (washers) must be installed in the number two cap and saddle. On four-cylinder engines, the thrust bearings (washers) must be installed in the number three (center) cap. Install the thrust washers with the grooved side facing out. Install the thrust washers so that one set is located in the block and the other set is with the main bearing cap. Clean the back sides of the lower main bearing inserts (without grooves) and lay them in the corresponding caps. Make sure the tab on the bearing insert fits into the recess in the block or main bearing cap.

❋❋ CAUTION:

Do not hammer the bearing insert into place and don't nick or gouge the bearing faces. DO NOT apply any lubrication at this time.

11.1 Checking crankshaft endplay with a dial indicator

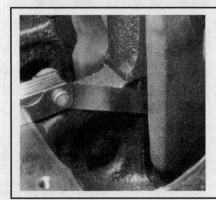

11.3 Checking crankshaft endplay with feeler gauges at the thrust bearing journal

11.17 Place the Plastigage onto the crankshaft bearing journal as shown

12 Clean the faces of the bearing inserts in the block and the crankshaft main bearing journals with a clean, lint-free cloth.

13 Check or clean the oil holes in the crankshaft, as any dirt here can go only one way - straight through the new bearings.

14 Once you're certain the crankshaft is clean, carefully lay it in position in the cylinder block.

15 Before the crankshaft can be permanently installed, the main bearing oil clearance must be checked.

16 Cut several strips of the appropriate size of Plastigage (they must be slightly shorter than the width of the main bearing journal).

17 Place one piece on each crankshaft main bearing journal, parallel with the journal axis (see illustration).

18 Clean the faces of the bearing inserts in the main bearing caps. Hold the bearing inserts in place and install the caps onto the crankshaft and cylinder block. DO NOT disturb the Plastigage. Make sure you install the main bearing cap with the arrow facing the front of the engine.

19 Apply clean engine oil to all bolt threads prior to installation, then install all bolts finger-tight. Tighten main bearing cap bolts in the sequence shown (see illustrations) progressing in two steps, to the torque listed in this Chapter's Specifications. DO NOT rotate the crankshaft at any time during this operation.

11.19a Main bearing cap bolt tightening sequence - four-cylinder models

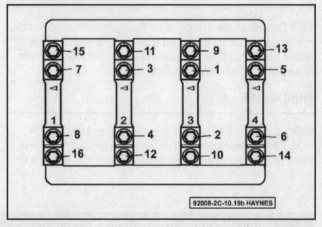

11.19b Main bearing cap bolt tightening sequence - V6 models

20 Remove the bolts in the reverse order of the tightening sequence and carefully lift the main bearing cap straight up and off the block. Do not disturb the Plastigage or rotate the crankshaft. If the main bearing cap is difficult to remove, tap it gently from side-to-side with a soft-face hammer to loosen it.

21 Compare the width of the crushed Plastigage on each journal

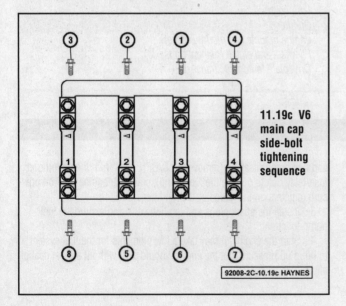

11.19c V6 main cap side-bolt tightening sequence

11.21 Use the scale on the Plastigage package to determine the bearing oil clearance - be sure to measure the widest part of the Plastigage and use the correct scale; it comes with both standard and metric scales

to the scale printed on the Plastigage envelope to determine the main bearing oil clearance (see illustration). A typical main bearing oil clearance should fall between 0.0015 to 0.0023-inch. Check with an automotive machine shop for the clearance specified for your engine.

22 If the clearance is not as specified, the bearing inserts may be the wrong size (which means different ones will be required). Before deciding if different inserts are needed, make sure that no dirt or oil was between the bearing inserts and the cap or block when the clearance was measured. If the Plastigage was wider at one end than the other, the crankshaft journal may be tapered. If the clearance still exceeds the limit specified, the bearing insert(s) will have to be replaced with an undersize bearing insert(s).

❋❋ CAUTION:

When installing a new crankshaft always install a standard bearing insert set.

23 Carefully scrape all traces of the Plastigage material off the main bearing journals and/or the bearing insert faces. Be sure to remove all residue from the oil holes. Use your fingernail or the edge of a plastic card - don't nick or scratch the bearing faces.

FINAL INSTALLATION

24 Carefully lift the crankshaft out of the cylinder block.
25 Clean the bearing insert faces in the cylinder block, then apply a thin, uniform layer of moly-base grease or engine assembly lube to each of the bearing surfaces. Be sure to coat the thrust faces as well as the journal face of the thrust bearing.

26 Make sure the crankshaft journals are clean, then lay the crankshaft back in place in the cylinder block.
27 Clean the bearing insert faces and then apply the same lubricant to them.
28 Hold the bearing inserts in place and install the main bearing caps on the crankshaft and cylinder block. Tap the bearing caps into place with a brass punch or a soft-face hammer.
29 Apply clean engine oil to the bolt threads, wipe off any excess oil and then install the bolts finger-tight.
30 Tighten the main bearing cap bolts in the indicated sequence (see illustration 11.19a or 11.19b), to 10 or 12 foot-pounds.
31 Push the crankshaft forward using a screwdriver or prybar to seat the thrust bearing. Once the crankshaft is pushed fully forward to seat the thrust bearing, leave the screwdriver in position so that force stays on the crankshaft until after all main bearing cap bolts have been tightened.
32 Tighten the main bearing cap bolts in two steps in the indicated sequence (see illustration 11.19a or 11.19b) and to the torque and angle listed in this Chapter's Specifications. On V6 engines, tighten the main cap side-bolts last (see illustration 11.19c).
33 Recheck crankshaft endplay with a feeler gauge or a dial indicator. The endplay should be correct if the crankshaft thrust faces aren't worn or damaged and if new bearings have been installed.
34 Rotate the crankshaft a number of times by hand to check for any obvious binding. It should rotate with a running torque of 50 in-lbs or less. If the running torque is too high, correct the problem at this time.
35 Install a new rear main oil seal (see Chapter 2A or 2B).

12 Engine overhaul - reassembly sequence

1 Before beginning engine reassembly, make sure you have all the necessary new parts, gaskets and seals as well as the following items on hand:

    Common hand tools
    A 1/2-inch drive torque wrench
    New engine oil
    Gasket sealant
    Thread locking compound

2 If you obtained a short block it will be necessary to install the cylinder head, the oil pump and pick-up tube, the oil pan, the water pump, the timing belt and timing cover, and the valve cover (see Chapter 2A or 2B). In order to save time and avoid problems, the external components must be installed in the following general order:

    Thermostat and housing cover
    Water pump
    Intake and exhaust manifolds
    Fuel injection components
    Emission control components
    Spark plugs
    Ignition coils
    Oil filter
    Engine mounts and mount brackets
    Flywheel and clutch (manual transaxle)
    Driveplate (automatic transaxle)

13 Initial start-up and break-in after overhaul

❋❋ WARNING:

Have a fire extinguisher handy when starting the engine for the first time.

1 Once the engine has been installed in the vehicle, double-check the engine oil and coolant levels.
2 With the spark plugs out of the engine and the ignition system and fuel pump disabled (remove the C/OPN relay from the underhood fuse/relay box to disable the fuel pump), crank the engine until oil pressure registers on the gauge or the light goes out.
3 Install the spark plugs and ignition coils and restore the fuel pump function.
4 Start the engine. It may take a few moments for the fuel system to build up pressure, but the engine should start without a great deal of effort.

5   After the engine starts, it should be allowed to warm up to normal operating temperature. While the engine is warming up, make a thorough check for fuel, oil and coolant leaks.

6   Shut the engine off and recheck the engine oil and coolant levels.

7   Drive the vehicle to an area with minimum traffic, accelerate from 30 to 50 mph, then allow the vehicle to slow to 30 mph with the throttle closed. Repeat the procedure 10 or 12 times. This will load the piston rings and cause them to seat properly against the cylinder walls. Check again for oil and coolant leaks.

8   Drive the vehicle gently for the first 500 miles (no sustained high speeds) and keep a constant check on the oil level. It is not unusual for an engine to use oil during the break-in period.

9   At approximately 500 to 600 miles, change the oil and filter.

10 For the next few hundred miles, drive the vehicle normally. Do not pamper it or abuse it.

11 After 2,000 miles, change the oil and filter again and consider the engine broken in.

GLOSSARY

B

Backlash - The amount of play between two parts. Usually refers to how much one gear can be moved back and forth without moving gear with which it's meshed.

Bearing Caps - The caps held in place by nuts or bolts which, in turn, hold the bearing surface. This space is for lubricating oil to enter.

Bearing clearance - The amount of space left between shaft and bearing surface. This space is for lubricating oil to enter.

Bearing crush - The additional height which is purposely manufactured into each bearing half to ensure complete contact of the bearing back with the housing bore when the engine is assembled.

Bearing knock - The noise created by movement of a part in a loose or worn bearing.

Blueprinting - Dismantling an engine and reassembling it to EXACT specifications.

Bore - An engine cylinder, or any cylindrical hole; also used to describe the process of enlarging or accurately refinishing a hole with a cutting tool, as to bore an engine cylinder. The bore size is the diameter of the hole.

Boring - Renewing the cylinders by cutting them out to a specified size. A boring bar is used to make the cut.

Bottom end - A term which refers collectively to the engine block, crankshaft, main bearings and the big ends of the connecting rods.

Break-in - The period of operation between installation of new or rebuilt parts and time in which parts are worn to the correct fit. Driving at reduced and varying speed for a specified mileage to permit parts to wear to the correct fit.

Bushing - A one-piece sleeve placed in a bore to serve as a bearing surface for shaft, piston pin, etc. Usually replaceable.

C

Camshaft - The shaft in the engine, on which a series of lobes are located for operating the valve mechanisms. The camshaft is driven by gears or sprockets and a timing chain. Usually referred to simply as the cam.

Carbon - Hard, or soft, black deposits found in combustion chamber, on plugs, under rings, on and under valve heads.

Cast iron - An alloy of iron and more than two percent carbon, used for engine blocks and heads because it's relatively inexpensive and easy to mold into complex shapes.

Chamfer - To bevel across (or a bevel on) the sharp edge of an object.

Chase - To repair damaged threads with a tap or die.

Combustion chamber - The space between the piston and the cylinder head, with the piston at top dead center, in which air-fuel mixture is burned.

Compression ratio - The relationship between cylinder volume (clearance volume) when the piston is at top dead center and cylinder volume when the piston is at bottom dead center.

Connecting rod - The rod that connects the crank on the crankshaft with the piston. Sometimes called a con rod.

Connecting rod cap - The part of the connecting rod assembly that attaches the rod to the crankpin.

Core plug - Soft metal plug used to plug the casting holes for the coolant passages in the block.

Crankcase - The lower part of the engine in which the crankshaft rotates; includes the lower section of the cylinder block and the oil pan.

Crank kit - A reground or reconditioned crankshaft and new main and connecting rod bearings.

Crankpin - The part of a crankshaft to which a connecting rod is attached.

Crankshaft - The main rotating member, or shaft, running the length of the crankcase, with offset throws to which the connecting rods are attached; changes the reciprocating motion of the pistons into rotating motion.

Cylinder sleeve - A replaceable sleeve, or liner, pressed into the cylinder block to form the cylinder bore.

D

Deburring - Removing the burrs (rough edges or areas) from a bearing.

Deglazer - A tool, rotated by an electric motor, used to remove glaze from cylinder walls so a new set of rings will seat.

E

Endplay - The amount of lengthwise movement between two parts. As applied to a crankshaft, the distance that the crankshaft can move forward and back in the cylinder block.

F

Face - A machinist's term that refers to removing metal from the end of a shaft or the face of a larger part, such as a flywheel.

Fatigue - A breakdown of material through a large number of loading and unloading cycles. The first signs are cracks followed shortly by breaks.

Feeler gauge - A thin strip of hardened steel, ground to an exact thickness, used to check clearances between parts.

Free height - The unloaded length or height of a spring.

Freeplay - The looseness in a linkage, or an assembly of parts, between the initial application of force and actual movement. Usually perceived as slop or slight delay.

Freeze plug - See Core plug.

G

Gallery - A large passage in the block that forms a reservoir for engine oil pressure.

Glaze - The very smooth, glassy finish that develops on cylinder walls while an engine is in service.

H

Heli-Coil - A rethreading device used when threads are worn or damaged. The device is installed in a retapped hole to reduce the thread size to the original size.

I

Installed height - The spring's measured length or height, as installed on the cylinder head. Installed height is measured from the spring seat to the underside of the spring retainer.

J

Journal - The surface of a rotating shaft which turns in a bearing.

K

Keeper - The split lock that holds the valve spring retainer in position on the valve stem.

Key - A small piece of metal inserted into matching grooves machined into two parts fitted together - such as a gear pressed onto a shaft - which prevents slippage between the two parts.

Knock - The heavy metallic engine sound, produced in the combustion chamber as a result of abnormal combustion - usually detonation. Knock is usually caused by a loose or worn bearing. Also referred to as detonation, pinging and spark knock. Connecting rod or main bearing knocks are created by too much oil clearance or insufficient lubrication.

L

Lands - The portions of metal between the piston ring grooves.

Lapping the valves - Grinding a valve face and its seat together with lapping compound.

Lash - The amount of free motion in a gear train, between gears, or in a mechanical assembly, that occurs before movement can begin. Usually refers to the lash in a valve train.

Lifter - The part that rides against the cam to transfer motion to the rest of the valve train.

M

Machining - The process of using a machine to remove metal from a metal part.

Main bearings - The plain, or babbitt, bearings that support the crankshaft.

Main bearing caps - The cast iron caps, bolted to the bottom of the block, that support the main bearings.

O

O.D. - Outside diameter.

Oil gallery - A pipe or drilled passageway in the engine used to carry engine oil from one area to another.

Oil ring - The lower ring, or rings, of a piston; designed to prevent excessive amounts of oil from working up the cylinder walls and into the combustion chamber. Also called an oil-control ring.

Oil seal - A seal which keeps oil from leaking out of a compartment. Usually refers to a dynamic seal around a rotating shaft or other moving part.

O-ring - A type of sealing ring made of a special rubberlike material; in use, the O-ring is compressed into a groove to provide the sealing action.

Overhaul - To completely disassemble a unit, clean and inspect all parts, reassemble it with the original or new parts and make all adjustments necessary for proper operation.

P

Pilot bearing - A small bearing installed in the center of the flywheel (or the rear end of the crankshaft) to support the front end of the input shaft of the transmission.

Pip mark - A little dot or indentation which indicates the top side of a compression ring.

Piston - The cylindrical part, attached to the connecting rod, that moves up and down in the cylinder as the crankshaft rotates. When the fuel charge is fired, the piston transfers the force of the explosion to the connecting rod, then to the crankshaft.

Piston pin (or wrist pin) - The cylindrical and usually hollow steel pin that passes through the piston. The piston pin fastens the piston to the upper end of the connecting rod.

Piston ring - The split ring fitted to the groove in a piston. The ring contacts the sides of the ring groove and also rubs against the cylinder wall, thus sealing space between piston and wall. There are two types of rings: Compression rings seal the compression pressure in the combustion chamber; oil rings scrape excessive oil off the cylinder wall.

Piston ring groove - The slots or grooves cut in piston heads to hold piston rings in position.

Piston skirt - The portion of the piston below the rings and the piston pin hole.

Plastigage - A thin strip of plastic thread, available in different sizes, used for measuring clearances. For example, a strip of plastigage is laid across a bearing journal and mashed as parts are assembled. Then parts are disassembled and the width of the strip is measured to determine clearance between journal and bearing. Commonly used to measure crankshaft main-bearing and connecting rod bearing clearances.

Press-fit - A tight fit between two parts that requires pressure to force the parts together. Also referred to as drive, or force, fit.

Prussian blue - A blue pigment; in solution, useful in determining the area of contact between two surfaces. Prussian blue is commonly used to determine the width and location of the contact area between the valve face and the valve seat.

R

Race (bearing) - The inner or outer ring that provides a contact surface for balls or rollers in bearing.

Ream - To size, enlarge or smooth a hole by using a round cutting tool with fluted edges.

Ring job - The process of reconditioning the cylinders and installing new rings.

Runout - Wobble. The amount a shaft rotates out-of-true.

S

Saddle - The upper main bearing seat.

Scored - Scratched or grooved, as a cylinder wall may be scored by abrasive particles moved up and down by the piston rings.

Scuffing - A type of wear in which there's a transfer of material between parts moving against each other; shows up as pits or grooves in the mating surfaces.

Seat - The surface upon which another part rests or seats. For example, the valve seat is the matched surface upon which the valve face rests. Also used to refer to wearing into a good fit; for example, piston rings seat after a few miles of driving.

Short block - An engine block complete with crankshaft and piston and, usually, camshaft assemblies.

Static balance - The balance of an object while it's stationary.

Step - The wear on the lower portion of a ring land caused by excessive side and back-clearance. The height of the step indicates the ring's extra side clearance and the length of the step projecting from the back wall of the groove represents the ring's back clearance.

Stroke - The distance the piston moves when traveling from top dead center to bottom dead center, or from bottom dead center to top dead center.

Stud - A metal rod with threads on both ends.

T

Tang - A lip on the end of a plain bearing used to align the bearing during assembly.

Tap - To cut threads in a hole. Also refers to the fluted tool used to cut threads.

Taper - A gradual reduction in the width of a shaft or hole; in an engine cylinder, taper usually takes the form of uneven wear, more pronounced at the top than at the bottom.

Throws - The offset portions of the crankshaft to which the connecting rods are affixed.

Thrust bearing - The main bearing that has thrust faces to prevent excessive end-play, or forward and backward movement of the crankshaft.

Thrust washer - A bronze or hardened steel washer placed between two moving parts. The washer prevents longitudinal movement and provides a bearing surface for thrust surfaces of parts.

Tolerance - The amount of variation permitted from an exact size of measurement. Actual amount from smallest acceptable dimension to largest acceptable dimension.

U

Umbrella - An oil deflector placed near the valve tip to throw oil from the valve stem area.

Undercut - A machined groove below the normal surface.

Undersize bearings - Smaller diameter bearings used with re-ground crankshaft journals.

V

Valve grinding - Refacing a valve in a valve-refacing machine.

Valve train - The valve-operating mechanism of an engine; includes all components from the camshaft to the valve.

Vibration damper - A cylindrical weight attached to the front of the crankshaft to minimize torsional vibration (the twist-untwist actions of the crankshaft caused by the cylinder firing impulses). Also called a harmonic balancer.

W

Water jacket - The spaces around the cylinders, between the inner and outer shells of the cylinder block or head, through which coolant circulates.

Web - A supporting structure across a cavity.

Woodruff key - A key with a radiused backside (viewed from the side).

Specifications

General

Displacement	
2AZ-FE four-cylinder	144 cubic inches (2.36 liters)
1MZ-FE V6	183 cubic inches (3.0 liters)
3MZ-FE V6	202 cubic inches (3.3 liters)
Cylinder compression pressure	
Four-cylinder	198 psi
V6	213 psi
Minimum compression pressure (all models)	142 psi
Variation between cylinders (all models)	14 psi
Oil pressure (all models)	
At curb idle	4.3 psi or more
At 3000 rpm	36 to 78 psi

Torque specifications

	Ft-lbs (unless otherwise indicated)	Nm

Four-cylinder models

Balance shaft carrier bolts		
Step 1	16	22
Step 2	Tighten an additional 90-degrees (1/4-turn)	
Connecting rod bearing cap bolts		
Step 1	18	25
Step 2	Tighten an additional 90-degrees (1/4-turn)	
Main bearing cap bolts (see illustration 11.19a)		
Step 1	15	20
Step 2	29	40
Step 3	Tighten an additional 90-degrees (1/4-turn)	

V6 models

Connecting rod bearing cap bolts		
Step 1	18	25
Step 2	Tighten an additional 90-degrees (1/4-turn)	
Main bearing cap bolts (see illustrations 11.19b and 11.19c)		
Step 1 (12-point bolts)	16	22
Step 2 (12-point bolts)	Tighten an additional 90-degrees (1/4-turn)	
Step 3 (side bolts)	20	27
Water inlet housing mounting bolts	71 in-lbs	8

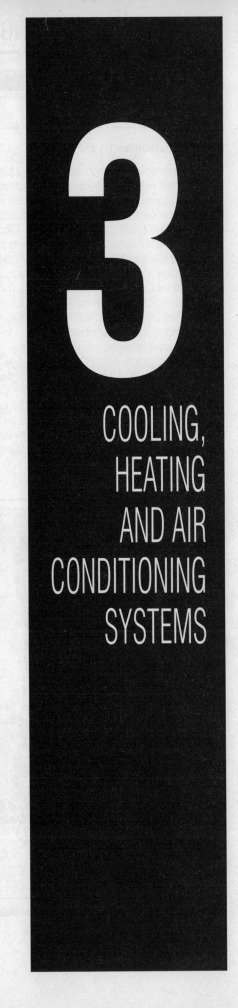

3

COOLING,
HEATING
AND AIR
CONDITIONING
SYSTEMS

Section

Reference to other Chapters

1 General information

ENGINE COOLING SYSTEM

▶ **Refer to illustrations 1.1 and 1.2**

All vehicles covered by this manual employ a pressurized engine cooling system with thermostatically controlled coolant circulation (see illustration). An impeller type water pump mounted on the front of the block pumps coolant through the engine. The coolant flows around each cylinder and toward the rear of the engine. Cast-in coolant passages direct coolant around the intake and exhaust ports, near the spark plug areas and in proximity to the exhaust valve guides.

A wax pellet-type thermostat is located in the thermostat housing near the front of the engine (see illustration). During warm up, the closed thermostat prevents coolant from circulating through the radiator. When the engine reaches normal operating temperature, the thermostat opens and allows hot coolant to travel through the radiator, where it is cooled before returning to the engine.

The cooling system is sealed by a pressure-type radiator cap. This raises the boiling point of the coolant, and the higher boiling point of the coolant increases the cooling efficiency of the radiator. If the system pressure exceeds the cap pressure relief value, the excess pressure in the system forces the spring-loaded valve inside the cap off its seat and allows the coolant to escape through the overflow tube into a coolant reservoir. When the system cools, the excess coolant is automatically drawn from the reservoir back into the radiator.

The coolant reservoir serves as both the point at which fresh cool-ant is added to the cooling system to maintain the proper fluid level and as a holding tank for overheated coolant.

This type of cooling system is known as a closed design because coolant that escapes past the pressure cap is saved and reused.

HEATING SYSTEM

The heating system consists of a blower fan and heater core located within the heater box, the inlet and outlet hoses connecting the heater core to the engine cooling system and the heater/air conditioning control head on the dashboard. Hot engine coolant is circulated through the heater core. When the heater mode is activated, a flap door opens to expose the heater box to the passenger compartment. A fan switch on the control head activates the blower motor, which forces air through the core, heating the air.

AIR CONDITIONING SYSTEM

The air conditioning system consists of a condenser mounted in front of the radiator, an evaporator mounted adjacent to the heater core under the dashboard, a compressor mounted on the engine, a receiver-drier which contains a high pressure relief valve and the plumbing connecting all of the above.

A blower fan forces the warmer air of the passenger compartment through the evaporator core (sort of a radiator-in-reverse), transferring the heat from the air to the refrigerant. The liquid refrigerant boils

1.1 Typical cooling system component locations (V6 shown, four-cylinder similar)

1	Coolant reservoir	4	Radiator	7	Thermostat (vicinity shown)
2	Upper radiator hose	5	Lower radiator hose	8	Fuse and relay box
3	Radiator cap	6	Cooling fans		

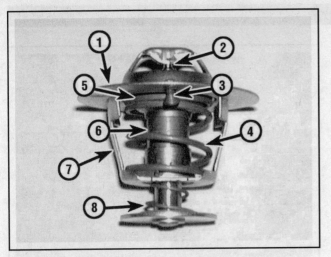

1.2 A typical thermostat

1	Flange	5	Valve seat
2	Piston	6	Valve
3	Jiggle valve	7	Frame
4	Main coil spring	8	Secondary coil spring

off into low-pressure vapor, taking the heat with it when it leaves the evaporator. The compressor keeps refrigerant circulating through the system, pumping the warmed coolant through the condenser where it is cooled and then circulated back to the evaporator.

Some models are equipped with an optional Automatic Air Conditioning system. With this system, you select the desired interior temperature with a knob on the controls, similar to setting the temperature on a home heating/cooling thermostat, and the system automatically adds the right blend of cool or warm air to maintain this temperature. The system has sensors that detect both the interior and outside temperature.

2   Antifreeze - general information

▶ Refer to illustration 2.4

✳✳ WARNING:

Do not allow antifreeze to come in contact with your skin or painted surfaces of the vehicle. Rinse off spills immediately with plenty of water. Antifreeze is highly toxic if ingested. Never leave antifreeze lying around in an open container or in puddles on the floor; children and pets are attracted by its sweet smell and may drink it. Check with local authorities on disposing of used antifreeze. Many communities have collection centers that will see that antifreeze is disposed of safely. Never dump used antifreeze on the ground or into drains.

➡Note: Non-toxic coolant is available at local auto parts stores. Although the coolant is non-toxic when fresh, proper disposal of used coolant is still required.

The cooling system should be filled with a water/ethylene-glycol based antifreeze solution, which will prevent freezing down to at least -20 degrees F, or lower if local climate requires it. It also provides protection against corrosion and increases the coolant boiling point.

The cooling system should be drained, flushed and refilled at least every other year (see Chapter 1). The use of antifreeze solutions for periods of longer than two years is likely to cause damage and encourage the formation of rust and scale in the system. If your tap water is 'hard,' i.e. contains a lot of dissolved minerals, use distilled water with the antifreeze.

Before adding antifreeze to the system, check all hose connections, because antifreeze tends to leak through very minute openings. Engines do not normally consume coolant. Therefore, if the level goes down,

2.4 An inexpensive hydrometer can be used to test the condition of your coolant

find the cause and correct it.

The exact mixture of antifreeze-to-water that you should use depends on the relative weather conditions. The mixture should contain at least 50 percent antifreeze, but should never contain more than 70-percent antifreeze. Consult the mixture ratio chart on the antifreeze container before adding coolant. Hydrometers are available at most auto parts stores to test the ratio of antifreeze to water (see illustration). Use antifreeze which meets the vehicle manufacturer's specifications.

3 Thermostat - check and replacement

❋❋ WARNING:

Do not attempt to remove the radiator cap, coolant or thermostat until the engine has cooled completely.

GENERAL CHECK

1 Before assuming the thermostat is responsible for a cooling system problem, check the coolant level (see Chapter 1), drivebelt tension (see Chapter 1) and temperature gauge (or light) operation.

2 If the engine takes a long time to warm up (as indicated by the temperature gauge or heater operation), the thermostat is probably stuck open. Replace the thermostat with a new one.

3 If the engine runs hot, use your hand to check the temperature of the lower radiator hose. If the hose is not hot, but the engine is, the thermostat is probably stuck in the closed position, preventing the coolant inside the engine from escaping to the radiator. Replace the thermostat.

❋❋ CAUTION:

Do not drive the vehicle without a thermostat. The computer may stay in open loop and emissions and fuel economy will suffer.

4 If the lower radiator hose is hot, it means that the coolant is flowing and the thermostat is open. Consult the *Troubleshooting* Section at the front of this manual for further diagnosis.

THERMOSTAT TEST

5 A more thorough test of the thermostat can only be made when it is removed from the vehicle (see below). If the thermostat remains in the open position at room temperature, it is faulty and must be replaced.

6 To test it fully, suspend the (closed) thermostat on a length of string or wire in a container of cold water, with a thermometer (cooking type that reads beyond 212-degrees F). A clear Pyrex cooking container is easiest to use.

7 Heat the water on a stove while observing the temperature and the thermostat. Neither should contact the sides of the container.

8 Note the temperature when the thermostat begins to open and

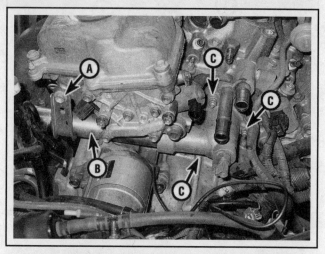

3.17 On V6 models, remove the water inlet pipe mounting fastener (A) and then the pipe (B). Move the hoses and wiring harnesses out of the way and then remove the thermostat housing mounting fasteners (C)

when it is fully open. Compare the temperatures to the Specifications in this Chapter. The number stamped into the thermostat is generally the fully-open temperature. Some manufacturers provide Specifications for the beginning-to-open temperature, the fully-open temperature, and sometimes the amount the valve should open.

9 If the thermostat doesn't open and close as specified, or sticks in any position, replace it.

REPLACEMENT

▶ **Refer to illustrations 3.17, 3.20 and 3.21**

10 Disconnect the cable from the negative terminal of the battery (see Chapter 5, Section 1).

11 Drain the cooling system (see Chapter 1).

12 On four-cylinder models, remove the alternator (see Chapter 5).

13 On V6 models, remove the air filter housing (see Chapter 4).

14 Follow the radiator hose to the thermostat housing cover or water

3.20 The thermostat seal fits around the edge of the thermostat

3.21 Note the position of the thermostat and how the seal is installed around it

inlet pipe and disconnect the hose.

15 On 2004 and later Lexus models, remove the starter (see Chapter 5).

16 On four-cylinder models, remove the thermostat housing cover mounting fasteners and then remove it from the cylinder block.

17 On V6 models, unbolt the water inlet pipe and remove the pipe and O-ring (see illustration).

18 Remove the fasteners and electrical connectors from the thermostat housing cover and detach the cover from the engine (see illustration 3.17). Be prepared for some coolant to spill as the gasket seal is broken.

19 Remove the thermostat, noting the direction in which it was

installed in the housing, and thoroughly clean the sealing surfaces.

20 Fit a new gasket onto the thermostat (see illustration). Make sure it is evenly fitted all the way around.

21 Install the thermostat and housing, positioning the jiggle pin in alignment with the highest mounting stud for the housing cover (see illustration).

22 Tighten the housing cover fasteners to the torque listed in this Chapter's Specifications and reinstall the remaining components in the reverse order of removal. On V6 models, use a new O-ring on the water inlet pipe and lubricate the O-ring with soapy water.

23 Refill the cooling system (see Chapter 1). Run the engine and check for leaks and proper operation.

4 Engine cooling fans - removal and installation

▶ Refer to illustrations 4.3a, 4.3b, 4.5 and 4.6

❄❄ WARNING 1:

Do not start this procedure until the engine is completely cool. Do not allow antifreeze to come in contact with your skin or painted surfaces of the vehicle. Rinse off spills immediately with plenty of water. Antifreeze is highly toxic if ingested. Never leave antifreeze lying around in an open container or in puddles on the floor; children and pets are attracted by its sweet smell and may drink it. Check with local authorities on disposing of used antifreeze. Many communities have collection centers, which will see that antifreeze is disposed of safely. Never dump used antifreeze on the ground or into drains.

❄❄ WARNING 2:

The models covered by this manual are equipped with Supplemental Restraint Systems (SRS), more commonly known as airbags. Always disarm the airbag system before working in the vicinity of any airbag system component to avoid the possibility of accidental deployment of the airbag, which could cause personal injury (see Chapter 12).

➡Note: On most models, the manufacturer's procedure for engine cooling fan removal involves removing the radiator and fan assembly together and then separating them afterwards. It is possible to remove the engine cooling fans without removing the radiator.

1 Disconnect the cable from the negative terminal of the battery (see Chapter 5, Section 1). Drain the cooling system (see Chapter 1), then detach the upper radiator hose from the radiator.

2 On 2004 and later Lexus models, remove the air intake duct (see Chapter 4), the plastic cover over the radiator support, and the hood release latch (see Chapter 11). Remove the six mounting bolts (two on each end and two in the middle) for the radiator support and then remove the support.

3 Disconnect the electrical connectors for both fan motors (see illustration).

➡Note: On 2004 and later Lexus models, disconnect the electrical connectors to the fan electronic control unit (ECU) and any other harness connectors attached to the fan shroud (see illustration).

4 Remove the fan/shroud mounting fasteners from the radiator, remove any hoses or wiring harnesses that may be attached to it and then lift out of the vehicle (see illustration 4.3a).

4.3a Engine cooling fan mounting details:

1 Fan motor connectors
2 Fan assembly mounting fasteners (top shown, bottom similar)

4.3b Disconnect these harnesses from the fan shroud

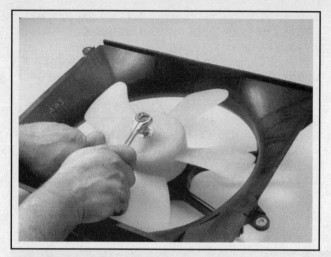

4.5 Remove the fan from the motor

4.6 Remove the screws and separate the motor from the shroud

5   Hold the fan blades and remove the fan retaining nut (see illustration).
6   Unbolt the fan motor from the shroud (see illustration).

7   Installation is the reverse of removal. Tighten the radiator support mounting bolts to the torque listed in this Chapter's Specifications. Refill the cooling system (see Chapter 1).

5   Radiator and coolant reservoir - removal and installation

✳✳ WARNING 1:

Do not start this procedure until the engine is completely cool. Do not allow antifreeze to come in contact with your skin or painted surfaces of the vehicle. Rinse off spills immediately with plenty of water. Antifreeze is highly toxic if ingested. Never leave antifreeze lying around in an open container or in puddles on the floor; children and pets are attracted by its sweet smell and may drink it. Check with local authorities on disposing of used antifreeze. Many communities have collection centers, which will see that antifreeze is disposed of safely. Never dump used antifreeze on the ground or into drains.

✳✳ WARNING 2:

The models covered by this manual are equipped with Supplemental Restraint Systems (SRS), more commonly known as airbags. Always disarm the airbag system before working in the vicinity of any airbag system component to avoid the possibility of accidental deployment of the airbag, which could cause personal injury (see Chapter 12).

➡Note: Non-toxic coolant is available at local auto parts stores. Although the coolant is non-toxic when fresh, proper disposal of used coolant is still required.

RADIATOR

Removal

▶ Refer to illustrations 5.7, 5.9 and 5.11

1   Disconnect the negative battery cable (see Chapter 5, Section 1).
2   On 2004 and later Lexus models, remove the air intake duct from above the radiator (see Chapter 4).

3   Remove the engine splash shield(s) from beneath the radiator and any covers for the engine, if equipped (see Chapter 2).
4   Drain the coolant into a container (see Chapter 1).
5   On 2003 and earlier Lexus models, remove the A/C condenser (see Section 15).
6   On 2004 and later Lexus models, remove the hood latch (see Chapter 11).
7   On 2004 and later Lexus models, remove the plastic cover over the radiator support (see illustration). Disconnect the horn electrical connectors, remove the mounting bolts for the radiator support and any brackets mounted to it and then remove the support.
8   Detach the upper and lower radiator hoses from the radiator and the reservoir hose from the radiator filler neck.

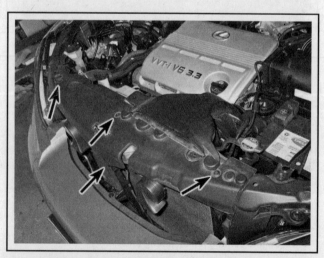

5.7 Radiator support cover mounting fasteners on 2004 and later Lexus models

5.9 Detach the automatic transaxle cooler lines

5.11 Radiator mounting bracket bolts

9 Disconnect the transaxle oil cooler lines from the radiator (see illustration). Place a drip pan to catch the fluid and cap the fittings.

10 Disconnect the electrical connectors for the cooling fans and detach any other wires or brackets from the fan shroud (see Section 4) and radiator.

11 Remove the mounting brackets for the radiator and then carefully remove the radiator and fan shroud as an assembly (see illustration).

12 Remove the cooling fans from the radiator (see Section 4).

13 With the radiator removed, it can be inspected for leaks, damage and internal blockage. If in need of repairs, have a professional radiator shop perform the work, as special techniques are required.

14 Bugs and dirt can be cleaned from the radiator with compressed air and a soft brush. Don't bend the cooling fins as this is done.

❊❊ WARNING:

Wear eye protection.

Installation

15 Installation is the reverse of the removal procedure. Be sure the rubber mounts are correctly in place.

16 After installation, fill the cooling system with the proper mixture of antifreeze and water. Refer to Chapter 1 if necessary.

17 Start the engine and check for leaks. Allow the engine to reach normal operating temperature, indicated by the upper radiator hose becoming hot. Recheck the coolant level and add more if required.

18 Be sure to check the automatic transaxle fluid level and add fluid as needed (see Chapter 1).

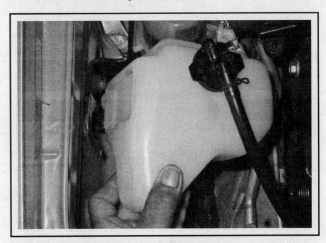

5.19 Pull the coolant reservoir up and out of its bracket to remove it

COOLANT RESERVOIR

▶ **Refer to illustration 5.19**

19 On most models, the coolant reservoir simply pulls up and out of the bracket on the fenderwell (see illustration).

➥**Note: On 2004 and later Lexus models, the reservoir is mounted to the engine cooling fan shroud.**

20 Pour the coolant into a container. Wash out and inspect the reservoir for cracks and chafing. Replace it if damaged.

21 Installation is the reverse of removal.

6   Water pump - check

1 A failure in the water pump can cause serious engine damage due to overheating.

2 Water pumps are equipped with weep or vent holes. If a failure occurs in the pump seal, coolant will leak from this hole. In most cases it will be necessary to use a flashlight to find the hole on the water pump by looking through the space behind the pulley just below the water pump shaft. A slight gray discoloration around the weep hole is normal, while dark brown stains indicate a problem.

3 If the water pump shaft bearings fail, there may be a howling sound at the front of the engine while it is running. Bearing wear can be felt if the water pump pulley is rocked up and down. Do not mistake drivebelt slippage, which causes a squealing sound, for water pump failure. Spray automotive drivebelt dressing on the belts to eliminate the belt as a possible cause of the noise.

7 Water pump - removal and installation

⁂ WARNING:

Do not start this procedure until the engine is completely cool. Do not allow antifreeze to come in contact with your skin or painted surfaces of the vehicle. Rinse off spills immediately with plenty of water. Antifreeze is highly toxic if ingested. Never leave antifreeze lying around in an open container or in puddles on the floor; children and pets are attracted by its sweet smell and may drink it. Check with local authorities on disposing of used antifreeze. Many communities have collection centers, which will see that antifreeze is disposed of safely. Never dump used antifreeze on the ground or into drains.

➡Note: Non-toxic coolant is available at local auto parts stores. Although the coolant is non-toxic when fresh, proper disposal of used coolant is still required.

FOUR-CYLINDER ENGINE

▶ **Refer to illustrations 7.4 and 7.5**

1   Disconnect the negative battery cable (see Chapter 5, Section 1) and drain the cooling system (see Chapter 1).
2   Refer to Chapter 2A and remove the splash shields on the right side of the engine and the inner fender well, the engine movement control rod and its bracket, then the #2 engine stay and bracket.
3   Remove the drivebelt (see Chapter 1) and the alternator (see Chapter 5).
4   Remove the water pump pulley, using a strap wrench or pin spanner wrench to hold the pulley while you loosen the bolts (see illustration).
5   Remove the four bolts and two nuts and pry the water pump off (see illustration). If necessary, tap the pump loose with a soft-face hammer.
6   To install, clean the water pump and block of any old gasket material or sealant, then clean with lacquer thinner.
7   Apply a new bead of sealant and install the water pump. Install the water pump bolts and nuts and tighten them to the torque listed in this Chapter's Specifications within five minutes.
8   Install the remaining parts in the reverse order of removal.

7.4 Using a pin spanner to hold the water pump pulley, remove the pulley mounting bolts

9   Refill the cooling system (see Chapter 1) and then run the engine and check for leaks and proper operation.

V6 ENGINE

▶ **Refer to illustrations 7.12, 7.13 and 7.15**

10 Disconnect the negative battery cable (see Chapter 5, Section 1) and drain the cooling system (see Chapter 1).
11 Remove the timing belt, camshaft sprockets (see Chapter 2B) and the number 2 idler pulley (above the water pump).
12 Remove the number 3 timing belt cover (see illustration).
13 Remove the water pump retaining bolts/nuts and separate the pump from the engine.

➡Note: There are two long studs that hold the engine mounting bracket and the water pump. The studs have a small hex-head that is used to remove the stud (see illustration). It may be necessary to remove one or both of these studs to remove the water pump. If the stud is stuck, remove it with locking pliers and replace it with a new one of the same length.

7.5 The four-cylinder water pump is retained by four bolts (A, one lower bolt not seen in this photo) and two nuts (B)

7.12 Release the three wiring harness clips, pull the harness back and remove the number 3 timing belt cover

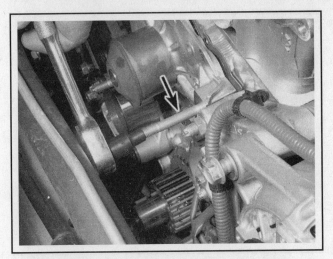

7.13 V6 water pumps have two long studs - one or both may have to be removed with a small wrench to allow water pump removal

7.15 On V6 models, a new metal/rubber gasket is used when installing the water pump

14 Thoroughly clean all sealing surfaces, removing all traces of old gaskets, sealer and O-rings. Remove any traces of oil with acetone or lacquer thinner and a clean rag.

15 These models use a metal and rubber gasket (see illustration).

16 Reinstall the pump and the remaining parts in the reverse order of removal. Tighten the water pump bolts/nuts in several steps to the torque listed in this Chapter's Specifications.

17 Refill the cooling system (see Chapter 1) and run the engine, checking for leaks and proper operation.

8 Coolant temperature indicator - check

※※ WARNING:

Wait until the engine is completely cool before beginning this procedure.

1 The coolant temperature indicator system consists of a temperature gauge on the dash and a sensor mounted on the engine. On all models, an Engine Coolant Temperature (ECT) sensor (see Chapter 6), which is an information sensor for the Powertrain Control Module (PCM), provides a signal to the PCM which controls and actuates the temperature gauge.

2 If an overheating indication has occurred, first check the coolant level in the system (see Chapter 1) and that the coolant mixture is correct (see Section 2). Also, refer to the *Troubleshooting* section at the beginning of this book before assuming that the temperature indicator is faulty.

3 Start the engine and warm it up for 10 minutes. If the temperature gauge has not moved from the C position, check the wiring harness connections going to the instrument cluster.

4 If there is a problem with the ECT sensor, it is very likely that the Service Engine Soon lamp will be illuminated and the sensor or circuit will need repair (see Chapter 6).

9 Blower motor - removal and installation

▶ **Refer to illustration 9.4 and 9.5**

※※ WARNING:

The models covered by this manual are equipped with Supplemental Restraint Systems (SRS), more commonly known as airbags. Always disarm the airbag system before working in the vicinity of any airbag system component to avoid the possibility of accidental deployment of the airbag, which could cause personal injury (see Chapter 12).

➡ **Note: On 2004 and later Highlander models, there is a rear heater assembly with a blower motor. The motor can be serviced in the same manner as the front motor. See Section 11 for removal and replacement of the rear blower motor.**

1 Disconnect the cable from the negative terminal of the battery (see Chapter 5, Section 1).

2 The blower unit is located under the dash and below the glove box.

3 Remove the right lower dash panel (just below the glove box), if

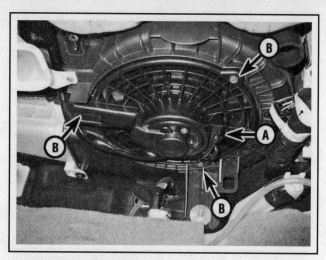

9.4 Blower motor details

A   *Blower motor connector*
B   *Blower motor mounting screws*

equipped (see Chapter 11).

4   To remove the blower, remove the blower unit retaining screws

9.5 Use pliers to release and remove the clip, then lift the blower fan off the motor shaft

and lower the unit from the housing (see illustration).

5   If the motor is being replaced, transfer the fan to the new motor prior to installation (see illustration).

6   Installation is the reverse of removal. Check for proper operation.

10 Heater and air conditioning control assembly - removal and installation

▶ Refer to illustrations 10.4, 10.5, 10.6 and 10.7

✳✳ WARNING:

The models covered by this manual are equipped with Supplemental Restraint Systems (SRS), more commonly known as airbags. Always disarm the airbag system before working in the vicinity of any airbag system component to avoid the possibility of accidental deployment of the airbag, which could cause personal injury (see Chapter 12).

1   Disconnect the negative cable from the battery (see Chapter 5,

Section 1).

2   On Highlander models, remove the center trim panel around the control assembly. On Lexus models, remove the center trim panel(s) from around the shift lever and that lead up to the control assembly (see Chapter 11).

3   On all models, the mounting fasteners will be exposed when the trim is removed. On 2003 and earlier Lexus models, there are two mounting fasteners at the bottom of the control assembly and audio unit. On 2004 and later Lexus models, there are four mounting fasteners. Once the fasteners are removed, there are retaining clips holding the assemblies in place.

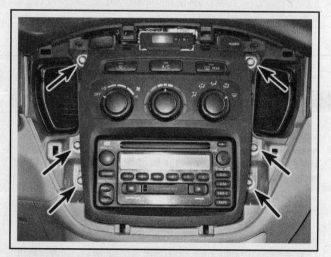

10.4 Control assembly mounting fasteners (2001 Highlander model shown, other models similar)

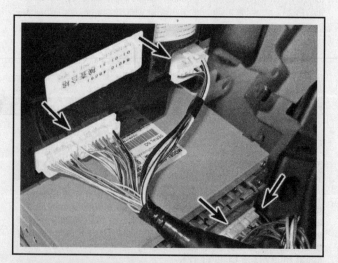

10.5 Unplug the connectors from the rear of the control assembly and audio unit

10.6 Mounting bracket fasteners

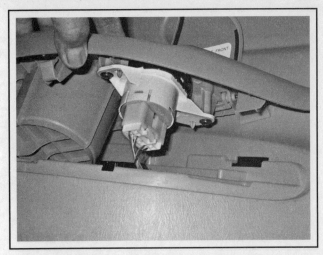

10.7 Carefully pry out the trim panel to access the control switch

4   On Highlander models, remove the mounting screws for the control assembly and the audio unit (see illustration).

5   On all models, pull the assembly out slightly and disconnect all electrical connectors for the control assembly and the audio unit (see illustration).

6   On Highlander models, detach the audio unit from the control assembly by removing the mounting bracket screws from the control

assembly (see illustration). On other models, the control unit can be separated in a similar fashion.

7   On models equipped with rear heating, there is an additional control switch mounted in the rear trim (see illustration).

8   Installation is the reverse of the removal procedure.

9   Run the engine and check for proper functioning of the heater and air conditioning.

11 Heater core - removal and installation

FRONT HEATER CORE

▶ Refer to illustrations 11.6a, 11.6b, 11.11, 11.12, 11.13, 11.14a, 11.14b, 11.16 and 11.17

❈❈ WARNING 1:

The models covered by this manual are equipped with Supplemental Restraint Systems (SRS), more commonly known as airbags. Always disarm the airbag system before working in the vicinity of any airbag system component to avoid the possibility of accidental deployment of the airbag, which could cause personal injury (see Chapter 12).

❈❈ WARNING 2:

Do not allow antifreeze to come in contact with your skin or painted surfaces of the vehicle. Rinse off spills immediately with plenty of water. Antifreeze is highly toxic if ingested. Never leave antifreeze lying around in an open container or in puddles on the floor; children and pets are attracted by its sweet smell and may drink it. Check with local authorities on disposing of used antifreeze. Many communities have collection center s which will see that antifreeze is disposed of safely. Never dump used antifreeze on the ground or into drains.

❈❈ WARNING 3:

Wait until the engine is completely cool before beginning this procedure.

➡Note 1: Non-toxic coolant is available at local auto parts stores. Although the coolant is non-toxic when fresh, proper disposal of used coolant is still required.

➡Note 2: Removal of the heater core is a difficult procedure for the home mechanic. There are numerous fasteners involved, some of which can be difficult to access, and we recommend that you have considerable mechanical experience before performing a heater core replacement.

1   Take the vehicle to a dealer service department or automotive air conditioning shop and have the air conditioning system discharged and the refrigerant recovered.

2   Disconnect the cable from the negative terminal of the battery (see Chapter 5, Section 1).

3   Refer to Chapter 1 and drain the cooling system.

4   The factory procedure for removal of the heater core involves removal of the heater/evaporator housing, which requires complete removal of the instrument panel and support tube.

5   Remove the floor console (if equipped) and instrument panel (see Chapter 11). On Lexus models, remove the cowl tray from the engine compartment (see Chapter 11).

6   From the engine side of the firewall, disconnect the refrigerant lines to the evaporator, then twist the coolant hoses off the heater core tubes (see illustrations).

7   If the hoses are stuck, cut the hoses off and cut the remaining hose from the metal heater core tubes.

➡Note: Not all models require a special tool for disconnecting the refrigerant lines.

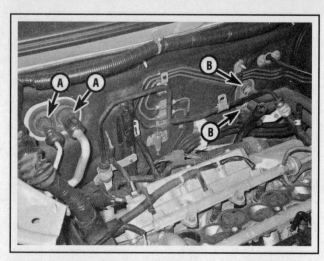

11.6a The location for the refrigerant lines (A) and heater hoses (B)

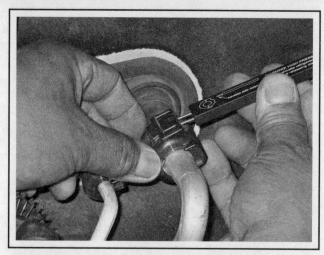

11.6b Using a special tool to release the clamp on the refrigerant lines

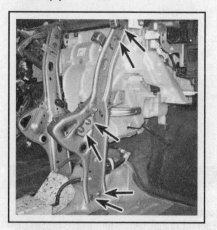

11.11 The cross-cowl tube brace and shift cable mounting fasteners

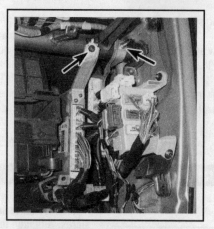

11.12 Mounting fasteners for the PCM and body control module on the cross-cowl tube

11.13 Mounting fasteners for the fuse and relay box and skid control module on the cross-cowl tube

8 Detach the steering column from the support tube (see Chapter 10).

9 Remove the shift lever (see Chapter 7).

10 Pull the carpet back enough to remove the floor air ducts coming from the heater/evaporator housing.

11 Detach the shift cable from the cross-cowl tube brace. Unscrew the mounting fasteners for each brace and remove them (see illustration).

12 On Highlander models, detach the PCM and the body control module from the right side of the cross-cowl tube (see illustration).

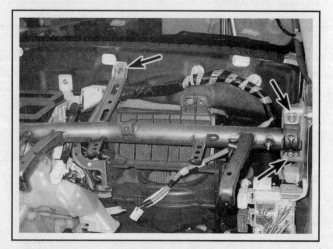

11.14a The cross-cowl mounting fasteners on the right side

11.14b The cross-cowl mounting fasteners on the left side

11.16 Mounting fasteners for the heater/evaporator housing (one hidden - vicinity given)

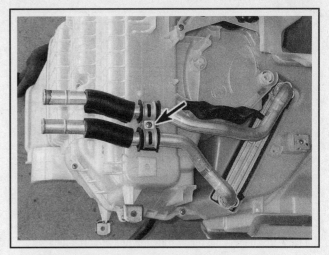

11.17 Heater core tube retainer and mounting fastener

13 On Highlander models, detach the instrument panel fuse and relay box and the skid control module from the left side of the cross-cowl tube (see illustration).

14 Remove the mounting bolts for the cross-cowl tube and carefully remove it (see illustrations).

➡**Note: Make certain that none of the wiring harnesses are damaged while removing the tube.**

15 Remove any small air ducts on the heater/evaporator housing.

16 Disconnect any wiring harnesses attached to the heater/evaporator housing and then remove the mounting fasteners for the housing and carefully pull it away from the firewall until the tubes for the heater core and evaporator are clear of the openings (see illustration).

17 On Highlander models, remove the retaining clamp securing the heater core tubes to the side of the housing and then slide the heater core out of the housing (see illustration). On Lexus models, it may be necessary to remove a few other components in order to remove the heater core.

18 Installation is the reverse order of removal. Use new O-rings to seal the evaporator refrigerant connections. Make sure all of the cross-cowl support tube mounting fasteners are tightened securely before

installing the instrument panel pad. Refill the cooling system (see Chapter 1).

REAR HEATER CORE

▶ **Refer to illustrations 11.24a, 11.24b, 11.25, 11.27 and 11.28**

➡**Note: This feature is only optional on 2004 and later Highlander models.**

19 Remove the right rear floor panel.

20 Pull the weatherstripping away from the bottom and right side of the rear door area.

21 Remove the trim from the back edge of the carpet.

22 Remove the small box that was covered by the right rear floor panel.

23 Remove the bottom trim and the weather-stripping for the right rear door.

24 Remove the various fastener covers and fasteners, then remove the right rear trim panel (see illustrations).

25 Disconnect the heater hoses from beneath the vehicle (see illustration).

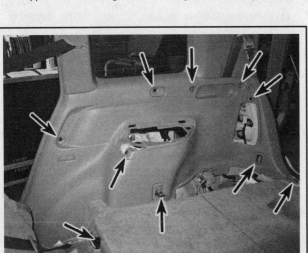

11.24a The right rear panel fastener locations

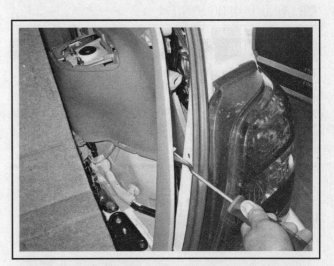

11.24b Separate the panel from the body by carefully prying with a trim panel tool

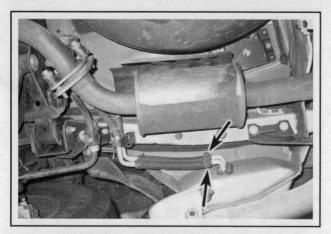

11.25 Disconnect the hoses from the heater core tubes

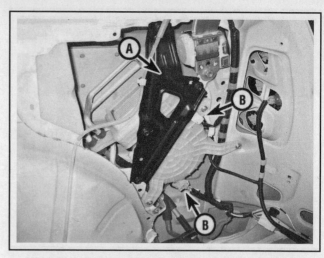

11.27 Remove the plate that covers the heater assembly (A) and disconnect the electrical connectors (B)

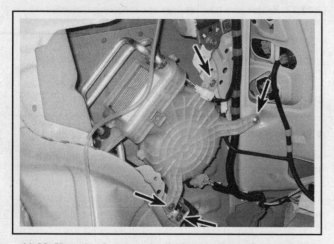

11.28 Mounting fasteners for the heater assembly

➡Note: Pinch the hoses off to avoid fluid loss before disconnecting them.

26 Remove the air ducts.

27 Remove the plate for the heater assembly and disconnect the electrical connectors (see illustration).

28 Remove the mounting fasteners for the heater assembly and then carefully remove it (see illustration).

29 Disassemble the heater assembly and remove the heater core.

30 Installation is the reverse order of removal. Be sure to check the coolant level and check the hose connections for any leaks.

12 Air conditioning and heating system - check and maintenance

AIR CONDITIONING SYSTEM

> ✷✷ **WARNING:**
>
> The air conditioning system is under high pressure. Do not loosen any hose fittings or remove any components until the system has been discharged. Air conditioning refrigerant must be properly discharged into an EPA-approved recovery/recycling unit by a dealer service department or an automotive air conditioning repair facility. Always wear eye protection when disconnecting air conditioning system fittings.

1 The following maintenance checks should be performed on a regular basis to ensure that the air conditioner continues to operate at peak efficiency.

  a) *Inspect the condition of the compressor drivebelt. If it is worn or deteriorated, replace it (see Chapter 1).*

  b) *Check the drivebelt tension and, if necessary, adjust it (see Chapter 1).*

  c) *Inspect the system hoses. Look for cracks, bubbles, hardening and deterioration. Inspect the hoses and all fittings for oil bubbles or seepage. If there is any evidence of wear, damage or leakage, replace the hose(s).*

  d) *Inspect the condenser fins for leaves, bugs and any other foreign material that may have embedded itself in the fins. Use a 'fin comb' or compressed air to remove debris from the condenser.*

  e) *Make sure the system has the correct refrigerant charge.*

2 It's a good idea to operate the system for about ten minutes at least once a month. This is particularly important during the winter months because long term non-use can cause hardening, and subsequent failure, of the seals.

3 Because of the complexity of the air conditioning system and the special equipment necessary to service it, in-depth troubleshooting and repairs are beyond the scope of this manual. However, simple component replacement procedures are provided in this Chapter.

4 The most common cause of poor cooling is simply a low system refrigerant charge. If a noticeable drop in system cooling ability occurs, one on the following quick checks will help you determine whether the refrigerant level is low.

12.8 Place an accurate thermometer in the center dash vent, turn the air conditioning on and check the output temperature

CHECK

▶ **Refer to illustration 12.8**

5 Warm the engine up to normal operating temperature.

6 Place the air conditioning temperature selector at the coldest setting and put the blower at the highest setting. Open the doors (to make sure the air conditioning system doesn't cycle off as soon as it cools the passenger compartment).

7 After the system reaches operating temperature, feel the two pipes connected to the evaporator at the firewall.

8 The pipe (thinner tubing) leading from the condenser outlet to the evaporator should be cold, and the evaporator outlet line (the thicker tubing that leads back to the compressor) should be slightly colder (3 to 10 degrees F). If the evaporator outlet is considerably warmer than the inlet, the system needs a charge. Insert a thermometer in the center air distribution duct (see illustration) while operating the air conditioning system - the temperature of the output air should be 35 to 40 degrees F below the ambient air temperature (down to approximately 40 degrees F). If the ambient (outside) air temperature is very high, say 110-degrees F, the duct air temperature may be as high as 60 degrees

12.12 Add R-134a only to the low side port - the procedure is easier if you wrap the can with a warm, wet towel to prevent icing

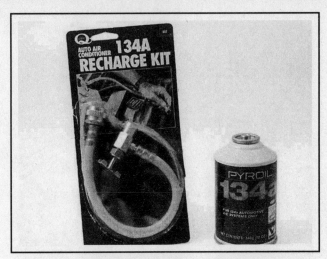

12.9 A basic charging kit for R-134a systems is available at most auto parts stores - it must say R-134a (not R-12) and so must the can of refrigerant

F, but generally the air conditioning is 35 to 40 degrees F cooler than the ambient air. If the air isn't as cold as it used to be, the system probably needs a charge. Further inspection or testing of the system is beyond the scope of the home mechanic and should be left to a professional.

ADDING REFRIGERANT

▶ **Refer to illustrations 12.9 and 12.12**

➡**Note: All models covered by this manual use the refrigerant R-134a. When recharging or replacing air conditioning components, use only refrigerant, refrigerant oil and seals compatible with this system. The seals and compressor oil used with older, conventional R-12 refrigerant are not compatible with the components in this system.**

9 Buy an automotive charging kit at an auto parts store. A charging kit includes a 12-ounce can of R-134a refrigerant, a tap valve and a short section of hose that can be attached between the tap valve and the system low side service valve (see illustration).

10 Connect the charging kit by following the manufacturer's instructions.

11 Back off the valve handle on the charging kit and screw the kit onto the refrigerant can, making sure first that the O-ring or rubber seal inside the threaded portion of the kit is in place.

❋❋❋ **WARNING:**

Wear protective eyewear when dealing with pressurized refrigerant cans.

12 Remove the dust cap from the low-side charging port and attach the quick-connect fitting on the kit hose (see illustration).

❋❋❋ **WARNING:**

DO NOT hook the charging kit hose to the system high side! The fittings on the charging kit are designed to fit only on the low side of the system.

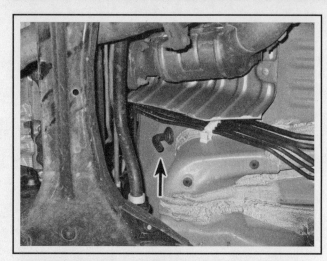

12.22 This drain hose from the heater/air conditioning unit should be kept clear to allow drainage of condensation (shown from under the vehicle)

13 Warm the engine to normal operating temperature and turn on the air conditioning. Keep the charging kit hose away from the fan and other moving parts. In some cases, if the refrigerant charge is low enough, the air conditioning system pressure switch may prevent the compressor from operating.

➡Note: The charging process requires that the compressor be running. If the clutch cycles off, you can switch the A/C controls to High and leave the vehicle's doors open to keep the clutch on the compressor working.

14 Turn the valve handle on the kit until the stem pierces the can, then back the handle out to release the refrigerant. You should be able to hear the rush of gas. Add refrigerant to the low side of the system until both the outlet and the evaporator inlet pipe feel about the same temperature. Allow stabilization time between each addition.

✻✻ WARNING:

Never add more than two cans of refrigerant to the system. The can may tend to frost up, slowing the procedure. Wet a shop towel with hot water and wrap it around the bottom of the can to keep it from frosting.

15 Put your thermometer back in the center register and check that the output air is getting colder.

16 When the can is empty, turn the valve handle to the closed position and release the connection from the low-side port. Reinstall the dust cap.

17 Remove the charging kit from the can and store the kit for future use with the piercing valve in the UP position, to prevent inadvertently piercing the can on the next use.

HEATING SYSTEMS

▶ **Refer to illustration 12.22**

18 If the air coming out of the heater vents isn't hot, the problem could stem from any of the following causes:
  a) *The thermostat is stuck open, preventing the engine coolant from warming up enough to carry heat to the heater core. Replace the thermostat (see Section 3).*

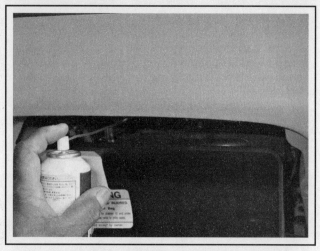

12.26 Remove the cabin air filter, then insert the disinfectant spray nozzle - be sure to support the nozzle so it doesn't get tangled up in the blower fan!

  b) *A heater hose is blocked, preventing the flow of coolant through the heater core. Feel both heater hoses at the firewall. They should be hot. If one of them is cold, there is an obstruction in one of the hoses or in the heater core, or the heater control valve is shut. Detach the hoses and back flush the heater core with a water hose. If the heater core is clear but circulation is impeded, remove the two hoses and flush them out with a garden hose.*
  c) *If flushing fails to remove the blockage from the heater core, the core must be replaced (see Section 11).*

19 If the blower motor speed does not correspond to the setting selected on the blower switch, the problem could be a bad fuse, circuit, blower relay, speed switch or blower resistor.

20 If there isn't any air coming out of the vents:
  a) *Turn the ignition ON and activate the fan control. Place your ear at the heating/air conditioning register (vent) and listen. Most motors are audible. Can you hear the motor running?*
  b) *If you can't (and have already verified that the blower switch and the blower motor resistor are good), the blower motor itself is probably bad (see Section 9).*

21 If the carpet under the heater core is damp, or if antifreeze vapor or steam is coming through the vents, the heater core is leaking. Remove it (see Section 11) and install a new unit (most radiator shops will not repair a leaking heater core).

22 Inspect the drain hose from the heater/evaporator, which exits the body under the floor (see illustration). If there is a humid mist coming from the system ducts, this hose may be plugged with leaves or road debris.

ELIMINATING AIR CONDITIONING ODORS

▶ **Refer to illustration 12.26**

23 Unpleasant odors that often develop in air conditioning systems are caused by the growth of a fungus, usually on the surface of the evaporator core. The warm, humid environment there is a perfect breeding ground for mildew to develop.

24 The evaporator core on most vehicles is difficult to access, and factory dealerships have a lengthy, expensive process for eliminating the fungus by opening up the evaporator case and using a powerful disinfectant and rinse on the core until the fungus is gone. You can ser-

vice your own system at home, but it takes something much stronger than basic household germ-killers or deodorizers.

25 Aerosol disinfectants for automotive air-conditioning systems are available in most auto parts stores, but remember when shopping for them that the most effective treatments are also the most expensive. The basic procedure for using these sprays is to start by running the system in the RECIRC mode for ten minutes with the blower on its highest speed. Use the highest heat mode to dry out the system and keep the compressor from engaging by disconnecting the wiring connector at the compressor (see Section 16).

26 The disinfectant can usually comes with a long spray hose. Remove the cabin air filter, point the nozzle inside the hole and spray according to the manufacturer's recommendations (see illustration). Follow the manufacturer's recommendations for the length of spray and waiting time between applications.

27 Once the evaporator has been cleaned, the best way to prevent the mildew from coming back again is to make sure your evaporator housing drain tube is clear (see illustration 12.23) and to run the defrost cycle briefly to dry the evaporator out after a long drive with the air conditioning on.

13 Air conditioning receiver/drier - removal and installation

▶ Refer to illustration 13.3

❋❋ WARNING:

The air conditioning system is under high pressure. Do not loosen any hose fittings or remove any components until the system has been discharged. Air conditioning refrigerant must be properly discharged into an EPA-approved recovery/recycling unit by a dealer service department or an automotive air conditioning repair facility. Always wear eye protection when disconnecting air conditioning system fittings.

1 Have the refrigerant discharged by an air conditioning technician.
2 Disconnect the negative battery cable (see Chapter 5, Section 1).
3 Using an Allen wrench, detach the end plug (see illustration) and remove the drier from the condenser with a pair of needle-nose pliers.

➡Note: **If the end plug is not accessible with the air conditioning condenser installed, the condenser may have to be removed.**

4 Use new O-rings when installing the new drier and tighten the end plug to the torque listed in this Chapter's Specifications. Be sure to lubricate the O-rings with R-134a compatible refrigerant oil.

13.3 After the system has been discharged, remove the Allen plug and pull the drier from the tube on the condenser

5 Installation is the reverse of removal.
6 Have the system evacuated, charged and leak tested by the shop that discharged it.

14 Air conditioning compressor - removal and installation

▶ Refer to illustrations 14.6

❋❋ WARNING:

The air conditioning system is under high pressure. Do not loosen any hose fittings or remove any components until the system has been discharged. Air conditioning refrigerant must be properly discharged into an EPA-approved recovery/recycling unit by a dealer service department or an automotive air conditioning repair facility. Always wear eye protection when disconnecting air conditioning system fittings.

❋❋ CAUTION:

The receiver/drier should be replaced whenever the compressor is replaced.

1 Have the refrigerant discharged by an automotive air conditioning technician.
2 Disconnect the negative cable from the battery (see Chapter 5, Section 1).
3 Remove the drivebelt from the compressor (see Chapter 1).
4 Remove the splash shield from below the engine compartment, if equipped (see Chapter 2)
5 On 2003 and earlier Lexus models, remove the alternator (see Chapter 5).
6 Detach the wiring connector and the refrigerant lines (see illustration).
7 Unbolt the compressor and lift it from the vehicle (see illustration 14.6).
8 If a new or rebuilt compressor is being installed, follow the

directions, which come with it regarding the proper level of oil prior to installation.

9   Installation is the reverse of removal. Replace any O-rings with new ones specifically made for the purpose and lubricate them with refrigerant oil.

10 Have the system evacuated, recharged and leak tested by the shop that discharged it.

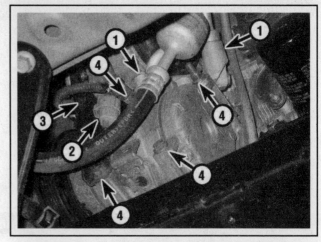

14.6 Compressor details - V6 model shown, other models are similar

1	Refrigerant line fittings	3	Mounting bracket
2	Electrical connector	4	Mounting fasteners

15 Air conditioning condenser - removal and installation

▶ **Refer to illustrations 15.4 and 15.5**

✳✳ WARNING:

The air conditioning system is under high pressure. Do not loosen any hose fittings or remove any components until the system has been discharged. Air conditioning refrigerant must be properly discharged into an EPA-approved recovery/recycling unit by a dealer service department or an automotive air conditioning repair facility. Always wear eye protection when disconnecting air conditioning system fittings.

✳✳ CAUTION:

The receiver/drier should be replaced whenever the condenser is replaced.

1   Have the refrigerant discharged by an air conditioning technician.

2   On Highland models, remove the radiator grill. On Lexus models, remove the front bumper (see Chapter 11).

3   On Lexus models, refer to Section 5 for removal of the radiator support. On Highlander models, remove the radiator upper mounts (see Section 5).

4   Disconnect the inlet and outlet fittings (see illustrations). Cap the open fittings immediately to keep moisture and dirt out of the system.

5   Remove the mounting fasteners (see illustration). Push the radiator back toward the engine, then push the condenser rearward until it's free of the radiator support and can be pulled up and out of the vehicle. On 2004 and later Lexus models, the condenser is removed towards the front of the vehicle.

6   Install the condenser, brackets and bolts, making sure the rubber cushions fit on the mounting points properly.

7   Reconnect the refrigerant lines, using new O-rings where needed.

8   Reinstall the remaining parts in the reverse order of removal.

9   Have the system evacuated, charged and leak tested by the shop that discharged it.

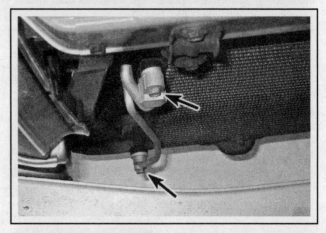

15.4 Condenser line fittings

15.5 Condenser mounting bracket fasteners

Specifications

General

Radiator cap pressure rating	13.4 to 17.6 psi (93 to 122 kPa)
Thermostat rating	
Opens	176 to 183-degrees F (80 to 84-degrees C)
Fully open	203-degrees F (95-degrees C)

Torque specifications	Ft-lbs (unless otherwise indicated)	Nm
Radiator support (Lexus)	62 in-lbs	10
Receiver/drier Allen plug	108 in-lbs	12
Thermostat housing bolts		
Four-cylinder engine	80 in-lbs	9
V6 engine	71 in-lbs	8
Water pump bolts/nuts		
Four-cylinder engine	80 in-lbs	9
V6 engine	71 in-lbs	8

Notes

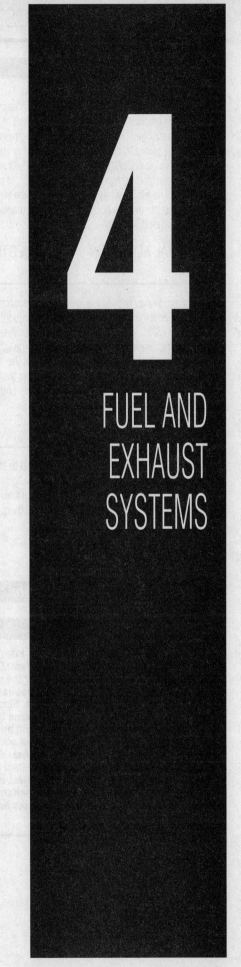

Section

Reference to other Chapters

4

FUEL AND EXHAUST SYSTEMS

1 General information

The fuel system consists of the fuel tank, the electric in-tank fuel pump, the fuel pressure regulator (an integral part of the pump), the fuel pulsation damper, the fuel rail, the fuel injectors and the fuel lines connecting the pump to the fuel rail. The injectors are turned on and off by the Powertrain Control Module (PCM). The fuel system electrical circuits are protected by the EFI fuse and the circuit opening relay (fuel pump relay). The air intake system consists of the air filter housing, the air intake duct, the throttle body and the intake manifold. (The intake manifold is covered in Chapter 2.)

SEQUENTIAL ELECTRONIC FUEL INJECTION (SFI) SYSTEM

Toyota and Lexus refer to the fuel injection systems used on these vehicles as Electronic Fuel Injection (EFI) or Sequential Electronic Fuel Injection (SFI). The system uses timed impulses to inject the fuel directly into the intake port of each cylinder in the ignition firing order. The Powertrain Control Module (PCM) controls the injectors. The PCM monitors various engine parameters and delivers the exact amount of fuel required into the intake ports. The throttle body controls the amount of air drawn into the engine.

FUEL PUMP AND LINES

Fuel is circulated from the in-tank fuel pump to the fuel rail through a fuel line running along the underside of the vehicle. Various sections of the fuel line are either rigid metal or nylon, or flexible fuel hose. The various sections of the fuel hose are connected by quick-connect fit-tings. An electric fuel pump/fuel level sending unit is located inside the fuel tank. The fuel pressure regulator is also located in the fuel tank and is an integral part of the fuel pump/fuel level sending unit.

The circuit opening relay (fuel pump relay) is equipped with a primary and secondary voltage circuit. The primary circuit is controlled by the PCM and the secondary circuit is linked directly to the EFI main relay from the ignition switch. With the ignition switch ON (engine not running), the PCM will ground the relay for two seconds. During cranking, the PCM grounds the fuel pump relay as long as the Camshaft Position (CMP) sensor sends its position signal. If there are no reference pulses, the fuel pump will shut off after two seconds.

EXHAUST SYSTEM

The exhaust system consists of the exhaust manifold(s), the exhaust pipes, the catalytic converters, a muffler, and a tail pipe. On 2001 through 2003 four-cylinder models, there are three catalytic converters: the two upstream catalysts are an integral part of the exhaust manifold, and the downstream catalyst is located underneath the vehicle. On 2004 and later four-cylinder models, there are two catalysts: one upstream catalyst, which is an integral part of the exhaust manifold and a down-stream catalyst, which is located under the vehicle. On all V6 models, there are three catalytic converters, an upstream catalyst just below each exhaust manifold and a downstream catalyst underneath the vehi-cle. On 1999 through 2003 models, the upstream catalysts are separate components that can be unbolted from the exhaust manifolds. On 2004 and later V6 models the upstream catalysts are integral component of the exhaust manifolds. For more information about the catalytic con-verters, refer to Chapter 6.

2 Fuel pressure relief procedure

✳✳ WARNING:

Gasoline is extremely flammable, so take extra precautions when you work on any part of the fuel system. Don't smoke or allow open flames or bare light bulbs near the work area, and don't work in a garage where a gas-type appliance (such as a water heater or a clothes dryer) is present. Since gasoline is carcinogenic, wear fuel resistant gloves when there's a possibil-ity of being exposed to fuel, and, if you spill any fuel on your skin, rinse it off immediately with soap and water. Mop up any spills immediately and do not store fuel-soaked rags where they could ignite. The fuel system is under constant pressure, so, if any fuel lines are to be disconnected, the fuel pressure in the system must be relieved first. When you perform any kind of work on the fuel system, wear safety glasses and have a Class B type fire extinguisher on hand.

1 Remove the gas cap to release any pressure in the fuel tank.

2 Remove the rear seat cushion (see Chapter 11).

3 Remove the fuel pump/sending unit floor service hole cover (see illustration 5.2).

4 Disconnect the fuel pump electrical connector (see illustra-tion 5.3).

5 Start the engine and allow it to run until it stops, then turn the ignition key to OFF.

6 Disconnect the cable from the negative battery terminal before working on the fuel system (see Chapter 5, Section 1).

7 The fuel system pressure is now relieved, and you can now safely open fuel line fittings anywhere in the system. But even though there is no longer any pressure in the system, it's still a good idea to wrap a shop rag around a fitting before opening it to soak up any fuel that leaks out.

3 Fuel pump/fuel pressure - check

GENERAL CHECKS

▶ **Refer to illustration 3.2**

1 Make sure that there is adequate fuel in the fuel tank.

2 .Verify the fuel pump actually runs. Have an assistant turn the ignition switch to ON - you should hear a brief whirring noise for about two seconds as the pump comes on and pressurizes the system.

➡**Note: The fuel pump is easy to hear through the fuel filler neck.**

If the fuel pump makes no sound, check the EFI fuse and the circuit opening (C/OPN) or EFI relay, both of which are located in the engine compartment fuse and relay box (see illustration). If the fuse and relay

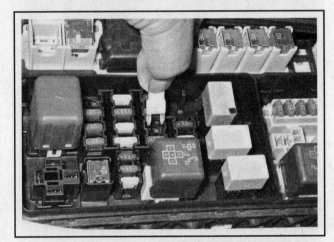

3.2 Location of the EFI fuse (2001 Highlander shown; check the fuse/relay box cover to verify the location of the EFI fuse on your vehicle)

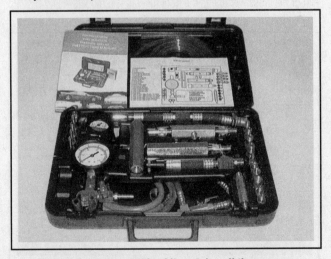

3.4 This fuel pressure testing kit contains all the necessary fittings and adapters, along with the fuel pressure gauge, to test most automotive fuel systems

are OK, check the wiring back to the fuel pump. If voltage is present at the fuel pump electrical connector for a couple of seconds when the ignition key is turned on, the fuel pump is defective. If no voltage is present, the PCM might be defective. Have the vehicle checked at a dealer service department.

FUEL PUMP PRESSURE CHECK

3 Relieve the fuel system pressure (see Section 2).

Four-cylinder models

▶ **Refer to illustrations 3.4 and 3.5**

4 To test the fuel pressure you will need a fuel pressure gauge capable of measuring the fuel pressure in the range listed in this Chapter's Specifications, and you'll need a T-fitting that fits the inside diameter of the fuel hoses. Fuel pressure gauge kits (see illustration) are available at many auto parts stores.

5 Disconnect the fuel supply line from the fuel rail (if you're unfamiliar with quick-connect fittings, see Section 4), then hook up your fuel pressure test rig between the fuel supply line and the fuel rail (see illustration). Make sure that the clamps are tight on the hoses.

V6 models

▶ **Refer to illustrations 3.6, 3.8 and 3.9a and 3.9b**

6 To test the fuel pressure you'll need a fuel pressure gauge that can measure fuel pressure in the range listed in this Chapter's Specifications, a fuel pressure test hose and an adapter with the same diameter and thread pitch as the banjo bolt that secures the crossover line

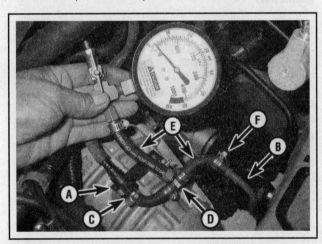

3.5 On four-cylinder models, tee the fuel pressure gauge into the fuel system between the fuel supply line and the fuel rail using a tee-fitting, some short sections of fuel hose and some hose clamps

A *Fuel rail quick-connect fitting*
B *Fuel supply line*
C *Use a short piece of fuel supply line tubing (available at most auto parts stores) to connect the hose to the quick-connect fitting*
D *Tee fitting*
E *Use short sections of fuel hose to connect the fuel pressure gauge to the tee fitting and the tee fitting to the fuel supply line and to the quick-connect fittings*
F *Secure everything with hose clamps*

3.6 To test the fuel pressure on V6 models you'll need a fuel pressure gauge, a hose and an adapter that will screw into the same hole in the end of the front fuel rail as the banjo bolt that secures the crossover line between the two fuel rails

between the two fuel rails (see illustration).

7   Remove the engine cover, if equipped (see *Intake manifold - removal and installation* in Chapter 2B).

8   Unscrew the banjo bolt that secures the crossover line to the left end of the front fuel rail (see illustration).

9   Using the same two sealing washers on both sides of the banjo fitting, screw the adapter into the fuel rail in place of the banjo bolt, then connect the fuel pressure test hose to the adapter (see illustrations).

All models

10   Turn off all accessories and turn the ignition switch key to ON. The fuel pump should run for about two seconds to pressurize the system. Note the reading on the gauge. After the pump stops running, the pressure should hold steady. After five minutes it should not drop below the minimum listed in this Chapter's Specifications.

11   Start the engine and let it idle at normal operating temperature. The pressure should remain the same. If all the pressure readings are within the limits listed in this Chapter's Specifications, the system is operating correctly.

12   If the fuel pressure is not within specifications, check the following:

a) *If the pressure is higher than specified, replace the fuel pressure regulator (see Section 6).*

b) *If the pressure is lower than specified, inspect the fuel line from the fuel tank to the fuel rail for a kinked or pinched line that would cause a restriction.*

c) *If the fuel lines are in good shape, inspect the fuel filter. Remove the fuel pump/fuel level sending unit (see Section 5) and inspect the condition of the fuel filter (see Section 6).*

d) *If the pressure is still low, most likely the fuel pressure regulator or the fuel pump is defective. In this situation, we recommend replacing both the fuel pressure regulator and the fuel pump (see Sections 5 and 6).*

e) *Another possibility of low fuel pressure is a leaking fuel injector, but that would most likely set a trouble code and turn on the CHECK ENGINE light (because the fuel mixture would be too rich).*

13   After testing is complete, relieve the fuel pressure (see Section 2) and remove the fuel pressure gauge.

3.8 To connect a fuel pressure gauge to a V6 engine, remove this banjo bolt from the left end of the front fuel rail. The banjo bolt connects the crossover tube (A) to the front fuel rail, which must remain connected to the fuel rail for this test

3.9a Screw the adapter into the front fuel rail of the V6 engine (make sure that there are still sealing washers on both sides of the banjo fitting) . . .

14   Reconnect the fuel supply line to the fuel rail (four-cylinder models) or the crossover line to the front fuel rail (V6 models).

15   Start the engine and check for leaks.

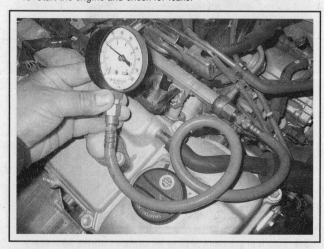

3.9b . . . then connect the fuel pressure test hose to the adapter

4   Fuel lines and fittings - general information

FUEL LINES

▶ **Refer to illustrations 4.4a, 4.4b, 4.4c and 4.4d**

1   Before servicing fuel lines or fittings, relieve the fuel pressure (see Section 2) and disconnect the cable from the negative battery terminal (see Chapter 5, Section 1).

2   The fuel line extends from the fuel tank to the engine compartment. The fuel line is secured to the underbody with plastic clips. Inspect these clips and the fuel line whenever you're working under the vehicle.

3   Look for leaks, kinks and dents in the fuel line. If you find any sign of damage, replace the fuel line immediately. Make sure that the metal fuel lines don't chafe against the pan or against any under-vehicle components. There must be a minimum of 1/4-inch clearance around the metal fuel lines to prevent chafing.

4   The fuel lines are secured to the underside of the vehicle by plastic clips. Some of these clips are bolted to the underside of the vehicle (see illustration). To detach a bolted clip, simply remove the retaining bolt. Most of the clips, however, are pushed onto threaded studs welded to the pan (see illustration). This type of clip is secured to the stud by ratchet teeth that grip the threads on the stud. To detach one of these clips, insert the tip of a large Phillips screwdriver up into the oblong hole in the clip to disengage the clip's teeth from the threads on the stud (see illustration). A third type of clip (see illustration) uses a welded stud and is retained the same way, but is much more difficult to remove because there is no oblong hole in the bottom of the clip through which you can insert a screwdriver. Instead, you'll have to use a pair of slotted screwdrivers to pry the retaining teeth away from the threads on the stud. It's virtually impossible to remove one of these clips without destroying it, so don't remove one unless you have to.

4.4b Most of the fuel line clips are pushed onto a threaded stud that's welded to the underside of the pan. To detach one of these clips, insert a large Phillips screwdriver up into the oblong hole in the underside of the clip . . .

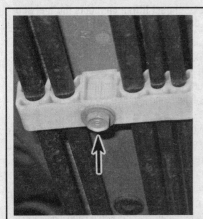

4.4a Some fuel line/EVAP line clips are bolted to the underside of the vehicle. To detach one of these clips, simply remove the bolt

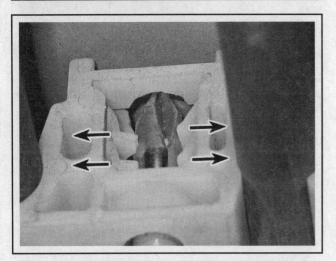

4.4c . . . to spread the clip's retaining teeth away from the stud threads so that you can pull off the clip

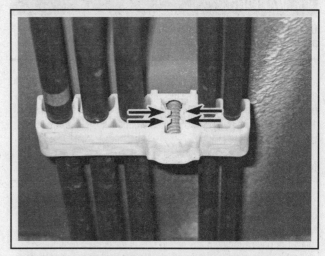

4.4d Some fuel/EVAP line clips are retained by a threaded stud but they don't have the oblong hole in the underside of the clip. These clips are difficult to remove but you can do it, using a pair of small standard screwdrivers to pry the retaining teeth away from the threads on the welded stud

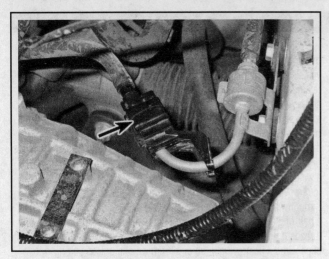

4.11 To remove the protective cover from the quick-connect fitting, simply push it off to the side

4.12 To disconnect a quick-connect fitting, depress the two buttons on opposite sides of the fitting, then pull the fitting off the end of the fuel line

5   To install a new fuel line clip:

a) *Push the fuel and EVAP lines into their guides in the clip.*
b) *Align the clip with the threaded stud.*
c) *Push the clip onto the threaded stud until it seats firmly against the underside of the vehicle.*

6   If you're replacing a damaged metal fuel line, replace the damaged line with factory replacement metal fuel line tubing or use equivalent grade steel tubing that meets the manufacturer's specifications. An unapproved line might fail when subjected to the high pressure of the fuel system. Don't use copper or aluminum tubing to replace steel tubing. These materials cannot withstand normal vehicle vibration.

7   Some fuel lines use banjo fittings with banjo bolts and sealing washers. Replace the two sealing washers every time that you open one of these fittings.

FUEL HOSES

❊❊ WARNING:

Use only original equipment replacement hoses or their equivalent. Unapproved hoses might fail when pressurized.

8   Never route flexible fuel hose within four inches of any part of the exhaust system or within ten inches of the catalytic converter. Rubber hoses must never be allowed to chafe against the frame. A minimum of 1/4-inch clearance must be maintained around a hose to prevent contact with the frame.

NYLON FUEL LINES

9   Some fuel lines are nylon, and use quick-connect fittings to connect them to the fuel pump, metal fuel lines and the fuel rail. These nylon fuel lines and quick-connect fittings cannot be serviced separately from one another. If any part of one of these lines or fittings is damaged, replace the entire assembly. Do not attempt to repair them.

10   Read through the following procedure for disconnecting and reconnecting a quick-connect fitting before trying to disconnect it.

QUICK-CONNECT FITTINGS

Button-type quick-connects

▶ **Refer to illustrations 4.11, 4.12 and 4.13**

➡**Note: This type of quick-connect fitting connects the fuel supply line to the pulsation damper on the fuel rail on V6 models.**

11   Some button-type quick-connect fittings are protected by a plastic cover (see illustration) that surrounds the fitting on three sides. To remove the cover, simply pull it off.

12   To disconnect a button-type quick-connect fitting, depress the two buttons on either side of the fitting and pull the fitting off the fuel line (see illustration).

13   Inspect the condition of the O-ring inside the fitting (see illustration). If it's damaged, replace it.

14   Installation is the reverse of removal. To reconnect a button-type quick-connect fitting, push it onto the fuel line (see illustration) until both retaining pawls lock with a clicking sound.

15   Before installing the plastic cover over the fitting, reconnect the cable to the negative battery terminal, start the engine and check for leaks.

Clamp-type quick-connects

▶ **Refer to illustrations 4.16a and 4.16b**

➡**Note: This type of quick-connect fitting connects the fuel supply line from the fuel tank to the fuel supply line that goes to the fuel rail.**

16   To unlock the clamp, press the locking tab and turn the clamp clockwise to unlock the fitting (see illustrations).

17   Pull the two halves of the fitting apart.

18   Installation is the reverse of removal. Make sure that the raised ridge on the fuel line clicks into the fitting.

19   To verify that the fitting is securely connected, try to pull the two sides of the fitting apart.

20   Reconnect the cable to the negative battery terminal, start the engine and check for leaks.

4.13 Inspect the O-ring inside the quick-connect fitting. If it's damaged, replace it

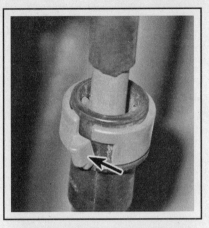

4.16a To unlock a clamp-type quick-connect fitting, push this locking tab to unlock the clamp . . .

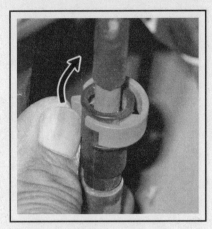

4.16b . . . then rotate the clamp in a clockwise direction to unlock the fitting

4.22 To disconnect a metal fuel line fitting, insert the special tool into the female side of the fitting, push it into the fitting until the four retainers are released, then pull the two sides of the fitting apart

METAL FUEL LINE CONNECTORS

▶ **Refer to illustration 4.22**

21 Four-cylinder models use a metal fitting to connect the fuel supply line to the fuel pulsation damper on the fuel rail. You will need to obtain a special tool, available at most auto parts stores, to disconnect one of these fittings.

22 Insert the special tool into the fuel line fitting (see illustration).

23 To reconnect a metal fuel line fitting, push the two halves of the fitting together until they snap together.

24 To verify that the fitting is securely connected, try to pull the two sides of the fitting apart.

25 Reconnect the cable to the negative battery terminal, start the engine and check for leaks.

---

5 Fuel pump/fuel level sending unit - removal and installation

▶ **Refer to illustrations 5.2, 5.3, 5.6a, 5.6b, 5.7, 5.8a and 5.8b**

❄❄ WARNING:

Gasoline is extremely flammable, so take extra precautions when you work on any part of the fuel system. See the Warning in Section 2.

➡**Note:** The removal procedure shown here depicts a typical pump used on 1999 through 2003 models. The pump used on 2004 and later models is slightly different.

1 Remove the rear seat (see Chapter 11).

2 Remove the fuel pump/sending unit floor service hole cover (see illustration).

3 Disconnect the fuel pump electrical connector (see illustration).

4 Relieve the fuel system pressure (see Section 2) and remove the fuel tank cap.

5 Disconnect the cable from the negative battery terminal (see Chapter 5, Section 1).

5.2 To remove the floor service hole cover, carefully pry it out with a screwdriver

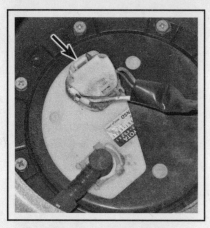

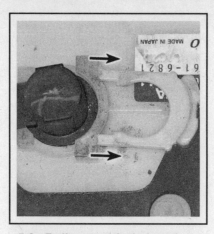

5.3 To disconnect the fuel pump/ fuel level sending unit electrical connector, depress this release tab (you might have to use a small pair of needle nose pliers - it's stiff) and pull up

5.6a To disconnect the fuel supply line fitting from the fuel pump outlet pipe, remove this clip . . .

5.6b . . . then pull the fitting straight up. Before reconnecting the fitting, inspect the O-ring inside the fitting. If the O-ring is damaged, replace it

6   Disconnect the fuel supply line from the top of the fuel pump unit (see illustrations).

7   Remove the retaining bolts (see illustration) from the fuel pump/ fuel level sending unit mounting flange.

8   Carefully lift the fuel pump/fuel level sending unit assembly from the fuel tank (see illustrations).

9   Inspect the fuel pump inlet strainer for contamination. If it's dirty, try cleaning the strainer with some clean solvent and an old toothbrush. If the strainer is too dirty to be cleaned while it's installed, remove it from the fuel pump/fuel level sending unit (see Section 6) and try cleaning it again. If you still can't clean it adequately, replace it.

10 Installation is the reverse of removal.

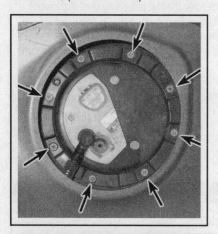

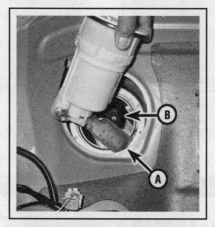

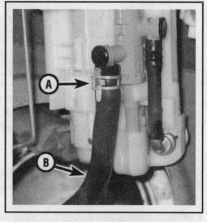

5.7 To detach the fuel pump/fuel level sending unit mounting flange from the fuel tank, remove these bolts

5.8a Angling the pump/sending unit assembly to protect the fuel inlet strainer (A) and the sending unit float (B), carefully lift the fuel pump/fuel level sending unit from the fuel tank . . .

5.8b . . . loosen and slide back this spring-type hose clamp (A) and disconnect the hose (B) from the underside of the pump

6   Fuel pump/fuel level sending unit - component replacement

◆ Refer to illustration 6.3, 6.4, 6.5, 6.6, 6.7, 6.8a, 6.8b, 6.9 and 6.10

✳✳ **WARNING:**

Gasoline is extremely flammable, so take extra precautions when you work on any part of the fuel system. See the Warning in Section 2.

➡ Note: Once you have removed the fuel pump/fuel level sending unit from the fuel tank, you can replace the entire assembly, or you can disassemble it and replace the fuel inlet strainer, the fuel pressure regulator, the fuel return jet tube, the fuel pump, the fuel filter or the fuel level sending unit. The disassembly procedure shown here depicts a typical fuel pump used on 1999 through 2003 models. The pump used on 2004 and later models is slightly different.

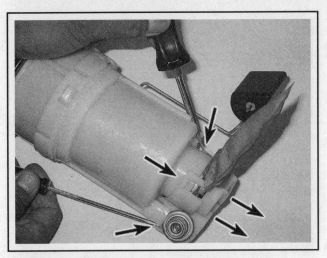

6.3 To remove the lower end cap, carefully disengage each of the locking tabs with a pair of screwdrivers, then slide it off

1   Remove the fuel pump/fuel level sending unit assembly from the fuel tank (see Section 5).

2   Drain any residual gasoline from the fuel pump/fuel level sending unit, then place the fuel pump/fuel level sending unit on a clean workbench. Make sure that the work area is well-ventilated, because there will be some gasoline evaporating from the pump/sending unit for awhile.

3   Carefully pry off the end cap from the lower end of the fuel pump/fuel level sending unit assembly (see illustration).

4   Remove the rubber insulator from the lower end of the fuel pump (see illustration).

5   Remove the fuel pump inlet strainer (see illustration) from the lower end of the fuel pump.

6   Remove the fuel pressure regulator (see illustration).

7   Remove the fuel return jet tube assembly (see illustration).

8   Remove the fuel pump from the fuel filter assembly (see illustrations).

9   Separate the fuel filter housing from the upper part of the fuel pump assembly (see illustration). If you're replacing the fuel filter, you must replace this housing. The fuel filter is not available separately from the housing.

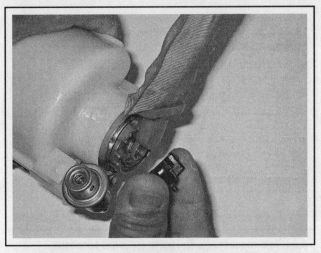

6.4 Remove and inspect this little rubber insulator from the lower end of the fuel pump. If it's damaged, replace it

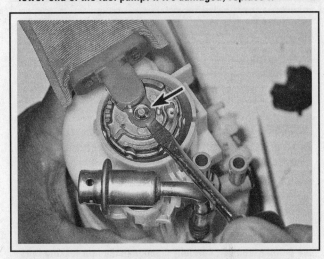

6.5 The inlet strainer is secured to the lower end of the fuel pump by this E-clip. To detach the strainer from the pump, insert a screwdriver between the strainer mounting flange and the pump and carefully pry the strainer and the E-clip loose (don't try to pry the E-clip off by itself - it's too difficult to slip a screwdriver under it)

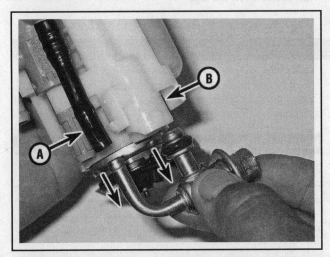

6.6 Disconnect the fuel pressure regulator from the fuel return jet tube hose (A) and from the fuel filter (B)

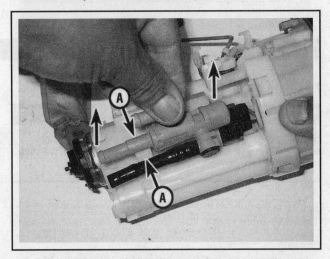

6.7 To detach the fuel return jet tube assembly, simply pull it out of this retainer clip (A)

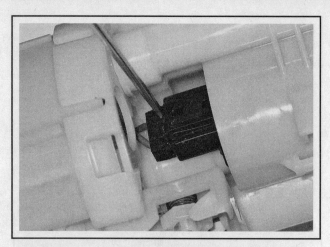

6.8a To disconnect the fuel pump electrical connector, depress this locking tab and pull out the connector

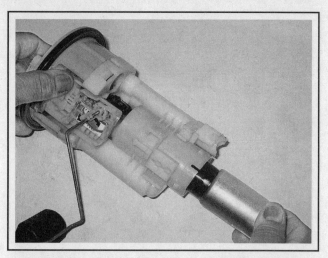

6.8b To remove the fuel pump from the fuel filter housing, pull it out the bottom

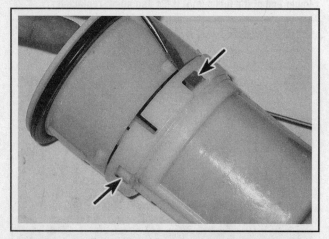

6.9 To separate the fuel filter housing from the upper part of the pump assembly, carefully disengage the four locking tabs from the filter housing and pull the two components apart

10 Detach the fuel level sending unit from the upper part of the fuel pump assembly (see illustration).

11 Reassembly is the reverse of disassembly.

12 Install the fuel pump/fuel level sending unit in the fuel tank (see Section 5).

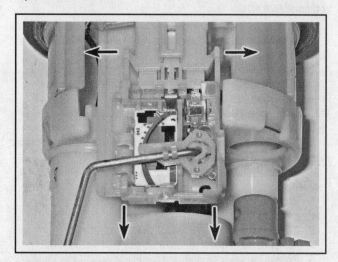

6.10 To detach the fuel level sending unit from the upper part of the fuel pump assembly, disengage these two locking tangs from the lugs on the upper part of the pump, then slide the sending unit straight down. When installing the sending unit, slide it into position until the two locking tangs click into position over the lugs

7 Fuel tank - removal and installation

▶ Refer to illustrations 7.8a, 7.8b, 7.9a, 7.9b, 7.10 and 7.12

❋❋ **WARNING:**

Gasoline is extremely flammable, so take extra precautions when you work on any part of the fuel system. See the Warning in Section 2.

1 Relieve the fuel system pressure (see Section 2). Remove the fuel filler cap to relieve fuel tank pressure.

2 Disconnect the cable from the negative battery terminal (see Chapter 5, Section 1).

3 Remove the fuel pump/fuel level sending unit assembly from the fuel tank (see Section 5).

4 If the tank is full or nearly full, siphon the fuel into an approved container with a siphoning kit. Siphoning kits are available at most auto parts stores.

❋❋ **WARNING:**

Do not start the siphoning action by mouth!

5 Raise the vehicle and place it securely on jackstands.

6 Remove the center part of the exhaust system (the downstream catalytic converter to the tailpipe behind the muffler - see Section 16).

7 On 4WD models, remove the driveshaft (see Chapter 10).

8 Remove the fuel tank protector(s) (see illustrations).

9 Remove the fuel tank heat shield (see illustrations).

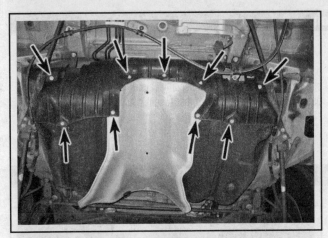

7.8a To detach the left and right fuel tank protectors from the fuel tank on a 2WD model, remove these fasteners (typical)

7.8b To detach the fuel tank protector from the fuel tank on a 4WD model, remove these fasteners (typical)

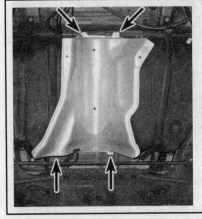

7.9a To detach the heat shield from the fuel tank on a 2WD model, remove these fasteners (typical)

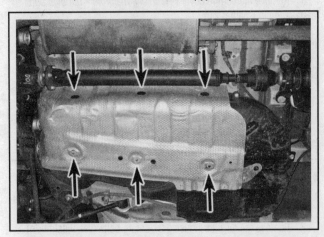

7.9b To detach the heat shield from the fuel tank on a 4WD model, remove these fasteners (typical)

10 Disconnect the fuel filler neck hose and the two smaller hoses next to it (see illustration).

11 Support the fuel tank securely.

12 Remove the bolts from the fuel tank retaining straps (see illustration).

13 Carefully lower the tank about six inches or so and verify that all electrical harnesses, fuel lines and EVAP system hoses are disconnected.

14 Carefully lower the fuel tank to the ground.

15 Installation is the reverse of removal. Be sure to tighten the fuel tank strap bolts securely.

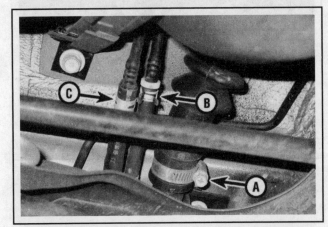

7.10 Loosen the screw-type hose clamp (A) and disconnect the fuel filler neck hose from the fuel tank filler neck pipe, then loosen the two spring-type clamps (B and C) and disconnect the two EVAP hoses

7.12 To detach the fuel tank from the vehicle, remove these four fuel tank strap bolts (2WD model shown, 4WD models similar)

8 Fuel tank cleaning and repair - general information

1 Any repairs to the fuel tank or filler neck should be done by a professional with experience in this critical and potentially dangerous work. Even after cleaning and flushing of the fuel system, explosive fumes can remain and ignite during repair of the tank.

2 If the fuel tank is removed from the vehicle, it should not be placed in an area where sparks or open flames could ignite the fumes coming out of the tank. Be especially careful inside garages where a gas-type appliance is located, because it could cause an explosion.

9 Air filter housing - removal and installation

AIR INTAKE DUCT AND RESONATORS

▸ **Refer to illustrations 9.2 and 9.4**

1 Remove the engine cover (see *Intake manifold - removal and installation* in Chapter 2).
2 Disconnect the Positive Crankcase Ventilation (PCV) fresh air inlet hose (see illustration) from the air intake duct.
3 Loosen the hose clamps at both ends of the air intake duct and remove the duct.
4 If you're replacing the air intake duct, remove the resonators (see illustration) and install them on the new intake duct.
5 Installation is the reverse of removal.

AIR FILTER HOUSING

▸ **Refer to illustration 9.7**

6 Remove the air filter housing cover and remove the filter element (see Chapter 1).
7 Remove the air filter housing mounting bolts (see illustration).
8 Remove the air filter housing.
9 Installation is the reverse of removal.

9.2 To remove the air intake duct, disconnect the PCV fresh air inlet hose (1) and loosen the hose clamps (2) at both ends of the duct, then pull the duct off the air filter housing and the throttle body (V6 model shown)

9.4 To remove the resonators, loosen the hose clamps and pull the resonators out of the air intake duct (V6 model shown)

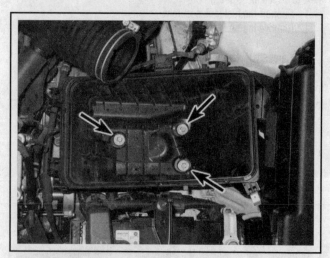

9.7 To detach the air filter housing, remove these three bolts (V6 model shown)

10 Accelerator cable - removal and installation

▶ Refer to illustrations 10.2, 10.3, 10.5 and 10.6

➡Note: This procedure applies to 1999 through 2003 models only. 2004 and later models do not use an accelerator cable.

1   Remove the engine cover (see *Intake manifold - removal and installation* in Chapter 2).

2   Loosen the accelerator cable locknut and back off the cable adjustment nut (see illustration), then disengage the cable from the cable bracket.

3   Disengage the accelerator cable from the throttle cam (see illustration).

4   Trace the accelerator cable back to the firewall, note the routing of the cable and detach any cable clips or guides.

5   Inside the vehicle, disengage the cable from the accelerator pedal (see illustration).

6   Remove the bolts and detach the cable from the firewall (see illustration).

7   Pull the cable through the firewall from the engine compartment side.

8   Installation is the reverse of removal.

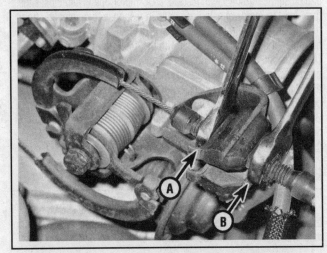

10.2 To disengage the accelerator cable from its cable bracket, loosen the locknut (A) and the adjusting nut (B), then slide the cable out of the bracket (V6 model shown, four-cylinder models similar)

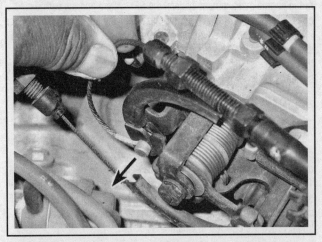

10.3 To disconnect the accelerator cable from the throttle cam, align the cable with the slot in the side of the cam, then slide the cable end plug out of the cam (V6 model shown, four-cylinder models similar)

10.5 To disconnect the cable from the accelerator pedal, pull the cable end (A) out of the pedal, then guide the cable out the slot on the left (B)

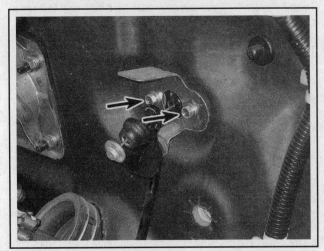

10.6 To detach the accelerator cable from the firewall, remove these two bolts

11 Sequential Electronic Fuel Injection (SFI) system - general information

The SFI system consists of the fuel tank, the electric in-tank fuel pump, the fuel pressure regulator (an integral component of the fuel pump), the EFI main relay, the circuit opening relay (fuel pump relay), the fuel rail and the fuel injectors. The induction system consists of the air filter housing, the air intake duct and the throttle body. The throttle bodies on all 1999 through 2003 models are conventional, cable-operated units. The throttle bodies on 2004 and later models are PCM-controlled electronic units. On these models, the PCM monitors the angle of the accelerator pedal with a sensor at the pedal assembly and directs a solenoid in the throttle body to alter the angle of the throttle valve in the throttle body in accordance with this data. Because they're all parts of the engine management system, the SFI system and the various emission control systems used on these vehicles are closely linked. For information about the emission control systems, see Chapter 6. The SFI system is easy to understand if you divide it into three sub-systems: the air intake system, the electronic control system and the fuel delivery system.

AIR INTAKE SYSTEM

The air intake system consists of the air filter housing, the air intake duct, the resonators, the throttle body and the intake manifold The resonators are oddly shaped appendages clamped to the air intake duct. They function as accumulators, or reservoirs, that help to quiet the flow of air through the air intake duct. The replacement procedures for all of these components are covered in this chapter except the intake manifolds, which are covered in Chapter 2A or 2B.

1999 through 2003 throttle bodies

The throttle body on 2001 through 2003 four-cylinder models is a conventional single-barrel design. The lower portion of the throttle body is heated by engine coolant to prevent icing in cold weather. The Throttle Position (TP) sensor is attached to the throttle shaft to monitor changes in the throttle opening. The Idle Speed Control (ISC) valve is attached to the underside of the throttle body. For information about the TP sensor and the ISC valve, refer to Chapter 6.

The throttle body on 1999 through 2003 V6 models is a two-barrel design that incorporates the three-stage Acoustic Control Induction System (ACIS). Like the single-barrel throttle body used on four-cylinder models, the two-barrel throttle body used on V6 models is also equipped with a Throttle Position (TP) sensor and an Idle Speed Control (ISC) valve. For more information about the ACIS, the TP sensor and the ISC valve, refer to Chapter 6.

When the engine is idling, the air/fuel ratio is controlled by the Idle Speed Control (ISC) system. The ISC system consists of the Engine Coolant Temperature (ECT) sensor, the Intake Air Temperature (IAT) sensor the Throttle Position (TP) sensor, the Mass Air Flow (MAF) sensor, various other sensors, the Powertrain Control Module (PCM) and the ISC valve. The ISC valve is activated by the PCM under certain conditions, such as cold temperature during cranking, or loads imposed on the engine by power steering demand at low vehicle speeds or by turn-

ing on the air conditioning. The ISC valve regulates the amount of airflow bypassing the throttle plate and into the intake manifold. For more information about the ISC valve and the ISC system, refer to Chapter 6.

2004 and later throttle bodies

The throttle bodies on 2004 and later four-cylinder and V6 models are equipped with the Electronic Throttle Control System-intelligent (ETCS-i) system. The single-barrel throttle bodies on these models are electronic units, i.e. they don't use an accelerator cable to open and close the throttle valve. Instead, a DC motor opens and closes the throttle valve inside one of these units in response to commands from the Powertrain Control Module (PCM). The PCM monitors the position, or angle, of the accelerator pedal with the Accelerator Pedal Position (APP) sensor, then commands the DC motor inside the throttle body to open or close the throttle plate accordingly. These throttle bodies have a Throttle Position (TP) sensor, but, unlike earlier units, it's an integral component of the throttle body, and not removable. There is no Idle Speed Control (ISC) valve on these units. The idle speed control function is handled by the PCM. For more information about the ETCS-i, the APP sensor and the PCM, refer to Chapter 6.

ELECTRONIC CONTROL SYSTEM

The information sensors, Powertrain Control Module (PCM) and output actuators, and the various emission control system employed on these vehicles, are described in Chapter 6.

FUEL DELIVERY SYSTEM

The fuel delivery system consists of the fuel pump, the fuel pressure regulator, the fuel filter, the fuel lines, the pulsation damper, the fuel rail and the fuel injectors. The in-tank fuel pump is an electric in-line type. Fuel is drawn through an inlet strainer into the pump, flows through the fuel pressure regulator, passes through the fuel filter and is delivered to the injectors. The fuel pressure regulator maintains a constant fuel pressure to the injectors.

The injectors are solenoid-actuated, constant stroke, pintle types consisting of a solenoid, plunger, needle valve and housing. When current is applied to the solenoid coil, the needle valve is raised off its seat and pressurized fuel squirts out of the injector body through the nozzle. The injection quantity is determined by the pulse width, i.e. the length of time that the pintle valve is open, which is controlled by the length of time during which current is supplied to the solenoid coil. Because injector timing determines opening and closing intervals, which in turn determines the air-fuel mixture ratio, injector timing must be precise.

The EFI main relay, located in the engine compartment relay/fuse box, supplies power to the fuel pump relay (circuit opening relay) from the ignition key. The PCM controls the grounding signal to the fuel pump in response to the starting and camshaft position signals at start-up.

12 Fuel injection system - check

▶ Refer to illustrations 12.7, 12.8 and 12.9

❋❋ WARNING:

Gasoline is extremely flammable, so take extra precautions when you work on any part of the fuel system. See the Warning in Section 2.

1 Check all electrical connectors, especially ground connections, for the system. Loose connectors and poor grounds can cause many engine control system problems.

2 Verify that the battery is fully charged because the Powertrain Control Module (PCM) and sensors cannot operate properly without adequate supply voltage.

3 Refer to Chapter 1 and check the air filter element. A dirty or partially blocked filter will reduce performance and economy.

4 Check fuel pump operation (see Section 3). If the fuel pump fuse is blown, replace it and see if it blows again. If it does, look for a short in the wiring harness to the fuel pump.

5 Inspect the vacuum hoses connected to the intake manifold for damage, deterioration and leakage.

6 Remove the air intake duct from the throttle body and check for dirt, carbon, varnish, or other residue in the throttle body, particularly around the throttle plate. If it's dirty, refer to Chapter 6 and troubleshoot the PCV and EGR systems for the cause of excessive varnish buildup.

7 With the engine running, place an automotive stethoscope against each injector, one at a time, and listen for a clicking sound that indicates operation (see illustration). If you don't have a stethoscope, you can place the tip of a long screwdriver against the injector and listen through the handle.

8 If an injector does not seem to be operating electrically (not clicking), purchase a special injector test light (sometimes called a

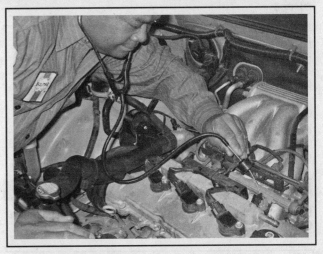

12.7 Using a stethoscope, listen to each fuel injector and verify that it's making a clicking sound when the engine is running

"noid" light) and install it into the injector wiring harness connector (see illustration). Start the engine and see if the noid light flashes. If it does, the injector is receiving proper voltage. If it doesn't flash, further diagnosis is necessary. You might want to have it checked by a dealership service department or other qualified repair shop.

9 With the engine off and the fuel injector electrical connectors disconnected, measure the resistance of each injector with an ohmmeter (see illustration). Refer to the Specifications at the end of this Chapter for the correct resistance.

10 Refer to Chapter 6 for other system checks.

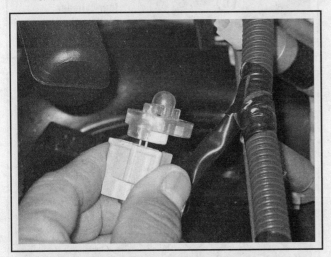

12.8 Plug the "noid" light into the electrical connector for each fuel injector and verify that it blinks with the engine running

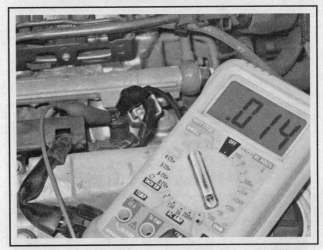

12.9 Turn the mode knob on your digital multimeter to the appropriate resistance range, touch the probes of the meter to the two terminals on each injector and measure the resistance, which should be within the range listed in this Chapter's Specifications

13   Throttle body - check, removal and installation

CHECK

▶ **Refer to illustration 13.2**

1   On 1999 through 2003 models, verify that the throttle linkage operates smoothly.

2   Remove the air intake duct from the throttle body and check for carbon and residue build-up. If it is dirty, clean it with aerosol carburetor cleaner (make sure the can specifically states that it is safe with oxygen sensor systems and catalytic converters) and a toothbrush (see illustration).

※ CAUTION:

Do not clean the Throttle Position (TP) sensor or the Idle Speed Control (ISC) valve with the solvent.

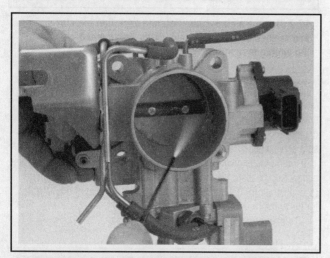

13.2 With the engine off, use aerosol carburetor cleaner, a toothbrush and a shop rag to clean the throttle body bore. Open the throttle plate so that you can clean behind it. (Make sure that the carb cleaner won't damage catalysts and oxygen sensors)

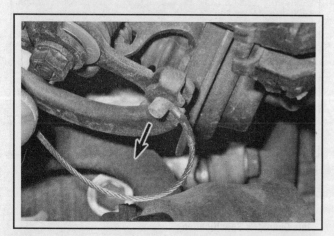

13.7b To disconnect the cruise control cable from the throttle cam, align the cable with the slot in the side of the cam, then slide the cable end plug out of the cam (V6 model shown, four-cylinder models similar)

REMOVAL AND INSTALLATION

※ WARNING:

Wait until the engine is completely cool before beginning this procedure.

2001 through 2003 models

▶ **Refer to illustrations 13.7a, 13.7b, 13.9, 13.13 and 13.14**

3   Disconnect the cable from the negative battery terminal (see Chapter 5, Section 1).

4   Remove the air intake duct (see Section 9).

5   Disconnect the electrical connector from the Throttle Position (TP) sensor.

6   Disconnect the electrical connector from the Idle Speed Control (ISC) valve.

7   Disconnect the cruise control cable from the throttle body (see illustrations).

8   Disconnect the accelerator cable from the throttle body (see illus-

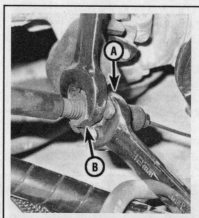

13.7a To disengage the cruise control cable from its cable bracket, loosen the locknut (A) and the adjusting nut (B), then slide the cable out of the bracket (V6 model shown, four-cylinder models similar)

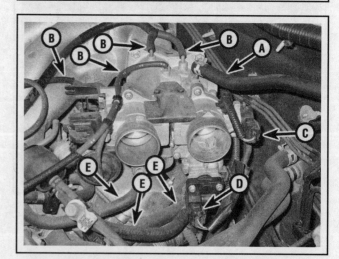

13.9 Before removing the throttle body, disconnect the power brake booster vacuum hose (A), the vacuum hoses (B), the Throttle Position (TP) sensor electrical connector (C), the Idle Speed Control (ISC) valve electrical connector (D) and the coolant hoses (E)

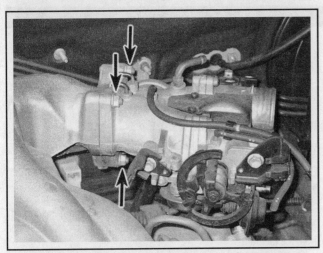

13.13 To detach the throttle body from the intake manifold, remove the three mounting nuts (V6 model shown, four-cylinder models similar)

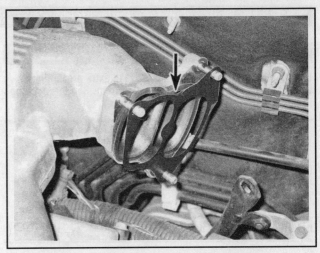

13.14 Remove and discard the old throttle body gasket (V6 model shown)

trations 10.2 and 10.3).

9   Disconnect the brake booster vacuum hose from the throttle body (see illustration).

10 Disconnect the vacuum hoses from the throttle body.

11 Clamp off the coolant hoses to the throttle body to minimize coolant loss. Plug the coolant hoses to prevent coolant leakage.

12 Disconnect the coolant hoses from the throttle body.

13 Remove the throttle body mounting bolts (four-cylinder models) or nuts (V6 models) (see illustration) and remove the throttle body.

14 Remove and discard the old throttle body gasket (see illustration).

15 Installation is the reverse of removal. Be sure to tighten the throttle body mounting nuts to the torque listed in this Chapter's Specifications. Check the coolant level, adding as necessary (see Chapter 1).

2004 and later models

▶ **Refer to illustrations 13.21 and 13.23**

16 Disconnect the cable from the negative battery terminal (see

Chapter 5, Section 1).

17 Remove the strut brace (see Chapter 10).

18 Remove the engine cover (see Intake manifold - removal and installation in Chapter 2A or 2B).

19 Remove the air intake duct (see Section 9).

20 Clamp off the coolant hoses to the throttle body to minimize coolant loss.

21 Disconnect the electrical connector from the throttle body and disengage the electrical harness from the clip on the fuel pipe bracket (see illustration).

22 Disengage the fuel hose from the fuel pipe bracket, then clearly label and disconnect the vacuum and coolant hoses from the throttle body (see illustration 13.21). Plug the coolant hoses to prevent coolant leakage.

23 Remove the throttle body mounting bolts (see illustration) and remove the throttle body.

24 Remove and discard the old throttle body gasket.

25 Installation is the reverse of removal. Be sure to use a new gasket and tighten the throttle body mounting bolts to the torque listed in this Chapter's Specifications. Check the coolant level, adding as necessary (see Chapter 1).

13.21 Before detaching the throttle body, disconnect the electrical connector (A), disengage the electrical harness from this clip (B), disengage the fuel hose (C) from the clip on the fuel pipe bracket and disconnect both coolant hoses (D) (four-cylinder model shown, V6 models similar)

13.23 To detach the throttle body from the intake manifold, remove these mounting bolts (four-cylinder model shown, V6 models similar)

14 Fuel pulsation damper - replacement

✳✳ WARNING:

Gasoline is extremely flammable, so take extra precautions when you work on any part of the fuel system. See the Warning in Section 2.

1   Relieve the fuel pressure (see Section 2).
2   Disconnect the cable from the negative battery terminal (see Chapter 5, Section 1).
3   Remove the engine cover (see *Intake manifold - removal and installation* in Chapter 2).

FOUR-CYLINDER MODELS

▶ **Refer to illustration 14.5**

4   Disconnect the fuel supply line from the fuel pulsation damper (see Section 4).

14.5 To detach the pulsation damper from the fuel rail on a four-cylinder model, remove these two bolts

5   Remove the two pulsation damper mounting bolts (see illustration) and remove the pulsation damper from the fuel rail.
6   Remove the old damper O-ring.
7   Installation is the reverse of removal. Be sure to use a new O-ring and tighten the pulsation damper mounting bolts to the torque listed in this Chapter's Specifications.

V6 MODELS

▶ **Refer to illustration 14.8**

8   On V6 models the pulsation damper (see illustration) is screwed onto the fuel rail for the rear cylinder bank. Simply unscrew the damper from the fuel rail.
9   Remove and discard the old sealing washers.
10 Installation is the reverse of removal. Be sure to install new sealing washers and tighten the damper to the torque listed in this Chapter's Specifications.

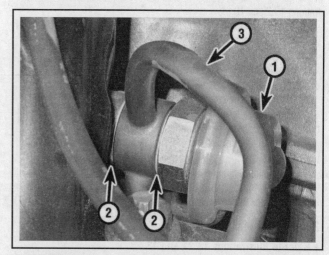

14.8 To remove the pulsation damper (1) from the rear fuel rail on a V6 model, simply unscrew it with a wrench. Discard the old sealing washers (2) between the pulsation damper and the crossover tube (3), and between the crossover tube and the fuel rail

15 Fuel rail and injectors - removal and installation

✳✳ WARNING:

Gasoline is extremely flammable, so take extra precautions when you work on any part of the fuel system. See the Warning in Section 2.

1   Relieve the fuel pressure (see Section 2).
2   Disconnect the cable from the negative battery terminal (see Chapter 5, Section 1).
3   Remove the engine cover, if equipped (see *Intake manifold - removal and installation* in Chapter 2).
4   Remove the air intake duct (see Section 9).

FOUR-CYLINDER MODELS

▶ **Refer to illustration 15.7**

5   Disconnect the electrical connectors from the fuel injectors and set the injector harness aside.
6   Disconnect the fuel supply line from the fuel pulsation damper (see Section 4).
7   Remove the fuel rail mounting bolts (see illustration).
8   Remove the fuel rail and the fuel injectors as a single assembly.
9   To remove the fuel injectors and replace the injector O-rings, refer to Steps 23 and 24.
10 Installation is the reverse of removal. Be sure to tighten the fuel rail mounting bolts to the torque listed in this Chapter's Specifications.

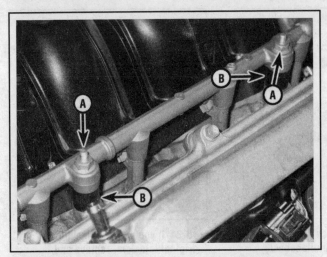

15.7 To detach the fuel rail from the intake manifold on four-cylinder models, remove these bolts (A). After removing the fuel rail and injectors, remove and inspect the condition of the spacers (B). If they're damaged, replace them

V6 MODELS

▶ Refer to illustrations 15.13, 15.20, 15.21, 15.22, 15.23 and 15.24

➥Note: On these models, the fuel rail is a two-piece assembly. The front fuel rail houses the fuel injectors for the cylinders in the front head. The rear fuel rail houses the fuel injectors for the cylinders in the rear head. We recommend replacing all injector O-rings even when only one injector O-ring or seal is leaking. On most fuel rails you have to remove the entire assembly anyway, so replace all of the O-rings/seals at one time to avoid having to remove the fuel rail later to replace another O-ring and/or seal. However, if the leak is occurring at only one front cylinder injector, please note that you do have the option to remove the front fuel rail without removing the rear fuel rail, which means that you won't have to remove the intake manifold. Conversely, you can also remove the rear fuel rail without removing the front fuel rail, though you will have to remove the intake manifold to do so. Finally, our recommendation is of that anytime any injector develops a leak it's a good idea to remove all of the injectors and replace all of the O-rings and/or seals.

15.13 Disconnect the electrical connectors (A) from the fuel injectors on the front cylinder head. To detach the Vacuum Switching Valve (VSV) mounting bracket from the intake manifold, remove these two nuts (B) (V6 models)

And removing the intake manifold isn't very difficult on these vehicles.

11 Remove the strut brace (see Chapter 10, Section 3).

12 Remove the intake manifold (see Chapter 2B).

13 Disconnect the electrical connectors from the injectors on the front cylinder head (see illustration).

14 Remove the Vacuum Switching Valve (VSV) mounting bracket from the intake manifold (see illustration 15.13) and set the bracket aside. Don't disconnect any of the vacuum hoses.

15 If you're going to remove the entire fuel rail assembly, disconnect the fuel supply line from the fuel rail (see Section 4).

16 If you're only going to remove the front fuel rail, disconnect the metal crossover tube that connects the front fuel rail to the rear fuel rail by removing the banjo bolt from the front fuel rail (see illustration 3.8). If you're going to remove the rear fuel rail, then remove the pulsation damper from the rear fuel rail (see illustration 14.8) and remove the crossover tube. Remove and discard the old sealing washers (use new ones during reassembly).

17 If you're just removing the front fuel rail and injectors, remove the front fuel rail mounting bolts (see illustration 15.21) and

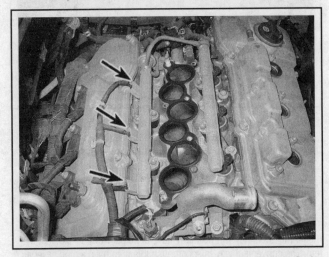

15.20 Disconnect the electrical connectors from the rear fuel injectors (V6 models)

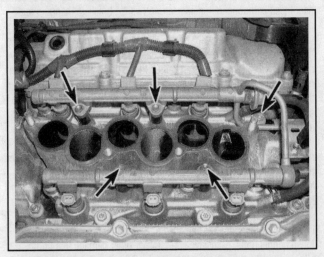

15.21 To detach the fuel rail assembly from the intake manifold, remove these bolts (V6 models)

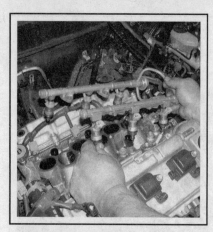

15.22 Remove the fuel rail and injectors as a single assembly

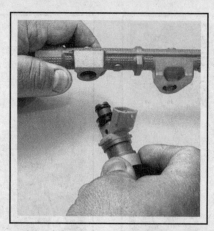

15.23 To remove each fuel injector from the fuel rail, simultaneously twist and pull it out of the fuel rail

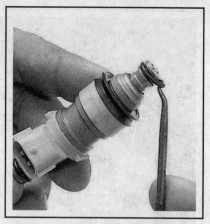

15.24 If you plan to reinstall the original injectors, remove and discard the old O-rings and grommets and replace them with new ones

remove the front fuel rail and injectors by lifting the assembly straight up.

18 To remove the fuel injectors and replace the injector O-rings, refer to Steps 23 and 24 below.

19 If you're going to remove the entire fuel rail assembly (or just the rear fuel rail), remove the intake manifold (see Chapter 2B).

20 Disconnect the electrical connectors from the rear fuel rail (see illustration).

21 Remove the fuel rail mounting bolts (see illustration).

22 Remove the fuel rail and the fuel injectors as a single assembly (see illustration).

23 Remove the fuel injectors from the fuel rail (see illustration).

24 If you intend to re-use the same injectors, replace the grommets and O-rings (see illustration).

25 Installation is the reverse of removal. Be sure to use new injector O-rings and tighten the fuel rail mounting bolts to the torque listed in this Chapter's Specifications.

16 Exhaust system servicing - general information

♦ Refer to illustrations 16.1a, 16.1b and 16.1c

✳✳ WARNING:

Inspect and repair exhaust system components only after the system components have cooled down.

1 The exhaust system consists of the exhaust manifold(s), the catalytic converters, the muffler, the tailpipe and the exhaust pipes that connect these components together, as well as the brackets, hangers and clamps that support and secure these components. The exhaust system is attached to the body with mounting brackets and rubber hangers (see illustration). If any of these parts are damaged or deteriorated, excessive noise and vibration will be transmitted to the body. The exhaust system is divided into various sections, which are bolted together. Some of these sections consist of several components welded together into one assembly. For example, the center part of the exhaust system, which is also the longest single section, includes the downstream catalytic converter and the pre-muffler (see illustrations). If either of these components is damaged, you will have to replace the entire assembly, or have the old catalyst or muffler cut off and a new unit welded in. You'll also have to remove the center exhaust pipe assembly in order to remove the fuel tank.

2 Conducting regular inspections of the exhaust system will keep it safe and quiet. Look for damaged or bent parts, open seams, holes, loose connections, excessive corrosion or other damage that could allow exhaust fumes to enter the vehicle. Do not repair deteriorated exhaust system components - replace them.

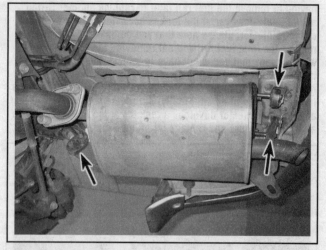

16.1a Each section of the exhaust system (this is the rear muffler section) is suspended by rubber exhaust hangers. Over time these hangers can become dried out, then crack or tear. Always inspect the exhaust hangers whenever you're servicing anything under the vehicle, and replace them immediately if they're damaged or worn out

3 If the exhaust system components are extremely corroded or rusted together, they will probably have to be cut from the exhaust system. The convenient way to accomplish this is to have a muffler repair shop remove the corroded sections with a cutting torch. If, however,

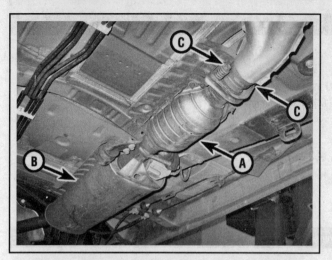

16.1b The center exhaust pipe assembly is the longest and biggest section of the exhaust system. It consists of the downstream catalytic converter (A) and the forward muffler (B). To unbolt the forward end of the center exhaust pipe assembly, remove these two bolts (C)

you want to save money by doing it yourself and you don't have an oxy/acetylene welding outfit with a cutting torch, simply cut off the old components with a hacksaw. If you have compressed air, you can also use special pneumatic cutting chisels. If you do decide to tackle the job at home, be sure to wear eye protection to protect your eyes from metal chips, and wear work gloves to protect your hands.

4   Here are some simple guidelines to apply when repairing the exhaust system:

a) *Work from the back to the front when removing exhaust system components.*

b) *Apply penetrating oil to the exhaust system component fasteners to make them easier to remove.*

c) *Use new gaskets, hangers and clamps when installing exhaust system components. Unless the fasteners are in good condition,*

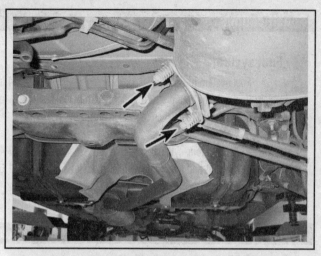

16.1c To unbolt the rear end of the center exhaust pipe assembly, remove these two bolts

it's also a good idea to replace them as well. Even if you're successful in unscrewing rusted fasteners, they will be even more difficult to unscrew the next time that you need to do so, and they might break the next time that they're loosened or tightened.

d) *Apply anti-seize compound to the threads of all exhaust system fasteners during reassembly. Be sure to allow sufficient clearance between newly installed parts and all points on the underbody to avoid overheating the floor pan and possibly damaging the interior carpet and insulation. Pay particularly close attention to the catalytic converter and its heat shield.*

✶✶ WARNING:

The catalytic converter operates at very high temperatures and takes a long time to cool. Wait until it's completely cool before attempting to remove the converter. Failure to do so could result in serious burns.

Specifications

Fuel system

Fuel system pressure	44 to 50 psi (303 to 343 kPa)
Fuel system hold pressure (after five minutes)	21 psi (147 kPa) minimum
Fuel injector resistance (approximate)	
All engines (except 2004 and later four-cylinder models)	13.4 to 14.2 ohms
2004 and later four-cylinder models	11.6 to 12.5 ohms

Torque specifications

	Ft-lbs (unless otherwise indicated)	Nm
Fuel rail mounting bolts or nuts		
Four-cylinder models	15	20
V6 models	84 in-lbs	10
Pulsation damper		
Four-cylinder models (mounting bolts)	80 in-lbs	9
V6 models (damper unscrews from fuel rail)	24	33
Fuel crossover pipe banjo bolt (V6 engine)	24	33
Throttle body mounting bolts/nuts		
Four-cylinder engines	22	30
V6 engines		
1999 through 2003	180 in-lbs	20
2004 on	96 in-lbs	11

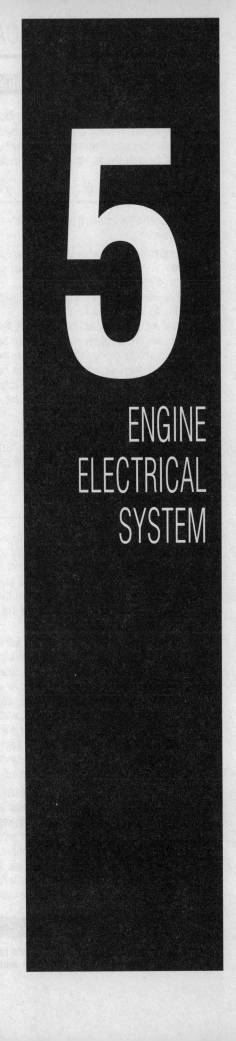

5

ENGINE
ELECTRICAL
SYSTEM

Section

Reference to other Chapters

1   General information, precautions and battery disconnection

The engine electrical systems include all ignition, charging and starting components. Because of their engine-related functions, these components are discussed separately from body electrical devices such as the lights, the instruments, etc. (which is included in Chapter 12).

PRECAUTIONS

Always observe the following precautions when working on the electrical system:

a) *Be extremely careful when servicing engine electrical components. They are easily damaged if checked, connected or handled improperly.*

b) *Never leave the ignition switched on for long periods of time when the engine is not running.*

c) *Never disconnect the battery cables while the engine is running.*

d) *Maintain correct polarity when connecting battery cables from another vehicle during jump starting - see the "Booster battery (jump) starting" section at the front of this manual.*

e) *Always disconnect the negative battery cable from the battery before working on the electrical system, but read the following battery disconnection procedure first.*

It's also a good idea to review the safety-related information regarding the engine electrical systems located in the "Safety first!" section at the front of this manual, before beginning any operation included in this Chapter.

BATTERY DISCONNECTION

Several systems - such as the alarm system, radio, power door locks or power windows - require continuous battery power, either to ensure their continued operation or to maintain control unit memory, such as the Powertrain Control Module (PCM), which will be lost if the battery is disconnected. Therefore, whenever the battery is to be disconnected, first note the following to ensure that there are no unforeseen consequences of this action:

a) *The engine management system's PCM will eventually lose some data stored in its memory when the battery is disconnected. This could include idling and operating values, detected fault codes and the system monitors required for emissions testing. Whenever the battery is disconnected, the computer may require a certain period of time to "re-learn" the operating values.*

b) *On any vehicle with power door locks, it is a wise precaution to remove the key from the ignition and to keep it with you, so that it does not get locked inside if the power door locks should engage accidentally when the battery is reconnected!*

Devices known as "memory-savers" can be used to avoid some of the above problems. Typically, a memory-saver is plugged into the cigarette lighter and connected to another voltage source. The vehicle battery is then disconnected from the electrical system, and the memory-saver provides current levels sufficient to maintain audio unit security codes and PCM memory values, and to power always-on circuits such as the clock and radio memory, all the while isolating the battery in the event that a short-circuit occurs while servicing the vehicle.

❊❊ WARNING 1:

Some of these devices allow a considerable amount of current to pass, which can mean that many of the vehicle's systems are still operational when the main battery is disconnected. So if you're using a memory-saver, make sure that a circuit is actually open before touching it.

❊❊ WARNING 2:

If you're working around an airbag, disconnect the battery and do not use a memory saver. If you use a memory-saver device to keep up power while working around an airbag, remember that the airbag is capable of accidental deployment.

The battery on all vehicles is located in the front left corner of the engine compartment. To disconnect the battery for service procedures requiring power to be cut from the vehicle, loosen the negative cable clamp nut and detach the negative cable from the negative battery post. Isolate the cable end to prevent it from accidentally coming into contact with the battery post.

INITIALIZE THE FOLLOWING SYSTEMS AFTER RECONNECTING OR REPLACING THE BATTERY, OR AFTER RECHARGING A LOW BATTERY

All models

The moon roof might not operate correctly, or at all, and the jam protection function will not operate correctly, after you reconnect or replace the battery or recharge a low battery. If this condition occurs, initialize the moon roof as follows. Push down and hold the slide control switch toward the TILT switch (Lexus) or TILT UP switch (Toyota). The moon roof will tilt up and down, then slide open and close. After the moon roof slides to its closed position, release the switch. Verify that the moon roof now functions normally. If it doesn't, have it checked by a dealer service department.

Lexus models

Adaptive front lighting system

The adaptive front lighting system might not operate correctly, or at all, after you reconnect or replace the battery or recharge a low battery. To initialize the system:

1) *Start the engine. The Adaptive Front Lighting System (AFS) OFF indicator light will start flashing.*

2) *Turn the steering wheel all the way to the left or right side, then push the AFS OFF switch twice.*

3) *Turn the steering wheel all the way to the opposite side, then push the AFS OFF switch twice.*

4) *Return the steering wheel to the center (straight ahead) position, then push the AFS OFF switch twice. The AFS OFF indicator light will go off.*

If the AFS OFF indicator light continues to flash, repeat the above four-step procedure.

Power back door system

The power back door system might not operate correctly, or at all, after you reconnect or replace the battery or recharge a low battery. To initialize the power back door system, simply close the back door completely, by hand. If the power back door still doesn't operate correctly after closing it by hand, have the system checked by a Lexus dealer service department or other qualified repair shop.

Power window system

The power window system might not operate correctly, or at all, and the jam protection function will not operate correctly, after you reconnect or replace the battery or recharge a low battery. At this time, the power window switch indicator lights will flash. To initialize the system:

1) *Push down the power window switch on each door and lower the windows half way.*

2) *Pull up the switch until the windows close and hold the switch for a second.*

Verify that the windows now open and close automatically. If they don't, have the system checked by a Lexus dealer or other qualified repair shop.

2   Battery - emergency jump starting

Refer to the *Booster battery (jump) starting* procedure at the front of this manual.

3   Battery - check and replacement

CHECK

▶ **Refer to illustrations 3.2 and 3.3**

1   Disconnect the negative battery cable, then the positive cable from the battery.

2   Check the battery state of charge. Visually inspect the indicator eye on the top of the battery; if the indicator eye is black in color, charge the battery as described in Chapter 1. Next perform an open voltage circuit test using a digital voltmeter (see illustration).

➡**Note: The battery's surface charge must be removed before accurate voltage measurements can be made. Turn on the high beams for ten seconds, then turn them off, and let the vehicle stand for two minutes. With the engine and all accessories turned off, touch the negative probe of the voltmeter to the negative terminal of the battery and the positive probe to the positive terminal of the battery. The battery voltage should be**

about 12.5 to 12.9 volts. If the battery is less than the specified voltage, charge the battery before proceeding to the next test. Do not proceed with the battery load test unless the battery charge is correct.

3   Perform a battery load test. An accurate check of the battery condition can only be performed with a load tester (available at most auto parts stores). This test evaluates the ability of the battery to operate the starter and other accessories during periods of heavy load (current draw). Install a special battery load-testing tool onto the terminals (see illustration). Load test the battery according to the manufacturer's instructions for the particular tool. This tool utilizes a carbon pile to increase the load demand (amperage draw) on the battery. Maintain the load on the battery for 15 seconds or less and observe that the battery voltage does not drop below 9.6 volts. If the battery condition is weak or defective, the tool will indicate this condition immediately.

➡**Note: Cold temperatures will cause the minimum voltage reading to drop slightly. Follow the chart given in the manufacturer's instructions to compensate for cold climates. Minimum load voltage for freezing temperatures (32-degrees F) should be approximately 9.1 volts.**

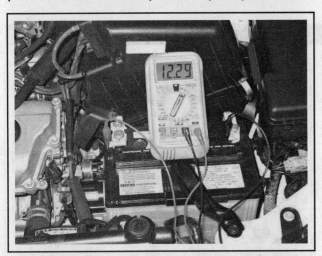

3.2 To test the open circuit voltage of the battery, touch the black probe of the voltmeter to the negative terminal and the red probe to the positive terminal of the battery. A fully charged battery should read between 12.5 and 12.9 volts, depending on its state of charge and on the ambient temperature

3.3 Some battery load testers are equipped with an ammeter that enables you to dial in the battery load (shown). Less expensive testers only have a load switch and voltmeter

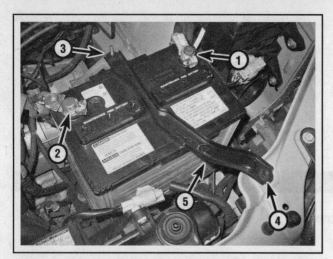

3.4 To remove the battery, disconnect the cable from the negative battery terminal (1), disconnect the cable from the positive terminal (2), then remove the nut (3) and bolt (4) and remove the hold-down clamp (5)

3.7 To detach the plastic battery tray from the tray support bracket underneath, simply pull these two locator pins from their holes in the bracket

REPLACEMENT

▶ **Refer to illustrations 3.4 and 3.7**

4   Disconnect the negative battery cable, then the positive cable from the battery (see illustration).

5   Remove the battery hold-down clamp (see illustration 3.4).

6   Lift out the battery. Be careful - it's heavy.

➡**Note: Battery straps and handlers are available at most auto parts stores for a reasonable price. They make it easier to remove and carry the battery.**

7   While the battery is out, remove the battery tray (see illustration) and inspect it for corrosion.

8   If there's corrosion on the battery tray, wash the tray thoroughly in clean water. If the tray is cracked or damaged, replace it. When installing the tray, make sure that you push the two locator pins down into their corresponding holes in the tray support bracket until they're fully seated.

9   If you are replacing the battery, make sure you get one that's identical, with the same dimensions, amperage rating, cold cranking rating, etc.

10   Installation is the reverse of removal.

4   Battery cables - check and replacement

▶ **Refer to illustrations 4.4a and 4.4b**

1   Periodically, inspect the entire length of each battery cable for damage, cracked or burned insulation and corrosion. Poor battery cable connections can cause starting problems and decreased engine performance.

2   Inspect the cable-to-terminal connections at the ends of the cables for cracks, loose wire strands and corrosion (see Chapter 1). The presence of white, fluffy deposits under the insulation at the cable terminal connection is a sign that the cable is corroded and should be replaced. Check the terminals for distortion, missing mounting bolts and corrosion.

3   When removing the cables, always disconnect the negative cable first and hook it up last or you might accidentally short the battery with the tool that you're using to loosen the cable clamps. Even if only the positive cable is being replaced, be sure to disconnect the negative cable from the battery first.

4   Disconnect the old cables from the battery, then trace each of them to their opposite ends and detach them from the starter solenoid

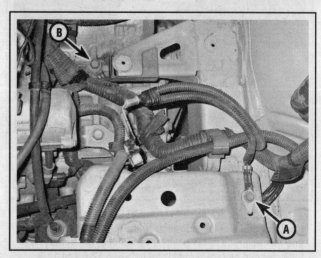

4.4a The battery ground cable is bolted to the battery tray support bracket (A) and to the transaxle (B)

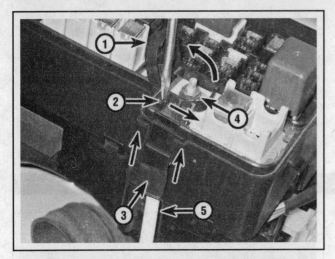

4.4b The battery starter cable is connected to the engine compartment fuse and relay box. To disconnect it, flip up the cover (1), pry the tab (2) loose with a screwdriver, lift up and remove the cable cover (3), then remove the nut (4) and disconnect the cable (5) from the terminal

and ground terminals (see illustrations). Note the routing of each cable to ensure correct installation. You'll have to remove the battery tray (see illustration 3.7) to access one of the ground cable bolts.

5 If you are replacing either or both of the old cables, take them with you when buying new cables. It is vitally important that you replace the cables with identical parts. Cables have characteristics that make them easy to identify. Positive cables are usually red and larger in cross-section; ground cables are usually black and smaller in cross section.

6 Clean the threads of the solenoid or ground connection with a wire brush to remove rust and corrosion. Apply a light coat of battery terminal corrosion inhibitor, or petroleum jelly, to the threads to prevent future corrosion.

7 Attach the cable to the solenoid or ground connection and tighten the mounting nut/bolt securely.

8 Before connecting a new cable to the battery, make sure that it reaches the battery post without having to be stretched.

9 Connect the positive cable first, followed by the negative cable.

5 Ignition system - general information and precautions

1 All models covered by this manual are equipped with a computer-controlled electronic ignition system. The ignition system includes the Camshaft Position (CMP) sensor, the Crankshaft Position (CKP) sensor, various other information sensors, the Powertrain Control Module (PCM), four or six coil-over-plug ignition coils with integral igniters, and the spark plugs. There is no distributor and there are no spark plug wires.

2 When working on the ignition system, take the following precautions:

a) Do not keep the ignition switch on for more than 10 seconds if the engine will not start.
b) Always connect a tachometer in accordance with the manufacturer's instructions. Some tachometers may be incompatible with this ignition system. Consult an auto parts counterperson before buying a tachometer for use with this vehicle.
c) Never allow the ignition coil terminals to touch ground. Grounding the coil could result in damage to the igniter and/or the ignition coil.
d) Do not disconnect the battery when the engine is running.

6 Ignition system - check

▶ **Refer to illustration 6.2**

☼☼ WARNING:

Because of the high voltage generated by the ignition system, be extremely cautious when servicing or checking ignition components. This not only includes the ignition coils and spark plugs but test equipment as well.

1 Relieve the fuel system pressure (see Chapter 4). Keep the fuel system disabled while performing the ignition system checks.

2 If the engine turns over but won't start, disconnect each ignition coil from its corresponding spark plug (see Section 7) and attach it to a calibrated ignition tester (see illustration). Calibrated ignition testers are available at most auto parts stores.

3 Crank the engine and watch the tester to see if bright blue, well-defined sparks occur.

4 If sparks occur, sufficient voltage is reaching the spark plug to fire it. Repeat this test at each spark plug to verify that all the ignition coils are functioning. If there is no spark at a plug, the ignition coil for that plug is probably bad. However, the plug itself might be fouled, so remove and check the plug as described in Chapter 1. If a number of

6.2 To use this type of calibrated ignition tester, unbolt and remove the ignition coil (don't disconnect the electrical connector), plug the tester into the boot at the lower end of the coil, clip the tester to a convenient ground and operate the starter with the ignition turned to ON. If there's enough power to fire the plug, sparks will be visible between the electrode tip and the tester body

plugs are fouled, it's a good idea to install a set of new spark plugs so that the plug gaps are uniform and the electrodes are all in the same condition.

5 If no sparks or intermittent sparks occur at all of the plugs, check for battery voltage to the ignition coils. If battery voltage is not present, check the ignition system fuse. The ignition fuse is located in the engine compartment fuse and relay box. It might be labeled as "IGN,"

as "AM" or as some other acronym. Refer to the fuse guide in your owner's manual for the name and number of the ignition fuse.

6 Use a code-reading tool or a scan tool to check for any stored Diagnostic Trouble Codes (DTCs) related to the ignition system, the Camshaft Position (CMP) sensor, the Crankshaft Position (CKP) sensor, other sensors or and/or the Powertrain Control Module (PCM).

7 Ignition coils - replacement

→Note: Toyota and Lexus do not publish primary or secondary resistance specifications for the ignition coil/igniter units used on these models.

1 Disconnect the cable from the negative battery terminal (see Sections 1 and 3).

2 Remove the engine cover (see *Intake manifold - removal and installation* in Chapter 2).

FOUR-CYLINDER MODELS

▸ Refer to illustrations 7.3 and 7.4

3 Disconnect the electrical connector from the ignition coil, then remove the mounting bolt (see illustration).

4 Pull the coil straight up and out of the valve cover (see illustration).

5 Installation is the reverse of removal.

V6 MODELS

▸ Refer to illustration 7.7

6 If you're going to remove/replace an ignition coil on the rear cylinder head, remove the upper intake manifold (see *Intake manifold - removal and installation* in Chapter 2).

7 Disconnect the electrical connector from the ignition coil, then

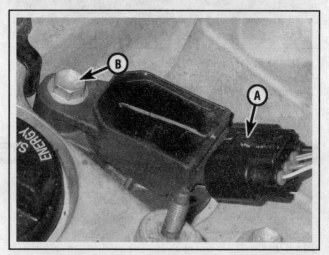

7.3 To detach an ignition coil from the valve cover on a four-cylinder engine, disconnect the electrical connector (A), then remove the coil mounting bolt (B)

remove the mounting bolt (see illustration).

8 Pull the coil straight up and out of the valve cover (see illustration 7.4).

9 Installation is the reverse of removal.

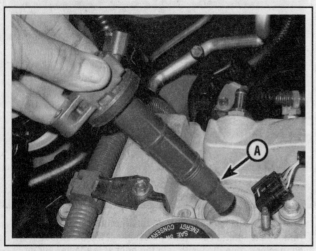

7.4 To remove an ignition coil from a four-cylinder engine, pull the coil straight up and out. Make sure the boot (A) is in good condition and fully seated before installing the coil

7.7 To detach an ignition coil from the valve cover on a V6 engine, depress the release tab (A) and disconnect the electrical connector, then remove the coil mounting bolt

8 Charging system - general information and precautions

The charging system includes the alternator, an internal voltage regulator, the discharge warning light, the battery, a fusible link and the wiring between all the components. The charging system supplies electrical power for the ignition system, the lights, the radio, etc. The alternator is driven by a drivebelt at the front of the engine. The voltage regulator limits the alternator's voltage to a preset value to prevent power surges and circuit overloads during peak voltage output.

The discharge warning light on the instrument cluster should come on when the ignition key is turned to START, then it should go off immediately. If it remains on, or if it comes on during vehicle operation, there is a malfunction in the charging system.

The charging system doesn't ordinarily require periodic maintenance. However, the drivebelt, battery and wires and connections should be inspected at the intervals outlined in Chapter 1. Be very careful when making electrical circuit connections to a vehicle equipped with an alternator and note the following:

a) When reconnecting wires to the alternator from the battery, be sure to note the polarity.
b) Before using arc-welding equipment to repair any part of the vehicle, disconnect the wires from the alternator and the battery terminals.
c) Never start the engine with a battery charger connected.
d) Always disconnect both battery leads before using a battery charger.
e) The alternator is driven by an engine drivebelt that could cause serious injury if your hand, hair or clothes become entangled in it with the engine running.
f) Because the alternator is connected directly to the battery, it could arc or cause a fire if overloaded or shorted out.
g) Wrap a plastic bag over the alternator and secure it with rubber bands before steam cleaning the engine.

9 Charging system - check

▶ **Refer to illustration 9.3**

1   If a malfunction occurs in the charging circuit, do not immediately assume that the alternator is causing the problem. First, check the following items:

a) Make sure the battery cable clamps are clean and tightly secured to the battery terminals.
b) Test the condition of the battery (see Section 2). If it does not pass all the tests, replace it.
c) Check the external alternator wiring and connections.
d) Check the drivebelt condition and tension (see Chapter 1).
e) Check the alternator mounting bolts for tightness.
f) Run the engine and check the alternator for abnormal noise.
g) Check the fusible link (see Chapter 12). If it's burned, determine the cause and repair the circuit.
h) Check the discharge warning indicator light on the instrument cluster. It should illuminate when the ignition key is turned to ON (engine not running). If it doesn't, check the circuit between the alternator and the discharge warning indicator light (see the Wiring Diagrams at the end of Chapter 12).
i) Check the fuses that are in series with the charging system circuit (see the Wiring Diagrams at the end of Chapter 12).

2   With the ignition key off, check the battery voltage with no accessories operating. It should be approximately 12.5 to 12.9 volts (see illustration 3.2). It may be slightly higher if the engine had been operating within the last hour.

3   Connect an ammeter to the charging system following the tool manufacturer's instructions. Start the engine, and check the battery voltage and amperage. It should now be approximately 13.2 to 14.8 volts (see illustration).

4   Load the battery by turning on the high beam headlights and the air conditioning system and place the blower fan on HIGH. Raise the engine speed to 2,000 rpm and check the voltage and amperage. If the

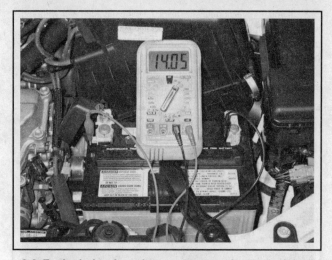

9.3 To check charging voltage, connect a voltmeter to the battery terminals and check the battery voltage with the engine running

charging system is working properly the voltage should stay above 13.5 volts and the amperage should be 30 amps or more (depending on the condition of the battery, it could be less than 30 amps).

5   If the voltage rises above 15.0 volts in either test, the regulator is defective.

6   If the indicated voltage reading is less than the specified charge voltage, the alternator is probably defective. Have the charging system checked at a dealer service department or other properly equipped repair facility.

➡ **Note: Many auto parts stores will bench test an alternator off the vehicle. Refer to your local auto parts store regarding their policy (many stores will perform this service free of charge).**

10 Alternator - removal and installation

REMOVAL

1   Disconnect the cable from the negative battery terminal (see Sections 1 and 3).

2   Remove the alternator drivebelt (see Chapter 1).

Four-cylinder models

▶ **Refer to illustration 10.4**

3   Disconnect the electrical connectors from the alternator.

4   Remove the two alternator mounting bolts (see illustration) and remove the alternator.

V6 models

▶ **Refer to illustrations 10.5 and 10.6**

5   Disconnect the electrical connectors from the alternator (see illustration).

6   Loosen the alternator pivot bolt and lock bolt, then loosen the adjustment bolt (see illustration).

7   Remove the lock bolt and the pivot bolt and remove the alternator.

INSTALLATION

8   If you're replacing the alternator, take the old alternator with you when purchasing a replacement unit. Make sure that the new/rebuilt unit is identical to the old alternator. Look at the terminals - they should be the same in number, size and locations as the terminals on the old alternator. Finally, look at the identification markings - they will be stamped in the housing or printed on a tag or plaque affixed to the

10.4 To detach the alternator from a four-cylinder engine, remove these mounting bolts

housing. Make sure that these numbers are the same on both alternators.

9   If the replacement alternator doesn't have a pulley installed, you might have to switch the pulley from the old unit to the replacement unit. When buying a new or rebuilt alternator, ask about the shop's policy regarding pulley swapping. Some shops will perform this service for free.

10   Installation is the reverse of removal. After the alternator is installed, adjust the drivebelt tension (see Chapter 1), then check the charging voltage to verify that the alternator is operating correctly (see Section 9).

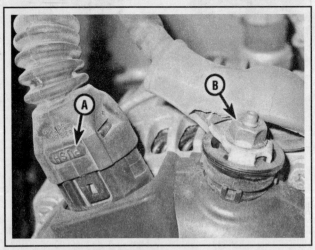

10.5 To disconnect the electrical connector from the alternator on a V6 model, push the release tab (A) in the area indicated and pull off the connector. To disconnect the battery cable from the B+ terminal, pull back the insulator and unscrew the nut (B)

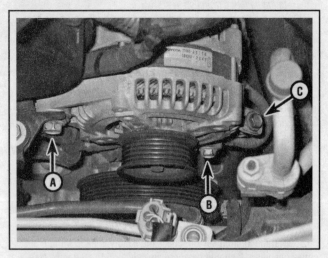

10.6 Alternator mounting details on a V6 engine

A   Pivot bolt   C   Adjustment bolt
B   Lock bolt

11 Starting system - general information and precautions

The starting system consists of the battery, the starter motor, the starter solenoid and the electrical circuit connecting the components. The solenoid is mounted directly on the starter motor. The solenoid/ starter motor assembly is installed on the upper part of the transaxle bellhousing.

When the ignition key is turned to the START position, the starter solenoid is actuated through the starter control circuit. The starter solenoid then connects the battery to the starter. The battery supplies the electrical energy to the starter motor, which does the actual work of cranking the engine.

Always observe the following precautions when working on the starting system:

a) *Excessive cranking of the starter motor can overheat it and cause serious damage. Never operate the starter motor for more than 15 seconds at a time without pausing to allow it to cool for at least two minutes.*
b) *The starter is connected directly to the battery and could arc or cause a fire if mishandled, overloaded or short-circuited.*
c) *Always detach the cable from the negative terminal of the battery before working on the starting system.*

12 Starter motor and circuit - check

▶ **Refer to illustrations 12.3 and 12.4**

1   If a malfunction occurs in the starting circuit, do not immediately assume that the starter is causing the problem. First, check the following items:

a) *Make sure the battery cable clamps, where they connect to the battery, are clean and tight.*
b) *Check the condition of the battery cables (see Section 4). Replace any defective battery cables with new parts.*
c) *Test the condition of the battery (see Section 3). If it does not pass all the tests, replace it with a new battery.*
d) *Check the starter solenoid wiring and connections. Refer to the wiring diagrams at the end of Chapter 12.*
e) *Check the starter mounting bolts for tightness.*
f) *Check the fusible link (see Chapter 12). If it's burned, determine the cause and repair the circuit.*
g) *Check the operation of the Park/Neutral switch. (Make sure the shift lever is in PARK or NEUTRAL). Refer to Chapter 7 for the Park/Neutral switch check and adjustment procedure. Refer to Chapter 12 wiring diagrams, if necessary, when performing circuit checks. These systems must operate correctly to provide battery voltage to the ignition solenoid.*

h) *Check the operation of the starter relay. The starter relay is located in the fuse/relay box inside the engine compartment. Refer to Chapter 12 for the relay testing procedure.*

2   If the starter does not actuate when the ignition switch is turned to the start position, check for battery voltage to the solenoid. This will determine if the solenoid is receiving the correct voltage signal from the ignition switch. Connect a test light or voltmeter to the starter solenoid positive terminal and while an assistant turns the ignition switch to the start position. If voltage is not available, refer to the wiring diagrams in Chapter 12 and check all the fuses and relays in series with the starting system. If voltage is available but the starter motor does not operate, remove the starter (see Section 13) and bench test it (see Step 4).

3   If the starter turns over slowly, check the starter cranking voltage and the current draw from the battery. This test must be performed with the starter assembly on the engine. Crank the engine over (for 10 seconds or less) and observe the battery voltage. It should not drop below 8.0 volts on manual transaxle models or 8.5 volts on automatic transaxle models. Also, observe the current draw using an ammeter (see illustration). It should not exceed 400 amps or drop below 250 amps. If the starter motor cranking amp values are not within the correct range, replace it with a new unit. There are several conditions that

12.3 To use an inductive ammeter, simply hold the ammeter over the positive or negative cable (whichever is more accessible)

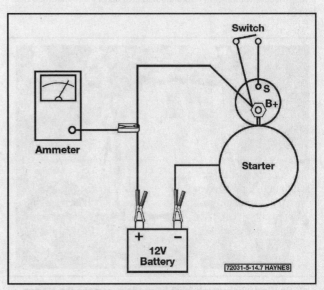

12.4 Starter motor bench testing details

may affect the starter cranking potential. The battery must be in good condition and the battery cold-cranking rating must not be under-rated for the particular application. Be sure to check the battery specifications carefully. The battery terminals and cables must be clean and not corroded. Also, in cases of extreme cold temperatures, make sure the battery and/or engine block is warmed before performing the tests.

4   If the starter is receiving voltage but does not activate, remove and check the starter/solenoid assembly on the bench (see illustration). Most likely the solenoid is defective. In some rare cases, the engine may be seized so be sure to try and rotate the crankshaft pulley (see Chapter 2A or 2B) before proceeding. With the starter/solenoid

assembly mounted in a vise on the bench, install one jumper cable from the negative battery terminal to the body of the starter. Install the other jumper cable from the positive battery terminal to the B+ terminal on the starter. Install a starter switch and apply battery voltage to the solenoid S terminal (for 10 seconds or less) and see if the solenoid plunger, shift lever and overrunning clutch extends and rotates the pinion drive. If the pinion drive extends but does not rotate, the solenoid is operating but the starter motor is defective. If there is no movement but the solenoid clicks, the solenoid and/or the starter motor is defective. If the solenoid plunger extends and rotates the pinion drive, the starter/solenoid assembly is working properly.

13 Starter motor - removal and installation

FOUR-CYLINDER MODELS

▶ **Refer to illustration 13.3**

1   Disconnect the cable from the negative battery terminal (see Sections 1 and 3).
2   On 2003 earlier models, remove the air filter housing and air intake duct (see Chapter 4). On 2004 and later models, remove the battery and the battery tray (see Section 3).
3   Detach the electrical connectors from the starter/solenoid assembly (see illustration).
4   Remove the starter motor mounting bolts (see illustration 13.3) and remove the starter.
5   Installation is the reverse of removal.

V6 MODELS

▶ **Refer to illustrations 13.8a, 13.8b, 13.8c and 13.9**

6   Disconnect the cable from the negative battery terminal (see Section 3).

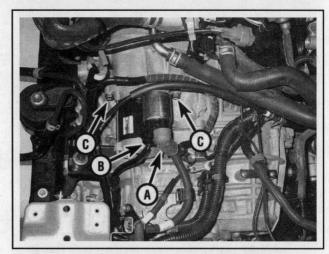

13.3 To remove the starter motor from a four-cylinder engine, disconnect the electrical connector (A) and the battery starter cable (B), then remove the starter mounting bolts (C)

13.8a On 1999 through 2003 V6 models, remove the starter mounting bolts and remove the starter . . .

13.8b . . . depress this release tab and disconnect the electrical connector from the starter solenoid . . .

13.8c . . . then remove this nut and disconnect the battery starter cable from the B+ terminal on the starter motor

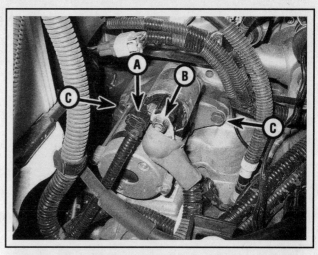

13.9 On 2004 and later V6 models, disconnect the electrical connector (A) and the battery starter cable (B), then remove the starter mounting bolts (C) and remove the starter

7   Remove the battery and the battery tray (see Section 3). Remove the air filter housing and the air intake duct (see Chapter 4).
8   On 1999 through 2003 models, remove the starter motor first, then disconnect the electrical connectors (see illustrations).
9   On 2004 and later models, disconnect the electrical connectors and remove the starter mounting bolts (see illustration), then remove

the starter.
10 Installation is the reverse of removal. Be sure to tighten the starter mounting bolts to the torque listed in this Chapter's Specifications.

Specifications

Charging system

Battery voltage	12.5 12.9 volts
Charging voltage	13.2 to 14.8 volts

Torque specifications

	Ft-lbs	Nm
Starter motor mounting bolts		
Four-cylinder models		
2001 through 2003	29	39
2004 on	27	37
V6 models		
1999 through 2003	31	42
2004 on	27	37

Notes

Section

Reference to other Chapters

Variable Valve Timing - intelligent (VVT-i) system - See Chapter 2

6

EMISSIONS AND ENGINE CONTROL SYSTEMS

1 General information

◆ **Refer to illustrations 1.6a and 1.6b**

To prevent pollution of the atmosphere from incompletely burned and evaporating gases, and to maintain good driveability and fuel economy, a number of emission control systems are incorporated. They include the:

Air Control Induction System (ACIS)
Catalytic converter
Electronic Throttle Control System - intelligent (ETCS-i) system
Evaporative Emissions Control (EVAP) system
Positive Crankcase Ventilation (PCV) system
Sequential Electronic Fuel Injection (SFI) system
Variable Valve Timing - intelligent (VVT-i) system

The Sections in this Chapter include general descriptions, general inspection procedures within the scope of the home mechanic and component replacement procedures, where possible, for the components used in each of the systems listed above. You'll also find a general description of each information sensor and output actuator used in the engine management system, including its function and location, and a brief description of how it works (see Section 2). We have also included replacement procedures for all sensors and actuators.

Before assuming that an emissions control system is malfunctioning, check the fuel and ignition systems carefully. Diagnosis of many emission control devices requires specialized tools, equipment and training. If a service procedure is beyond the scope of this book, or your ability, consult a dealer service department or other repair shop.

But keep in mind that the most frequent cause of emissions problems is simply a loose connector, a broken wire or a disconnected vacuum hose, so always check the hoses, wiring and connections first.

Emission control systems are not, however, particularly difficult to maintain and repair. You can quickly and easily perform many checks and do most of the regular maintenance at home with common tune-up and hand tools.

➡**Note: Because of a Federally mandated warranty which covers the emissions control system components, check with your dealer about warranty coverage before working on any emissions-related systems. Once the warranty has expired, you may wish to perform some of the component checks and/or replacement procedures in this Chapter to save money.**

Pay close attention to any special precautions outlined in this Chapter. It should be noted that the illustrations of the various systems might not exactly match the system installed on your vehicle because of changes made by the manufacturer during production or from year-to-year.

A Vehicle Emissions Control Information (VECI) label (see illustration) is attached to the underside of the hood. This label contains important emissions specifications and adjustment information. Another label, the vacuum hose routing diagram (see illustration), provides a vacuum hose schematic with emissions components identified. When servicing the engine or emissions systems, always refer to the VECI label and the vacuum hose routing diagram on your vehicle for up-to-date information.

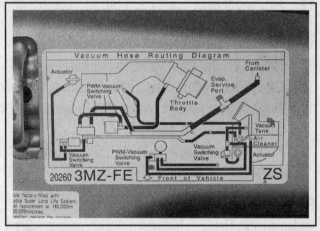

1.6a The Vehicle Emission Control Information (VECI) label contains such essential information as the types of emission control systems installed on the engine and certain tune-up specifications

1.6b A typical vacuum hose routing diagram for a V6 model

2 On Board Diagnostic (OBD) system and trouble codes

SCAN TOOL INFORMATION

◆ **Refer to illustrations 2.2a and 2.2b**

1 Hand-held scanners are handy for analyzing the engine management systems used on late-model vehicles. Because extracting the Diagnostic Trouble Codes (DTCs) from an engine management system

is now the first step in troubleshooting many computer-controlled systems and components, even the most basic generic code readers are capable of accessing a computer's DTCs (see illustration). More powerful scan tools can also perform many of the diagnostics once associated with expensive factory scan tools. If you're planning to obtain a generic scan tool for your vehicle, make sure that it's compatible with OBD-II systems. If you don't plan to purchase a code reader or scan

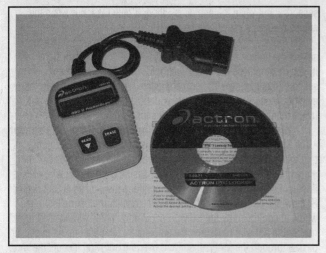

2.2a Simple code readers are an economical way to extract trouble codes when the CHECK ENGINE light comes on

tool and don't have access to one, you can have the codes extracted by a dealer service department or by an independent repair shop.

2   With the advent of the Federally mandated emission control system known as On-Board Diagnostics-II (OBD-II), specially designed scanners were developed. Several tool manufacturers have released OBD-II scan tools for the home mechanic (see illustration).

OBD-II SYSTEM GENERAL DESCRIPTION

3   All models are equipped with the second generation On-Board Diagnostic (OBD-II) system. The OBD-II system includes a computer known as the Powertrain Control Module (PCM), information sensors that monitor various functions of the engine and send data to the PCM, and output actuators that carry out the PCM's various control commands. The OBD-II system also incorporates a series of diagnostic monitors that detect and identify fuel injection and emission control system faults and store the information in the computer memory. The system also tests sensors and output actuators, diagnoses drive cycles, freezes data and clears codes.

4   This powerful diagnostic computer must be accessed with an

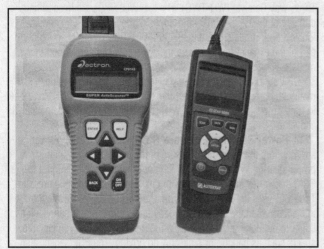

2.2b Scanners like these from Actron and AutoXray are powerful diagnostic aids - they can tell you just about anything that you want to know about your engine management system

OBD-II scan tool and the 16 pin Data Link Connector (DLC) located under the driver's dash area. The PCM is the brain of the electronically controlled fuel and emissions system. It receives data from a number of sensors and other electronic components (switches, relays, etc.). Based on the information it receives, the PCM generates output signals to control various relays, solenoids (i.e. fuel injectors) and other actuators. The PCM is specifically calibrated to optimize the emissions, fuel economy and driveability of the vehicle.

5   It isn't a good idea to attempt diagnosis or replacement of the PCM or emission control components at home while the vehicle is under warranty. Because of a Federally mandated warranty which covers the emissions system components and because any owner-induced damage to the PCM, the sensors and/or the control devices may void this warranty, take the vehicle to a dealer service department if the PCM or a system component malfunctions.

INFORMATION SENSORS

➡Note: The following list provides a brief description of the function, operation and location of each of the important information sensors. We use the standard sensor terminology recommended by the Society of Automotive Engineers (SAE). Where Toyota uses other names for some sensors we provide the Toyota name as well.

6   **Accelerator Pedal Position (APP) sensor** - All 2004 and later models covered by this manual are equipped with an Electronic Throttle Control System - intelligent (ETCS-i). The ETCS-i system dispenses with a conventional accelerator cable by using an electronic throttle body instead of a conventional cable-operated throttle body. The Powertrain Control Module (PCM) controls the position of the throttle plate with a solenoid that's located in the throttle body. The PCM's commands are based on the inputs that it receives from the APP sensor, which is located at the top of the accelerator pedal assembly. Its electrical output signal, which is proportional to the angle of the accelerator pedal, is used by the ETCS-i system to determine the corresponding opening angle of the throttle plate inside the throttle body.

7   **Camshaft Position (CMP) sensor** - A CMP sensor monitors the position of the camshaft and tells the Powertrain Control Module (PCM) when the piston in the No. 1 cylinder is on its compression stroke. The PCM uses the CMP sensor signal to synchronize the sequential firing of the fuel injectors. On four-cylinder engines, there is one CMP sensor, and it's located on the left end of the cylinder head, near the left end of the intake camshaft. On V6 engines, there are two CMP sensors, one for each cylinder head. They're located at the left end of the cylinder heads, near the left end of each intake camshaft. On V6 engines, Lexus and Toyota refer to the CMP sensors as VVT sensors. For more information about the Variable Valve Timing-intelligent (VVT-i) system, refer to Section 19.

The CMP sensor is a variable reluctance (pick-up coil) type sensor that generates an analog (sine wave) signal pulse each time that a boss on the timing rotor passes by the sensor. On four-cylinder engines, the timing rotor is mounted on the left end of the intake camshaft. On V6 engines, the timing rotors are located on the backside of the VVT-i controllers, which are mounted on the left ends of the intake camshafts. There are three bosses on the timing rotor. As the timing rotor rotates with the camshaft, the CMP sensor generates a voltage output each time one of the bosses no the rotor passes by the sensor. The intake camshaft rotates once for each two rotations of the crankshaft, so there are three voltage outputs for every two rotations of the crankshaft. The PCM uses these voltage outputs to determine the position of the intake camshaft.

8 **Crankshaft Position (CKP) sensor** - Like the CMP sensor, the CKP sensor is a variable reluctance (pick-up coil) type sensor that generates an analog (sine wave) signal pulse each time that a boss on the timing rotor passes by the sensor. On four-cylinder engines, the CKP sensor is located on the front of the timing chain cover, next to the crankshaft damper pulley, and the timing rotor is mounted on the front of the crankshaft timing sprocket. On V6 engines, the CKP sensor is located on the front of the engine, next to the crankshaft damper pulley, and the timing rotor is mounted on the back of the crankshaft timing belt sprocket. All timing rotors have 36 evenly spaced teeth, with two missing, i.e. there are 34 actual teeth and a blank space for the other two. As the crankshaft rotates, each tooth on the timing rotor moves past the CKP sensor. As each tooth moves past the sensor, the magnetic flux in the sensor coil changes because the air gap between the sensor and the timing rotor changes. This change in magnetic flux induces a voltage pulse in the sensor, and this pulse is sent to the PCM as an output signal. And the blank space with the missing teeth? The PCM uses the missing teeth to determine Top Dead Center (TDC).

The CKP sensor is the primary sensor that provides ignition information to the PCM. The PCM uses the CKP sensor to determine crankshaft position (which piston will be at TDC next) and crank speed (rpm), both of which it needs to synchronize the ignition system.

9 **Engine Coolant Temperature (ECT) sensor** - The ECT sensor measures the temperature of the engine coolant. The ECT sensor is a thermistor, i.e. its resistance decreases as the temperature increases, and its resistance increases as the temperature decreases. This type of thermistor is also referred to as a Negative Temperature Coefficient (NTC) thermistor. This variable resistance produces an analogous voltage drop across the sensor terminals, thus providing an electrical signal to the PCM that accurately reflects the engine coolant temperature.

The ECT sensor is a critical sensor because it tells the PCM when the engine is warmed up sufficiently to go into closed loop operation. And once the engine is in closed loop, the PCM also uses the ECT sensor to control fuel injector pulse width and ignition timing, and it uses the ECT sensor signal to determine when to purge the EVAP system.

On four-cylinder engines, the ECT sensor is located at the left (driver's) end of the cylinder head. On V6 models, the ECT sensor is on the right (timing belt) end of the engine, on the water outlet housing.

10 **Intake Air Temperature (IAT) sensor** - The IAT sensor is used by the PCM to calculate air density, which is one of the variables that it must know in order to calculate injector pulse width and adjust ignition timing (to prevent spark knock when air intake temperature is high). Like the ECT sensor, the IAT sensor is a Negative Temperature Coefficient (NTC) type thermistor, whose resistance decreases as the temperature increases. The IAT sensor is an integral component of the Mass Air Flow (MAF) sensor, which is located on the air filter housing. For more information about the MAF sensor, see paragraph 12.

11 **Knock sensor** - The knock sensor monitors engine vibration caused by detonation. Basically, a knock sensor converts engine vibration to an electrical signal. When the knock sensor detects a knock in one of the cylinders, it signals the PCM so that the PCM can retard ignition timing accordingly. The knock sensor contains a piezoelectric material, a certain type of piezoresistive crystal, that has the ability to produce a voltage when subjected to a mechanical stress. The piezoelectric crystal in the knock sensor vibrates constantly and produces an output signal that's proportional to the intensity of the vibration. As the intensity of the vibration increases, so does the voltage of the output signal. When the intensity of the crystal's vibration reaches a specified threshold, the PCM stores that value in its memory and retards ignition timing in all cylinders (the PCM does not selectively retard timing only at the affected cylinder). The PCM doesn't respond to the knock

sensor's input when the engine is idling; it only responds when the engine reaches a specified speed.

On four-cylinder engines, the knock sensor is located on the back-side of the engine block, under the intake manifold runners. To access it, you must remove the intake manifold. On V6 engines, the knock sensors (there are two of them) are located in the valley between the cylinder heads. On these models, you must remove the upper and lower intake manifolds to access the knock sensors.

12 **Mass Air Flow/Intake Air Temperature (MAF/IAT) sensor** - The MAF sensor is the principal means by which the PCM monitors intake airflow. It uses a hot-wire sensing element to measure the amount of air entering the engine. Air passing over the hot wire causes it to cool down. The hot wire's temperature is maintained at 392 degrees F. above the ambient temperature by electrical current supplied to the wire and controlled by the PCM. A constantly "cold" wire located right next to the hot wire measures the ambient air temperature. As intake air passes through the MAF sensor and over the hot wire, it cools the wire, and the control system immediately corrects the temperature back to its constant value. The current required to maintain the specified constant temperature value is used by the PCM as an indicator of airflow.

The functions of the MAF and the Intake Air Temperature (IAT) sensors are combined into one assembly on all models. For more information about the IAT sensor, see paragraph 10. The MAF/IAT sensor is located on top of the air filter housing.

13 **Oxygen sensors** - Oxygen sensors generate a voltage signal that varies in accordance with the amount of oxygen in the exhaust stream. The PCM uses the data from the upstream oxygen sensor to calculate the injector pulse width. The downstream oxygen sensor monitors the oxygen content of the exhaust gases as they exit the catalytic converters. This information is used by the PCM to predict catalyst deterioration and/or failure. One job of the catalytic converter is to store excess oxygen. As long as the catalyst is functioning correctly, the downstream sensor should show little activity because there should be little oxygen exiting the catalyst. But as the catalyst deteriorates, its ability to store oxygen is compromised. When the output signal from the downstream sensor starts to look like the output signal from the upstream sensor, the PCM stores a DTC and turns on the MIL to let you know that it's time to replace the catalyst.

On 2001 through 2003 four-cylinder models, there are four oxygen sensors: one upstream and one downstream sensor for each Warm-Up Three-Way Catalyst (WU-TWC). The two WU-TWCs are integral components of the exhaust manifold on these models. On 2004 and later four-cylinder models, there is just one WU-TWC (it too is an integral part of the exhaust manifold), so there are only two sensors on these models - one upstream and one downstream.

On V6 models there are four oxygen sensors: one in each exhaust manifold, above the Warm-Up Three-Way Catalyst (WU-TWC), and one in each downpipe, between the WU-TWCs and the downstream catalytic converter.

14 **Power Steering Pressure (PSP) sensor** - The PSP sensor monitors the hydraulic pressure of the power steering fluid in the power steering system. The PSP sensor provides a voltage input to the PCM that varies in accordance with changes in the hydraulic pressure. The PCM uses the input signal from the PSP sensor to elevate the idle speed when the engine is already under some other load, such as the air conditioning compressor, while maneuvering the vehicle at low speed, such as parking or stop-and-go driving. The PSP sensor also signals the PCM to adjust the air assist injector solenoid valve during high-load situations such as parking. The PSP sensor is located at the power steering pump. On four-cylinder models, it's screwed directly into the pump. On V6 models, it's screwed into the top of the banjo

bolt that connects the power steering pressure hose banjo fitting to the pump.

15 **Throttle Position (TP) sensor** - The TP sensor, which is located on the throttle body, is a rotary potentiometer, which is a type of variable resistor, that produces a variable voltage signal in proportion to the opening angle of the throttle plate. The PCM sends 5 volts to the TP sensor. As the plate opens and closes, the resistance of the TP sensor changes with it, altering the signal back to the PCM. The output voltage of the TP sensor is about 0.6 volt at idle (closed throttle plate) to 4.5 volts at wide-open throttle. This variable signal enables the PCM to calculate the position (opening angle) of the throttle plate. The PCM uses the TP sensor input, along with other sensor inputs, to adjust fuel injector pulse-width and ignition timing.

A TP sensor is used on all 1999 through 2003 models. On these models you can replace the TP sensors. All 2004 and later models are equipped with an Electronic Throttle Control-intelligent (ETCS-i) system. On models with the ETCS-i system, the TP sensor is an integral part of the electronic throttle body and is not serviceable separately from the throttle body.

16 **Transmission Range (TR) sensor** - Like the Park/Neutral Position (PNP) switch that it replaces, the TR sensor prevents you from starting the engine unless the automatic transaxle is in Park or Neutral, and it activates the back-up lights when you put the shift lever in Reverse. Unlike a PNP or inhibitor switch, however, the TR sensor also tells the PCM what gear the transaxle is in. The PCM uses this information to determine what gear the transaxle should be in based on the load, engine speed, vehicle speed, etc. and to determine when to upshift and downshift the transaxle. The TR sensor is mounted on the front of the transaxle.

17 **Transmission speed sensors** - There are two speed sensors on all transaxles: the **Input Turbine Speed Sensor**, or **Input Shaft RPM Sensor**, and the **Counter Gear Speed Sensor**, or **Counter Gear RPM Sensor**. The speed sensors are variable reluctance (pick-up coil) type sensors that generate an analog (sine wave) signal pulse each time that a boss on a timing rotor passes by the sensor. Both sensors are located on top of the transaxle. The PCM uses the signal from the Input Turbine Speed Sensor to monitor input turbine or input shaft speed. And it uses the signal from the Counter Gear Speed Sensor to monitor counter gear or output shaft speed. The PCM constantly compares these two speeds to its map (program) for the transaxle in order to determine shift scheduling, Torque Converter Clutch (TCC) engagement scheduling and optimal hydraulic pressure for various hydraulically-controlled components inside the transaxle. Both of the speed sensors are located on top of the transaxle. The Input Turbine Speed Sensor is unit located on the left (driver's) end of the transaxle; the Counter Gear Speed Sensor is located to the right of the Input Turbine Speed Sensor, closer to the bellhousing.

18 **Vapor pressure sensor** - The vapor pressure sensor is a component of the Evaporative Emission Control (EVAP) system. It's located on top of the fuel tank. The vapor pressure sensor monitors the pressure of fuel vapors inside the tank. When the vapor pressure exceeds the upper threshold, the vapor pressure sensor signals the PCM, which opens the Vacuum Switching Valve (VSV) for the pressure switching valve, allowing the fuel vapors to migrate to the EVAP canister, where they are stored until they're purged.

POWERTRAIN CONTROL MODULE (PCM)

19 Based on the information that it receives from the information sensors described above, the PCM adjusts fuel injector pulse width, idle speed, ignition spark advance, ignition coil dwell, EVAP canister

purge operation and a lot of other things. It does so by controlling the output actuators. The following list provides a brief description of the function, location and operation of each of the important output actuators.

OUTPUT ACTUATORS

20 **Camshaft timing oil control valve** - The camshaft timing oil control valve is a component of the Variable Valve Timing-intelligent (VVT-i) system, which alters valve timing in response to engine operating conditions in order to improve torque, increase fuel economy and decrease emissions. The intake valves are advanced or retarded by a hydraulic device known as the VVT-i controller, which is mounted on the end of each intake camshaft. The camshaft timing oil control valve houses a spool valve that controls the flow of oil to the advance or retard side of the controller on each intake cam. A coil inside the oil control valve controls the spool valve. There is one camshaft timing oil control valve on four-cylinder engines, and it's located on the right (timing chain) rear corner of the cylinder head, adjacent to the VVT-i controller, which is installed on the right end of the intake camshaft. There are two camshaft timing oil control valves on V6 engines, and they're located on the left (driver's) end of the cylinder heads, adjacent to the VVT-i controllers, which are installed on the left ends of the intake camshafts. For more information about the VVT-i system, refer to Chapter 2A (four-cylinder engine) or 2B (V6 engines).

21 **Canister closed valve** - The canister closed valve is a component of the Evaporative Emission Control (EVAP) system. The canister closed valve opens and closes the EVAP system's fresh air line (between the air filter housing and the EVAP canister) in response to signals from the Powertrain Control Module (PCM). When commanded by the PCM, the canister closed valve also closes the EVAP canister for monitor testing. The canister closed valve is located in the engine compartment, on the backside of the air filter housing.

➡**Note: On earlier models, Lexus and Toyota refer to this device as the Vacuum Switching Valve (VSV) for Canister Closed Valve (CCV). However, in this book we refer to it simply as the canister closed valve.**

22 **Canister purge valve** - The canister purge valve is a component of the Evaporative Emission Control (EVAP) system. When the engine is cold or still warming up, no captive fuel vapors are allowed to escape from the EVAP canister. After the engine is warmed up, the PCM energizes the canister purge valve, which regulates the flow of these vapors from the canister to the intake manifold. The rate of vapor flow is regulated by the purge valve in response to commands from the PCM, which controls the duty cycle of the valve. On V6 models, the EVAP canister purge valve is located in the engine compartment, at the right (timing belt) end of the engine. For more information about the EVAP system, see Section 18.

➡**Note: On earlier models, the purge valve is referred to by Toyota and Lexus as the Vacuum Switching Valve for the Evaporative Emission Control system, or VSV for EVAP. However, in this book we refer to it as the canister purge valve, or simply, the purge valve.**

23 **Electronic Throttle Body** - 2004 and later models are equipped with the Electronic Throttle Control System-intelligent (ETCS-i). These vehicles do not have a conventional accelerator cable-actuated throttle body. Instead, they use an electronic throttle body that is controlled by the Powertrain Control Module (PCM). The throttle plate inside the throttle body is opened and closed by a PCM-controlled throttle motor. There is no cruise control cable and no Idle Air Control (IAC) valve. Cruise control and idle speed are handled electron-

ically by the PCM. The electronic throttle body has a Throttle Position (TP) sensor, but it's an integral part of the throttle body and cannot be replaced separately from the throttle body assembly. The PCM determines the correct throttle plate angle by processing the input signal from the Accelerator Pedal Position (APP) sensor, which is located at the upper end of the accelerator pedal (see Accelerator Pedal Position sensor in paragraph 6).

24 **Fuel injectors** - The PCM opens the fuel injectors sequentially (in firing order sequence). The PCM also controls the injector pulse width, which is the interval of time during which each injector is open. The pulse width of an injector (measured in milliseconds) determines the amount of fuel delivered. For more information on the fuel system and the fuel injectors, including injector replacement, refer to Chapter 4.

25 **Fuel pump relay** - Lexus and Toyota also refer to the fuel pump relay as the circuit opening relay, because it controls the circuits for other components in the fuel system besides the fuel pump. When grounded by the PCM, the fuel pump relay provides battery voltage to the fuel pump (and to other fuel system components). On Lexus models, the fuel pump relay is usually located in the fuse and relay box in the engine compartment. On Toyota models, it's usually located in relay box No. 1, which is located in the dash, behind the lower finish panel. The location of this relay may change depending on model and year, so if you can't find it, refer to the fuse and relay guide in your owner's manual.

26 **Idle Air Control (IAC) valve** - The PCM-controlled IAC valve, which is located on the throttle body on all 1999 through 2003 models, regulates the flow of air that bypasses the throttle plate when the engine is idling. The PCM opens and closes the IAC valve in response to loads - air conditioning and power steering loads, for example - to keep the engine idle speed at its target rpm. The IAC valve also increases the idle speed during the early stages of the warm-up period and functions as a dashpot when the throttle plate is abruptly closed during sudden deceleration conditions. There is no IAC valve on 2004 and later vehicles. These models are equipped with electronic throttle bodies that do no require an IAC valve to control idle speed.

27 **Ignition coil(s)** - All ignition coils used by the vehicles covered in this manual are the coil-over-plug type. Each spark plug has its own coil, mounted directly over the plug. There are no spark plug wires, and there is no separate ignition control module. The Powertrain Control Module (PCM) handles this function. For more information about the ignition coils, refer to Chapter 5.

28 **Intake air control valve(s)** - The intake air control valves are components of the Acoustic Control Induction System (ACIS). The ACIS system improves torque throughout the operating range of V6 engines, particularly in the low-speed range, by varying the effective length of the intake manifold. (For more information about the ACIS, refer to Section 15).

On 1999 through 2003 V6 models, the ACIS uses two intake air control valves: one is mounted inside the throttle body and the other is mounted inside the air intake plenum. A vacuum-operated actuator activates each intake air control valve, and a PCM-controlled Vacuum Switching Valve (VSV) operates each actuator.

The ACIS is simplified on 2004 and later V6 models: there is only one intake air control valve, which is located on the air intake plenum. Like earlier units, it's activated by a vacuum-operated actuator, which in

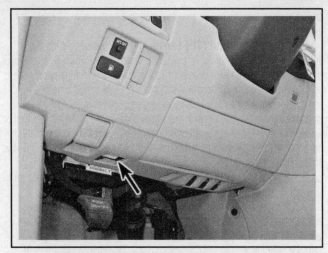

2.32 The 16-pin Data Link Connector (DLC) is located under the left side of the dash

turn is operated by a PCM-controlled VSV.

29 **Vacuum Switching Valve (VSV)** for the pressure switching valve - The VSV for the pressure switching valve is a component of the Evaporative Emission Control (EVAP) system on 1999 through 2003 models. It's located underneath the vehicle, at the EVAP canister, which is located behind the rear suspension crossmember. Gasoline inside the fuel tank is constantly emitting hydrocarbon vapors on a warm day. As these vapors build up inside the tank the pressure goes up. When the pressure inside the tank exceeds atmospheric pressure, the PCM-controlled pressure switching valve opens, allowing the fuel vapors to migrate to the EVAP canister, where they are stored until they're purged.

30 **Vapor pressure sensor** - The vapor pressure sensor, which is a component of the Evaporative Emission Control (EVAP) system, monitors the pressure inside the fuel tank and sends a signal to the Powertrain Control Module (PCM). When the pressure exceeds the upper threshold, the PCM commands the canister closed valve to send the excessive vapors to the EVAP canister. The vapor pressure sensor is located on top of the fuel tank.

OBTAINING TROUBLE CODES

▶ **Refer to illustration 2.32**

31 The PCM illuminates the CHECK ENGINE light (also called the Malfunction Indicator Light) on the dash if it recognizes a component fault for two consecutive drive cycles. It will continue to turn on the light until the PCM does not detect any malfunction for three or more consecutive drive cycles.

32 To extract the Diagnostic Trouble Codes (DTCs) you will need an OBD-II scan tool (see illustration 2.2). Plug the scan tool into the PCM's data link connector (see illustration), which is located under the left end of the dash.

33 Plug the scan tool into the 16-pin data link connector (DLC), and then follow the instructions included with the scan tool to extract all the diagnostic codes.

DIAGNOSTIC TROUBLE CODES

Trouble code	Code identification
P0010	Camshaft Position (CMP) sensor C "A" actuator
P0011	Camshaft Position (CMP) sensor "A" - timing over-advanced or system performance (Bank 1)
P0012	Camshaft Position (CMP) sensor "A" - timing over-retarded or system performance (Bank 1)
P0016	Crankshaft Position (CKP) sensor-Camshaft Position (CMP) sensor correlation (Bank 1, Sensor 1)
P0031	Oxygen sensor heater control circuit, low voltage (Bank 1, Sensor 1)
P0032	Oxygen sensor heater control circuit, high voltage (Bank 1, Sensor 1)
P0037	Oxygen sensor heater control circuit, low voltage (Bank 1, Sensor 2)
P0038	Oxygen sensor heater control circuit, high voltage (Bank 1, Sensor 2)
P0100	Mass Air Flow (MAF) sensor circuit fault
P0101	Mass Air Flow (MAF) sensor, range or performance problem
P0102	Mass Air Flow (MAF) sensor, low input voltage
P0103	Mass Air Flow (MAF) sensor, high input voltage
P0110	Intake Air Temperature (IAT) sensor, circuit fault
P0112	Intake Air Temperature (IAT) sensor circuit, low input voltage
P0113	Intake Air Temperature (IAT) sensor circuit, high input voltage
P0115	Engine Coolant Temperature (ECT) sensor, circuit fault
P0116	Engine Coolant Temperature (ECT) sensor, range or performance problem
P0117	Engine Coolant Temperature (ECT) sensor circuit, low input voltage
P0118	Engine Coolant Temperature (ECT) sensor circuit, high input voltage
P0120	Throttle Position (TP) sensor or Accelerator Pedal Position (APP) sensor, circuit fault
P0121	Throttle Position (TP) sensor or Accelerator Pedal Position (APP) sensor circuit, range or performance problem
P0122	Throttle Position (TP) sensor or Accelerator Pedal Position (APP) sensor circuit, low input voltage
P0123	Throttle Position (TP) sensor or Accelerator Pedal Position (APP) sensor circuit, high input voltage
P0125	Insufficient coolant temperature for closed loop fuel control
P0128	Thermostat malfunction
P0136	Oxygen sensor circuit fault (Bank 1, Sensor 2)
P0137	Oxygen sensor circuit, low voltage (Bank 1, Sensor 2)
P0138	Oxygen sensor circuit, high voltage (Bank 1, Sensor 2)
P0141	Oxygen sensor heater circuit fault (Bank 1, Sensor 2)
P0156	Oxygen sensor circuit fault (Bank 2, Sensor 2)

DIAGNOSTIC TROUBLE CODES (CONTINUED)

Trouble code	Code identification
P0161	Oxygen sensor heater circuit fault (Bank 2, Sensor 2)
P0171	Fuel injection system, fuel trim too lean (Bank 1)
P0172	Fuel injection system fuel trim too rich (Bank 1)
P0174	Fuel injection system fuel trim too lean (Bank 2)
P0175	Fuel injection system fuel trim too rich (Bank 2)
P0300	Random or multiple cylinder misfire detected
P0301	Cylinder no. 1 misfire detected
P0302	Cylinder no. 2 misfire detected
P0303	Cylinder no. 3 misfire detected
P0304	Cylinder no. 4 misfire detected
P0305	Cylinder no. 5 misfire detected
P0306	Cylinder no. 6 misfire detected
P0325	Knock sensor No. 1, circuit fault (Bank 1, or single sensor)
P0327	Knock sensor No. 1 circuit, low input voltage (Bank 1, or single sensor)
P0328	Knock sensor No. 1 circuit, high input voltage (Bank 1, or single sensor)
P0330	Knock sensor No. 2, circuit fault (Bank 2)
P0335	Crankshaft Position (CKP) sensor, circuit fault
P0339	Crankshaft Position (CKP) sensor, circuit intermittent
P0340	Camshaft Position (CMP) sensor circuit fault (Bank 1, or single sensor)
P0341	Camshaft Position (CMP) sensor circuit, range or performance problem (Bank 1, or single sensor)
P0351	Ignition coil A primary or secondary circuit malfunction
P0352	Ignition coil B primary or secondary circuit malfunction
P0353	Ignition coil C primary or secondary circuit malfunction
P0354	Ignition coil D primary or secondary circuit malfunction
P0420	Catalyst system efficiency below threshold (Bank 1)
P0430	Catalyst system efficiency below threshold (Bank 2)
P043E	Evaporative Emission Control (EVAP) system, reference orifice clogged up
P043F	Evaporative Emission Control (EVAP) system, reference orifice high flow
P0440	Evaporative Emission Control (EVAP) system malfunction
P0441	Evaporative Emission Control (EVAP) system, incorrect purge flow

Trouble code	Code identification
P0442	Evaporative Emission Control (EVAP) system malfunction or small leak detected
P0446	Evaporative Emission Control (EVAP) system, vent control malfunction
P0450	Evaporative Emission Control (EVAP) system, pressure sensor malfunction
P0451	Evaporative Emission Control (EVAP) system, pressure sensor range or performance problem
P0452	Evaporative Emission Control (EVAP) system, pressure sensor/switch, low input voltage
P0453	Evaporative Emission Control (EVAP) system, pressure sensor/switch, high input voltage
P0455	Evaporative Emission Control (EVAP) system, gross leak detected
P0456	Evaporative Emission Control (EVAP) system, small leak detected
P0500	Vehicle Speed Sensor (VSS) malfunction
P0503	Vehicle Speed Sensor (VSS), intermittent, erratic or high voltage
P0504	Brake switch correlation
P0505	Idle control system malfunction
P0560	System voltage
P0604	Internal control module Random Access Memory (RAM) error
P0606	Powertrain Control Module (PCM) processor malfunction
P0607	Powertrain Control Module (PCM) performance problem
P0617	Starter relay circuit, high voltage
P0657	Actuator supply voltage, open circuit
P0705	Transmission Range (TR) sensor circuit malfunction (PRNDL input)
P0710	Transaxle fluid temperature sensor, circuit fault
P0711	Transaxle fluid temperature sensor circuit, range or performance problem
P0712	Transaxle fluid temperature sensor circuit, low input voltage
P0713	Transaxle fluid temperature sensor circuit, high input voltage
P0717	Input/turbine speed sensor circuit, no signal
P0724	Brake switch circuit, high voltage
P0741	Torque converter clutch circuit, performance problem or stuck in OFF position
P0746	Pressure control solenoid performance problem (shift solenoid valve SL1)
P0748	Pressure control solenoid, electrical problem (shift solenoid valve SL1)
P0750	Transaxle shift solenoid A malfunction, or stuck open or closed
P0753	Transaxle shift solenoid A, electrical malfunction or circuit fault
P0755	Transaxle shift solenoid B malfunction, or stuck open or closed

DIAGNOSTIC TROUBLE CODES (CONTINUED)

Trouble code	Code identification
P0758	Transaxle shift solenoid B, electrical malfunction or circuit fault
P0765	Transaxle shift solenoid D malfunction, or stuck open or closed
P0766	Transaxle shift solenoid D, performance problem
P0768	Transaxle shift solenoid D, electrical malfunction or circuit fault
P0770	Transaxle shift solenoid E malfunction, or stuck open or closed
P0773	Transaxle shift solenoid E, electrical malfunction or circuit fault
P0776	Transaxle pressure control solenoid B, performance problem (shift solenoid valve SL2)
P0778	Transaxle pressure control solenoid B, electrical malfunction (shift solenoid valve SL2)
P0793	Transaxle intermediate shaft speed sensor A circuit, no signal
P0982	Transaxle shift solenoid D, low control voltage
P0983	transaxle shift solenoid D, high control voltage

3   Accelerator Pedal Position (APP) sensor - replacement

→Note: The APP sensor is located at the upper end of the accelerator pedal assembly on 2004 and later models. The APP sensor and the accelerator pedal are removed as a single assembly.

1   Disconnect the cable from the negative battery terminal (see Chapter 5, Section 1).

2   Disconnect the APP sensor electrical connector.

3   Remove the accelerator pedal/APP sensor assembly mounting bolts and remove the pedal/APP sensor assembly.

4   Installation is the reverse of removal. Be sure to tighten the accelerator pedal assembly mounting bolts securely.

4   Camshaft Position (CMP) sensor - replacement

FOUR-CYLINDER MODELS

▶ Refer to illustration 4.3

→Note: The CMP sensor is located on the left end of the cylinder head.

1   Disconnect the cable from the negative battery terminal (see Chapter 5, Section 1).

2   Remove the air intake duct, if necessary, to provide more room to work (see *Air filter housing - removal and installation* in Chapter 4).

3   Disconnect the electrical connector from the CMP sensor (see illustration).

4   Remove the CMP sensor mounting bolt and remove the sensor. If you're going to install the same sensor, check the condition of the O-ring. Replace the O-ring if it's damaged.

5   Installation is the reverse of removal.

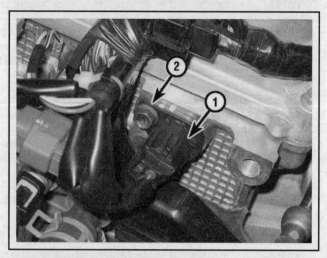

4.3 To remove the CMP sensor from the cylinder head on a four-cylinder engine, disconnect the electrical connector (1) and remove the sensor mounting bolt (2)

4.8 To disconnect the electrical connector from the CMP sensor for the front cylinder head on V6 engines, depress the release tab and pull off the connector

V6 MODELS

➡Note: There are two CMP sensors, one for each intake camshaft (Lexus and Toyota call them VVT-i sensors because their signals are used by the PCM to determine the optimal intake camshaft valve timing). The CMP sensors are located on the left (driver's) ends of the front and rear cylinder heads.

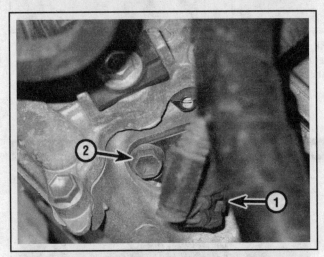

4.12 The CMP sensor the rear cylinder head is accessible through the left wheelhousing after you remove the wheel. To remove the sensor, depress the release tab (1) and disconnect the electrical connector, then remove the sensor mounting bolt (2)

4.9 To detach the CMP sensor from the front cylinder head on a V6 engine, remove this bolt

6   Disconnect the cable from the negative battery terminal (see Chapter 5, Section 1).

7   Remove the engine cover (see *Intake manifold - removal and installation* in Chapter 2B).

CMP sensor for front cylinder head

▶ Refer to illustrations 4.8 and 4.9

➡Note: The CMP sensor for the front cylinder head is on the left front corner of the head, right below the valve cover and right above the starter motor.

8   Disconnect the electrical connector from the CMP sensor (see illustration).

9   Remove the CMP sensor mounting bolt (see illustration) and remove the sensor.

10 Installation is the reverse of removal.

CMP sensor for rear cylinder head

▶ Refer to illustration 4.12

➡Note: The CMP sensor for the rear cylinder head is on the left rear corner of the head. It's difficult, but not impossible, to access.

11 Loosen the wheel lug nuts for the front left wheel. Raise the front of the vehicle and place it securely on jackstands. Remove the left front wheel.

12 Turn the steering wheel all the way to the right to get the brake rotor out of your way, then, looking through the area between the upper rear part of the transaxle bellhousing and the stabilizer bar, locate the CMP sensor (see illustration).

13 Disconnect the electrical connector from the CMP sensor.

14 Remove the CMP sensor mounting bolt and remove the sensor.

15 Installation is the reverse of removal.

5   Crankshaft Position (CKP) sensor - replacement

FOUR-CYLINDER MODELS

▶ **Refer to illustrations 5.2 and 5.4**

➡**Note: The CKP sensor is located on the front of the engine, on the timing chain cover, next to the crankshaft damper pulley.**

1   Disconnect the cable from the negative battery terminal (see Chapter 5, Section 1).

2   Disconnect the electrical connector (see illustration).

3   Raise the front of the vehicle and support it securely on jackstands.

4   Remove the CKP sensor mounting bolt (see illustration) and remove the CKP sensor.

5   Installation is the reverse of removal.

V6 MODELS

▶ **Refer to illustration 5.7**

6   Disconnect the cable from the negative battery terminal (see Chapter 5, Section 1).

7   Disconnect the electrical connector from the CKP sensor (see illustration).

8   Remove the CKP sensor mounting bolt and remove the CKP sensor.

9   Installation is the reverse of removal.

5.2 On four-cylinder engines, the electrical connector for the CKP sensor is located to the right of the exhaust manifold, just above the alternator

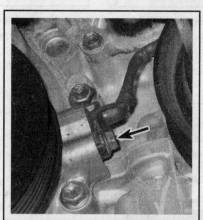

5.4 To detach the CKP sensor from the timing chain cover on a four-cylinder engine, remove this bolt

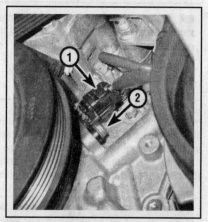

5.7 To remove the CKP sensor from a V6 engine, depress this release tab (1) and disconnect the electrical connector from the sensor, then remove the sensor mounting bolt (2)

6   Engine Coolant Temperature (ECT) sensor - replacement

✳✳ **WARNING:**

Wait until the engine is completely cool before beginning this procedure.

FOUR-CYLINDER MODELS

▶ **Refer to illustrations 6.2 and 6.4**

✳✳ **CAUTION:**

Handle the ECT sensor with care. Damage to the ECT sensor will affect the operation of the entire fuel injection system.

➡**Note: The ECT sensor is located at the left (driver's) end of the cylinder head.**

1   Disconnect the cable from the negative battery terminal (see Chapter 5, Section 1). Partially drain the cooling system (see Chapter 1).

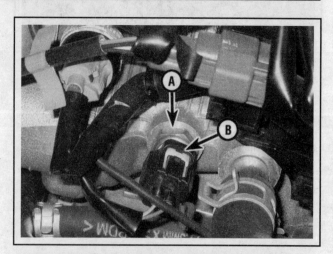

6.2 To remove the ECT sensor (A) from the left end of the cylinder head on a four-cylinder engine, disconnect the electrical connector (B) and unscrew the sensor from the head

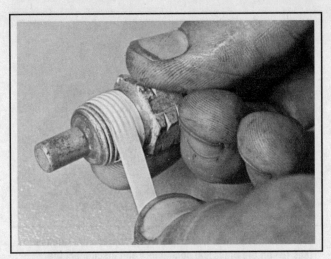

6.4 To prevent leaks, wrap the threads of the ECT sensor with Teflon tape

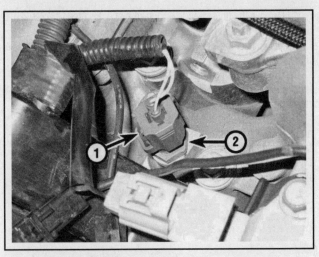

6.7 To remove the ECT sensor from the water outlet casting on a V6 engine, disconnect the electrical connector (1), then unscrew the sensor (2)

2   Disconnect the electrical connector from the ECT sensor (see illustration).
3   Unscrew the ECT sensor from the cylinder head.
4   Seal the threads of the ECT sensor with Teflon tape (see illustration).
5   Installation is the reverse of removal. Be sure to tighten the ECT sensor securely. Refill the cooling system (see Chapter 1).

V6 MODELS

▶ Refer to illustration 6.7

✳✳ CAUTION:

Handle the ECT sensor with care. A damaged ECT sensor will affect the operation of the entire fuel injection system.

➡Note: The ECT sensor is located at the right (timing belt) end of the engine, on the water outlet casting.

6   Disconnect the cable from the negative battery terminal (see Chapter 5, Section 1). Partially drain the cooling system (see Chapter 1).
7   Disconnect the electrical connector from the ECT sensor (see illustration).
8   Unscrew the ECT sensor from the water outlet casting.
9   Seal the threads of the ECT sensor with Teflon tape (see illustration 6.4).
10 Installation is the reverse of removal. Be sure to tighten the ECT sensor securely. Refill the cooling system (see Chapter 1).

7   Knock sensor - replacement

✳✳ WARNING:

Wait until the engine is completely cool before beginning this procedure.

FOUR-CYLINDER MODELS

➡Note: The knock sensor is located on the backside of the engine block. To access it, you have to remove the intake manifold.

1   Disconnect the cable from the negative battery terminal (see Chapter 5, Section 1).
2   Remove the intake manifold (see Chapter 2A).
3   Disconnect the knock sensor electrical connector.

4   Remove the knock sensor retaining nut and remove the knock sensor.
5   When installing the knock sensor, angle the electrical terminal of the sensor slightly below the horizontal. Installation is otherwise the reverse of removal. Be sure to tighten the knock sensor retaining nut to the torque listed in this Chapter's Specifications.

V6 MODELS

▶ Refer to illustration 7.8

➡Note: The knock sensors are located in the valley between the cylinder heads. To access them, you have to remove the intake manifold.

6   Disconnect the cable from the negative battery terminal (see Chapter 5, Section 1).

7   Remove the upper and lower intake manifolds (see Chapter 2B) and the water transfer hose (see illustration 5.23 in Chapter 2B).

8   Disconnect the electrical connectors from the knock sensors (see illustration).

9   On 1999 through 2003 models unscrew the knock sensors (see illustration 7.8). On 2004 and later models, remove the knock sensor retaining nuts.

10   When installing the knock sensors on 2004 and later models, angle the electrical terminals of the sensors slightly below the horizontal. Installation is otherwise the reverse of removal. Be sure to tighten the knock sensors or the knock sensor retaining nuts to the torque listed in this Chapter's Specifications.

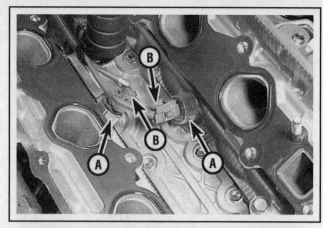

7.8 To remove the knock sensors (A), depress the release tabs (B) on the electrical connectors and disconnect the connectors, then unscrew the knock sensors (1999 through 2003 model shown; 2004 and later models are secured by a retaining nut)

8   Mass Air Flow/Intake Air Temperature (MAF/IAT) sensor - replacement

▸ Refer to illustration 8.2

➡Note: The MAF/IAT sensor is located on top of the air filter housing.

1   Disconnect the cable from the negative battery terminal (see Chapter 5, Section 1).

2   Disconnect the electrical connector from the MAF sensor (see illustration).

3   Remove the MAF sensor mounting screws and remove the MAF sensor.

4   Remove and discard the old MAF sensor O-ring.

5   Installation is the reverse of removal. Be sure to use a new O-ring.

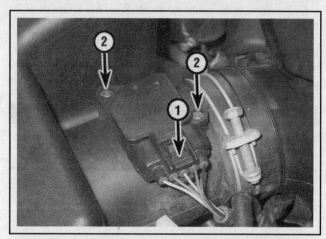

8.2 To disconnect the electrical connector from the MAF sensor, depress the release tab (1) and pull off the connector, then remove the sensor mounting screws (2) and remove the sensor from the air filter housing (V6 model shown, four-cylinder models similar)

9   Oxygen sensors - replacement

GENERAL INFORMATION

1   Use special care when servicing an oxygen sensor:
• Oxygen sensors have a permanently attached pigtail and electrical connector that can't be removed from the sensor. Damage to or removal of the pigtail or the electrical connector will ruin the sensor.
• Keep grease, dirt and other contaminants away from the electrical connector and the oxygen sensor.
• Do not use cleaning solvents of any kind on an oxygen sensor.
• Do not drop or roughly handle an oxygen sensor.

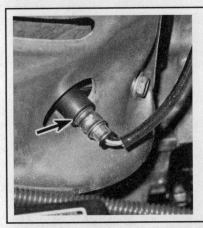

9.3 On four-cylinder engines, the upstream oxygen sensors are located in the exhaust manifold and are accessible through holes in the heat shield (2004 and later model with one upstream oxygen sensor shown)

REPLACEMENT

➡Note: Because it is installed in the exhaust manifold or catalytic converter, both of which contract when cool, an oxygen sensor might be very difficult to loosen when the engine is cold. Rather than risk damage to the sensor, start and run the engine for a minute or two, then shut it off. Be careful not to burn yourself during the following procedure.

Four-cylinder models

Upstream oxygen sensor

▶ Refer to illustration 9.3

➡Note: On 2003 and earlier models, there are two upstream oxygen sensors, one for each of the two Warm-Up Three-Way Catalysts (WU-TWCs) that are integral parts of the exhaust manifold. The upstream oxygen sensors are located in the upper ends of the WU-TWCs and are accessible through the exhaust manifold heat shield. 2004 and later models use a single WU-TWC so there is just one upstream oxygen sensor, which is also located at the upper end of the single WU-TWC and accessible through the exhaust manifold heat shield.

2 Disconnect the cable from the negative battery terminal (see Chapter 5, Section 1).

3 Locate the upstream oxygen sensor in the exhaust manifold heat shield (see illustration), then trace the sensor electrical lead to the connector and disconnect the connector.

4 Carefully unscrew the upstream oxygen sensor from the exhaust manifold. You'll have to use an oxygen sensor socket (see illustration 9.9) to unscrew it, unless you remove the exhaust manifold heat shield.

5 Installation is the reverse of removal. Be sure to coat the threads of the oxygen sensor with anti-seize compound, then tighten it to the torque listed in this Chapter's Specifications.

Downstream oxygen sensor

▶ Refer to illustration 9.9

➡Note: On 2001 through 2003 models, there are two downstream oxygen sensors, one for each of the two Warm-Up Three-Way Catalysts (WU-TWCs) that are integral parts of the exhaust manifold. The downstream sensors are located right at the lower ends of the WU-TWCs. 2004 and later models use a single WU-TWC so there is just one downstream sensor, which is also located on the front exhaust pipe, which is the pipe that connects the WU-TWC to the downstream catalyst.

6 Disconnect the cable from the negative battery terminal (see Chapter 5, Section 1).

9.9 After disconnecting the electrical connector for the downstream oxygen sensor, carefully unscrew the sensor with an oxygen sensor socket (shown) or with a large wrench (oxygen sensor sockets are handy where you don't have room to maneuver a wrench)

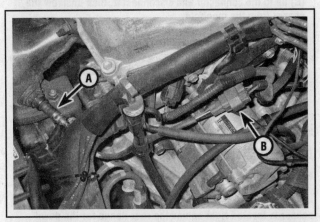

9.12a The upstream oxygen sensor (A) for the front cylinder head on a V6 engine is located on the front exhaust manifold. To find the electrical connector, trace the electrical harness to the connector (B) and disconnect it, then unscrew the sensor from the manifold

7 Raise the vehicle and place it securely on jackstands.

8 Locate the downstream oxygen sensor in the front exhaust pipe, then trace the sensor electrical lead to the connector and disconnect the connector.

9 Unscrew the oxygen sensor (see illustration).

10 Installation is the reverse of removal. Be sure to coat the threads of the oxygen sensor with anti-seize compound and tighten it to the torque listed in this Chapter's Specifications.

V6 models

Upstream oxygen sensors

▶ Refer to illustrations 9.12a and 9.12b

➡Note: The upstream oxygen sensors are located on the exhaust manifolds, above the Warm-Up Three-Way Catalysts (WU-TWCs).

11 Disconnect the cable from the negative battery terminal (see Chapter 5, Section 1).

12 Disconnect the upstream oxygen sensor electrical connector (see illustrations).

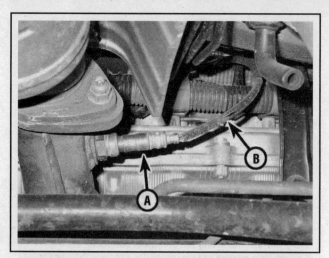

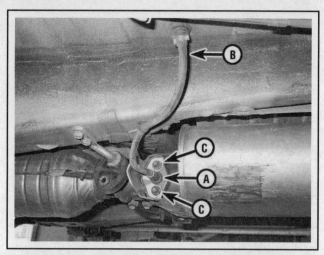

9.12b The upstream oxygen sensor (A) for the rear cylinder head on a V6 engine is located on the rear exhaust manifold. To find the electrical connector, trace the electrical harness (B) up to the connector and disconnect it, then unscrew the sensor from the manifold

9.18 On 1999 through 2003 V6 models, the downstream oxygen sensor (A) is located right behind the flange for the downstream catalyst and just ahead of the first muffler. To replace the sensor, remove the back seat and the carpeting, disconnect the electrical connector from the sensor harness (B), then remove the two sensor mounting nuts (C) and remove the sensor

13 If you're replacing the upstream oxygen sensor for the rear exhaust manifold, raise the front of the vehicle and place it securely on jackstands.

14 Unscrew the oxygen sensor from the exhaust manifold.

15 Installation is the reverse of removal. Be sure to coat the threads of the oxygen sensor with anti-seize compound and tighten it to the torque listed in this Chapter's Specifications.

Downstream oxygen sensor(s)

▶ Refer to illustration 9.18

➡Note: On 1999 through 2003 models, the downstream oxygen sensor is located right behind the rear flange for the downstream catalyst, just ahead of the first muffler. On 2004 and later models, there are two downstream oxygen sensors. The downstream sensor for the front exhaust manifold is located in the front exhaust pipe, which connects the front exhaust manifold to the Y-shaped pipe that connects both the front and rear

exhaust manifolds to the center exhaust pipe. The downstream oxygen sensor for the rear exhaust manifold is also in this Y-shaped pipe (there is no connecting pipe between the rear exhaust manifold and the center exhaust pipe).

16 Disconnect the cable from the negative battery terminal (see Chapter 5, Section 1).

17 Raise the vehicle and support it securely on jackstands.

18 Locate the downstream oxygen sensor (see illustration) and trace the sensor electrical harness to the sensor connector. On some models, you will have to remove the back seat and the carpeting (see Chapter 11) to access the connector for the downstream sensor.

19 Unscrew the oxygen sensor.

20 Installation is the reverse of removal. Be sure to coat the threads of the oxygen sensor with anti-seize compound and tighten it to the torque listed in this Chapter's Specifications.

---

10 Power Steering Pressure (PSP) sensor - replacement.

❊❊ **WARNING:**

Wait until the power steering fluid has cooled completely before beginning this procedure.

1 Disconnect the cable from the negative battery terminal (see Chapter 5, Section 1).

2 Raise the front of the vehicle and support it securely on jackstands.

FOUR-CYLINDER MODELS

➡Note: The PSP sensor is located on the power steering pump. It's screwed directly into the pump.

3 Locate the PSP sensor on the power steering pump, then disconnect the sensor electrical connector.

4 Place a drain pain under the pump, then unscrew the PSP sensor from the pump. Be prepared for some power steering fluid to leak out.

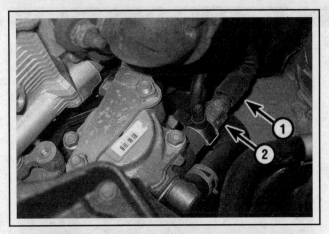

10.6 On V6 models, the Power Steering Pressure (PSP) sensor is screwed into the banjo fitting for the high-pressure line connection at the power steering pump. To remove the PSP sensor, disconnect the electrical connector (1), then unscrew the sensor from the banjo fitting (2)

5   Installation is the reverse of removal. After you're done, check the power steering fluid level (see Chapter 1) and add fluid as necessary.

V6 MODELS

▸ **Refer to illustration 10.6**

➡**Note: The PSP sensor is located at the power steering pump. It's screwed into the top of the banjo bolt that connects the power steering pressure hose to the pump.**

6   Disconnect the electrical connector from the PSP sensor (see illustration).

7   Place a drain pan under the power steering pump, then unscrew the PSP sensor from the banjo bolt. Be prepared for some power steering fluid to leak out.

8   Installation is the reverse of removal. After you're done, check the power steering fluid level (see Chapter 1) and add fluid as necessary.

11 Throttle Position (TP) sensor - replacement

▸ **Refer to illustrations 11.3a, 11.3b and 11.5**

➡**Note: On 1999 through 2003 models, the TP sensor is located on the throttle body. (On 2004 and later models, the TP sensor is an integral part of the electronic throttle body, and is not removable.)**

11.3a To disconnect the electrical connector from the TP sensor on a four-cylinder model, depress the release tab (1) and pull off the connector, then remove the two sensor mounting screws (2) and remove the sensor from the throttle body

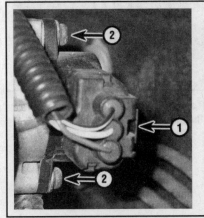

11.3b To disconnect the electrical connector from the TP sensor on a V6 model, depress the release tab (1) and pull off the connector, then remove the two sensor mounting screws (2) and remove the sensor from the throttle body.

1   Disconnect the cable from the negative battery terminal (see Chapter 5, Section 1).

2   Remove the engine cover, if necessary (see *Intake manifold - removal and installation* in Chapter 2).

3   Disconnect the TP sensor electrical connector (see illustrations).

4   Remove the TP sensor mounting screws and remove the sensor from the throttle body.

➡**Note: On V6 models there's not a lot of room between the TP sensor and the firewall, so if you're unable to loosen the sensor screws, remove the throttle body (see Chapter 4) and try again.**

5   Install the TP sensor 30 to 60-degrees counterclockwise from its installed position (see illustration), then rotate it clockwise until the mounting holes in the TP sensor are aligned with the mounting holes in the throttle body.

6   Installation is otherwise the reverse of removal.

11.5 Install the TP sensor 30 to 60-degrees counterclockwise from its installed position, then rotate it clockwise until the mounting holes in the TP sensor are aligned with the mounting holes in the throttle body

12 Transmission Range (TR) sensor - replacement

REMOVAL

▶ **Refer to illustrations 12.3 and 12.6**

➡**Note: The TR sensor is located on the front of the transaxle.**

1 Disconnect the cable from the negative battery terminal (see Chapter 5, Section 1).

2 Raise the vehicle and place it securely on jackstands.

3 Disconnect the electrical connector from the TR sensor (see illustration).

4 Remove the nut that secures the shift control cable to the control shaft lever and disconnect the cable from the lever.

5 Remove the nut and washer that secures the control shaft lever to the manual valve shaft and remove the control shaft lever from the manual valve shaft.

6 Pry open the lockplate tabs (see illustration) and remove the manual valve shaft nut.

7 Remove the TR sensor mounting bolts and remove the TR sensor.

INSTALLATION

▶ **Refer to illustration 12.11**

8 To install the TR sensor, slide it onto the manual valve shaft, then loosely install the mounting bolts.

9 Install a new lockplate and the manual valve shaft nut on the manual valve shaft.

10 Install the control shaft lever on the manual valve shaft, turn it counterclockwise through the gears - it will click as it changes to the next gear - until it stops, then turn it clockwise two clicks. It's now in the Neutral position.

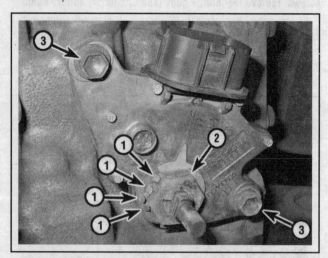

12.6 To detach the TR sensor from the transaxle:

1 *Pry open the lock plate tabs*
2 *Remove the manual valve shaft nut*
3 *Remove the TR sensor mounting bolts and remove the sensor*

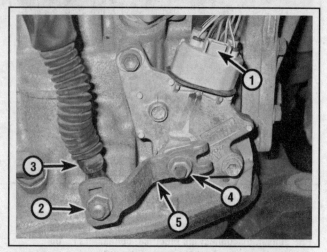

12.3 First, unplug the electrical connector and remove the control shaft lever from the TR sensor:

1 *Depress this release tab and disconnect the electrical connector from the TR sensor*
2 *Remove this nut . . .*
3 *. . . and disconnect the shift control cable from the control shaft lever*
4 *Remove this nut . . .*
5 *. . . and remove the control shaft lever from the manual valve shaft*

11 Align the pointer on the lockplate with the Neutral basic line (see illustration).

12 Install a new lockplate, install the manual valve shaft nut, tighten it securely, then bend the lockplate tabs up against the nut.

13 The remainder of installation is the reverse of removal.

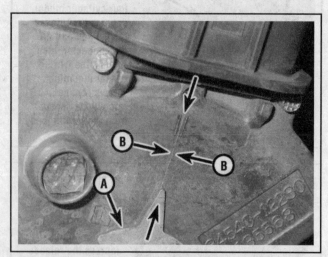

12.11 To adjust the TR sensor, align the pointer on the lockplate (A) with the neutral basic line (B) on the transaxle case

13 Transmission speed sensors - replacement

▶ Refer to illustrations 13.3, 13.4a, 13.4b and 13.5

➡Note: The transmission speed sensors are located on top of the transaxle. The Input Turbine Speed Sensor is the unit located on the left (driver's) end of the transaxle; the Counter Gear Speed Sensor is located to the right of the Input Turbine Speed Sensor, closer to the bellhousing.

1   Remove the battery (see Chapter 5).

2   Remove the air intake duct (see *Air filter housing - removal and installation* in Chapter 4).

3   Locate the speed sensor that you want to replace (see illustration).

4   To replace a sensor, disconnect the electrical connector, then remove the sensor mounting bolt (see illustrations).

5   Remove the old sensor O-ring (see illustration) and discard it.

6   Installation is the reverse of removal.

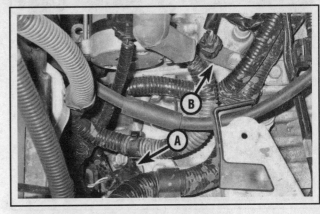

13.3 Transmission speed sensor locations:

A   Input Turbine Speed Sensor
B   Counter Gear Speed Sensor

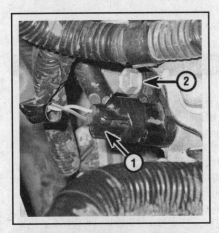

13.4a To replace the Input Turbine Speed Sensor, depress the release tab (1) and disconnect the electrical connector, then remove the sensor mounting bolt (2)

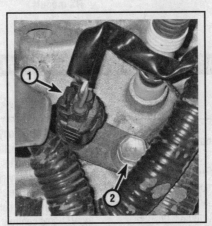

13.4b To replace the Counter Gear Speed Sensor, depress this release tab (1) and disconnect the electrical connector, then remove the sensor mounting bolt (2)

13.5 Remove and discard the old transmission speed sensor O-ring. Always use a new O-ring when installing either speed sensor

14 Powertrain Control Module (PCM) - replacement

▶ Refer to illustrations 14.3 and 14.4

✳✳ WARNING:

The models covered by this manual are equipped with Supplemental Restraint systems (SRS), more commonly known as airbags. Always disable the airbag system before working in the vicinity of any airbag system component to avoid the possibility of accidental deployment of the airbag, which could cause personal injury (see Chapter 12).

✳✳ CAUTION:

To avoid electrostatic discharge damage to the PCM, handle the PCM only by its case. Do not touch the electrical terminals during removal and installation. If available, ground yourself to the vehicle with an anti-static ground strap, available at computer supply stores.

➡Note: The PCM is located inside the passenger compartment behind the glove box, to the right of the heater blower case.

1   Disconnect the cable from the negative battery terminal (see Chapter 5, Section 1).

2   Remove the glove box assembly (see Chapter 11).

3   Disconnect the electrical connectors from the PCM (see illustration).

4   Remove the upper and lower PCM mounting bracket nuts (see illustration) and carefully remove the PCM.

✳✳ CAUTION:

Avoid any static electricity damage to the computer by grounding yourself to the body before touching the PCM and using a special anti-static pad to store the PCM on once it is removed.

5   Installation is the reverse of removal.

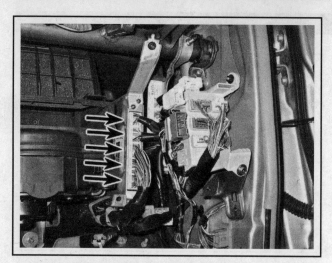

14.3 To remove the PCM, disconnect the electrical connectors . . .

14.4 . . . then remove the upper and lower mounting bracket nuts (Toyota Highlander PCM shown, mounting bracket and location of upper and lower nuts slightly different on Lexus RX models)

15 Air Control Induction System (ACIS) - description and component replacement

DESCRIPTION

➡Note: ACIS is used on V6 models only. It is not used on four-cylinder engines.

1 When an intake valve is closed, incoming air rushing down the intake runner slams against the closed valve, is compressed by its own inertia force and bounces back up the intake runner and into the plenum, then bounces back down the intake runner, where it hits the closed intake valve, etc. At some combinations of load and engine speed the timing of this uncontrolled pressure wave coincides with the opening of the intake valve, producing increased intake air volume and improving torque. But at most combinations of load and speed, it doesn't, because the length of the intake manifold is fixed. The Air Control Induction System (ACIS) takes advantage of this phenomenon by altering the effective length of the intake manifold to control the pulsing of this pressure wave. The result is improved torque and power throughout the operating range of V6 engines, particularly high-load torque at lower engine speeds. By improving the torque and power at low engine speeds the ACIS improves driveability and fuel economy.

1999 through 2003 models

2 On 1999 through 2003 models, the ACIS consists of a two-barrel throttle body and a pair of intake air control valves. One intake air control valve is mounted inside the throttle body and the other is mounted inside the air intake plenum. A vacuum-operated actuator activates each intake air control valve, and a PCM-controlled Vacuum Switching Valve (VSV) operates each actuator. This setup allows three different intake manifold lengths.

3 During heavy engine loads in the low-speed range, the intake pulses are extremely long. The PCM turns on both VSVs, which allows

intake vacuum to operate the two actuators, which closes both of the intake air control valves. When both valves are closed, the air intake duct, the throttle body and the entire intake manifold (plenum and intake runners) function as one big long intake manifold to take advantage of these long intake pulses.

4 During heavy engine loads in the medium-speed range, the intake pulses are relatively long. The PCM turns on the VSV that controls the intake air control valve in the intake manifold, and it turns off the VSV that controls the air control valve inside the throttle body. Now the intake manifold and the air intake plenum function as the intake manifold.

5 During idling, light loads and no-load high speed operation, the intake pulsations are short. The PCM turns off both VSVs, which allows both intake air control valves to open, reducing the effective intake manifold length to just the intake runners themselves.

2004 and later models

6 The ACIS is simplified on 2004 and later V6 models: there is only one intake air control valve, which is located on the air intake plenum. Like earlier units, it's activated by a vacuum-operated actuator, which in turn is operated by a PCM-controlled VSV. On these models there are just two effective intake manifold lengths.

7 During heavy engine loads at low engine speeds, the PCM turns on the VSV, which closes the intake air control valve, creating an effective intake manifold length that includes the air intake duct, the throttle body and the entire intake manifold (including the plenum and the intake runners).

8 During idling, light loads and no-load high speed operation, the intake pulsations are short. The PCM turns off the VSV, which opens the intake air control valve, reducing the effective intake manifold length to just the intake runners themselves.

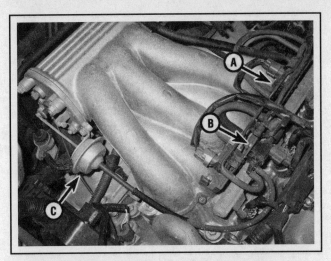

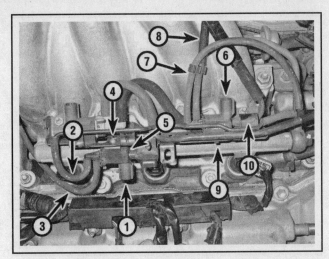

15.12 On 1999 through 2003 V6 models, the Vacuum Switching Valves (VSVs) for the Acoustic Control Induction System (ACIS) are located on this mounting bracket on the front of the intake manifold

A   *VSV for actuator on throttle body*
B   *VSV for actuator on intake manifold*
C   *Actuator for intake manifold (throttle body actuator, which is on the underside of the throttle body, not visible in this photo)*

9   On these models, there is no provision for altering the effective intake manifold length during heavy engine loads in the medium-speed range. But all of these models are equipped with the Electronic Throttle Control System-intelligent (ETCS-i), which provides much finer control of the throttle plate inside the throttle body than conventional induction systems that utilize an accelerator cable.

10   Lexus models are also equipped with a special air intake control system that alters the effective volume of the air filter housing (see next Section).

COMPONENT REPLACEMENT

Vacuum Switching Valves (VSVs)

▶ **Refer to illustrations 15.12 and 15.13**

➡**Note: This procedure applies to either VSV. The photos accompanying this section depict the VSVs used on 1999 through 2003 models. 2004 and later models use one VSV, but it's located on a similar mounting bracket in the same location as the two VSVs shown here.**

11   Remove the engine cover (see *Intake manifold - removal and installation* in Chapter 2B).

12   The vacuum switching valves are located on a mounting bracket located on the front of the intake manifold (see illustration).

13   Disconnect the electrical connector from the VSV (see illustration).

14   Clearly label the vacuum hoses, then disconnect them from the VSV.

15   Remove the VSV mounting screw and remove the VSV.

16   Installation is the reverse of removal.

15.13 Vacuum Switching Valve (VSV) removal and installation details:

1   *Disconnect the electrical connector from the VSV for the intake manifold actuator*
2   *Disconnect the vacuum source hose (from intake manifold)*
3   *Disconnect the vacuum signal hose (to intake manifold actuator)*
4   *Remove the VSV mounting screw*
5   *Remove the VSV from the mounting bracket*
6   *Disconnect the electrical connector from the VSV for the throttle body actuator*
7   *Disconnect the vacuum source hose (from intake manifold)*
8   *Disconnect the vacuum signal hose (to throttle body actuator)*
9   *Remove the VSV mounting screw*
10   *Remove the VSV from the mounting bracket*

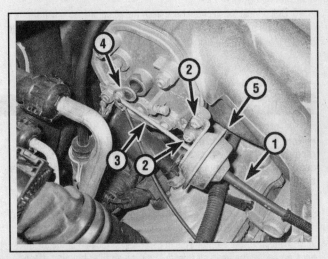

15.18 Intake manifold actuator removal and installation details:

1	*Vacuum source hose (from VSV)*	3	*Actuator rod*
2	*Actuator mounting bolts*	4	*Actuator lever*
		5	*Actuator*

15.19 To disconnect the intake manifold actuator rod from the actuator lever, remove this E-clip with a small screwdriver

15.23 To detach the intake air control valve actuator from the throttle body, remove these two screws

Intake air control valve actuators

Intake manifold actuator

▶ **Refer to illustrations 15.18 and 15.19**

➡**Note: The intake manifold actuator is located on the right end of the intake manifold. The photos accompanying this section depict the actuator used on 1999 through 2003 models. 2004 and later models also use a similar device.**

17 Remove the engine cover (see *Intake manifold - removal and installation* in Chapter 2B).

18 Disconnect the vacuum hose from the actuator (see illustration).

19 Pry off the E-clip (see illustration) that secures the actuator rod to the actuator lever, then disconnect the rod from the lever.

20 Remove the actuator mounting bolts and remove the actuator.

21 Installation is the reverse of removal.

Throttle body actuator

▶ **Refer to illustration 15.23**

➡**Note: The throttle body actuator is located on the underside of the throttle body. This actuator is used only on 1999 through 2003 models. The electronic throttle bodies on 2004 and later models don't use this actuator.**

22 Remove the throttle body (see Chapter 4).

23 Remove the actuator mounting screws (see illustration) and remove the actuator from the throttle body.

24 Installation is the reverse of removal.

16 Air intake control system - description

➡**Note: This system is used on 2004 and later Lexus RX330 models. At the time of publication, there was no information available regarding the availability of replacement parts for this system.**

1 The air intake control system reduces intake noise in the low-speed range and increases power in the high-speed range. It consists of a Vacuum Switching Valve (VSV), an actuator and an air intake control valve. The air intake control system assembly is located at the air filter housing end of the fresh air inlet duct for the air filter housing. (Don't confuse the fresh air inlet duct with the air intake duct, which connects the air filter housing to the throttle body. The fresh air inlet duct brings fresh air into the air filter housing.) VSV and the actuator are located on top of the air intake duct and the air intake control valve is located

directly below them, inside the duct. The fresh air inlet duct is divided into two chambers: an upper and a lower chamber.

2 When the engine is operating in the low-speed to mid-speed range, the PCM energizes the VSV, which operates the actuator, which moves the air intake control valve to close off one chamber inside the fresh air inlet duct. This reduces the cross-sectional area of the fresh air inlet duct and reduces intake noise.

3 When the engine is operating in the high-speed range, the PCM de-energizes the VSV, which deactivates the actuator, which opens the air intake control valve, allowing fresh air to enter both chambers of the air inlet duct. This increases the cross-sectional area of the fresh air inlet duct and improves inlet efficiency.

17 Idle Air Control (IAC) valve - replacement

FOUR-CYLINDER MODELS

➡Note: The IAC valve is located on the underside of the throttle body.

1 Disconnect the cable from the negative battery terminal (see Chapter 5, Section 1).

2 Remove the throttle body (see Chapter 4).

3 Remove the IAC valve mounting screws (see illustration 17.8) and remove the IAC valve.

4 Remove and discard the old IAC valve gasket.

5 Installation is the reverse of removal. Be sure to use a new gasket and tighten the IAC valve mounting screws securely.

V6 MODELS

▶ Refer to illustration 17.8

➡Note: The IAC valve is located on the underside of the throttle body.

6 Disconnect the cable from the negative battery terminal (see Chapter 5, Section 1).

7 Remove the throttle body (see Chapter 4).

8 Remove the IAC valve mounting screws (see illustration) and remove the IAC valve.

9 Remove and discard the old IAC valve gasket.

17.8 To detach the Idle Air Control (IAC) valve from the throttle body on a V6 model, remove these four screws. Be sure to remove and discard the old IAC valve gasket. Always use a new gasket when installing the IAC valve to prevent leaks

10 Installation is the reverse of removal. Be sure to use a new gasket and tighten the IAC valve mounting screws securely.

18 Catalytic converters - description, check and component replacement

➡Note 1: Because of a Federally mandated extended warranty which covers emissions-related components such as the catalytic converter, check with a dealer service department before replacing the converter at your own expense.

➡Note 2: The front catalytic converter is incorporated into the exhaust manifold. Refer to Chapter 2A or 2B for the exhaust manifold replacement procedure.

GENERAL DESCRIPTION

1 The catalytic converter is an emission control device installed in the exhaust system that reduces pollutants from the exhaust gas stream. There are two types of converters: The oxidation catalyst reduces the levels of hydrocarbon (HC) and carbon monoxide (CO) by adding oxygen to the exhaust stream to produce water vapor (H_2O) and carbon dioxide (CO_2). The reduction catalyst lowers the levels of oxides of nitrogen (NOx) by removing oxygen from the exhaust gases to produce nitrogen (N) and oxygen. These two types of catalysts are combined into a three-way catalyst that reduces all three pollutants.

2 The amount of oxygen entering the catalyst is critical to its operation because without oxygen it cannot convert harmful pollutants into harmless compounds. The catalyst is most efficient at capturing and storing oxygen when it converts the exhaust gases of an intake charge that's mixed at the ideal (stoichiometric) air/fuel ratio of 14.7:1. If the air/fuel ratio is leaner than stoichiometric for an extended period of time, the catalyst will store even more oxygen. But if the air/fuel ratio is richer than stoichiometric for any length of time, the oxygen content in the catalyst can become totally depleted. If this condition occurs, the catalyst will not convert anything!

3 Because the catalyst's ability to store oxygen is such an important factor in its operation, it can also be considered a factor in the catalyst's eventual inability to do its job. The PCM monitors the oxygen content going into and coming out of the catalyst by comparing the voltage signals from the upstream and downstream oxygen sensors. When the catalyst is functioning correctly, there is very little oxygen to monitor at the outlet end of the catalyst because it's capturing, storing and releasing oxygen as needed to convert HC, CO and NOx into more benign substances. If the catalyst isn't doing its job, the downstream oxygen sensor tells the PCM that the oxygen content in the catalyzed exhaust gases is going up. When the amount of oxygen exiting the catalyst reaches a specified threshold, the PCM stores a Diagnostic Trouble Code (DTC) and turns on the Malfunction Indicator Light (MIL), also known as the CHECK ENGINE light.

CHECK

4 The test equipment for a catalytic converter is expensive and highly sophisticated. If you suspect that the converter on your vehicle is malfunctioning, take it to a dealer or authorized emissions inspection facility for diagnosis and repair.

5 Whenever the vehicle is raised for servicing of underbody components, check the converter for leaks, corrosion, dents and other damage. Check the welds/flange bolts that attach the front and rear ends of the converter to the exhaust system. If damage is discovered, the

converter should be replaced.

6   Although catalytic converters don't break too often, they can become plugged. The easiest way to check for a restricted converter is to use a vacuum gauge to diagnose the effect of a blocked exhaust on intake vacuum.

   a) Connect a vacuum gauge to an intake manifold vacuum source (see Chapter 2C).
   b) Warm the engine to operating temperature, place the transaxle in Park (automatic) or Neutral (manual) and apply the parking brake.
   c) Note and record the vacuum reading at idle.
   d) Quickly open the throttle to near full throttle and release it shut. Note and record the vacuum reading.
   e) Perform the test three more times, recording the reading after each test.
   f) If the reading after the fourth test is more than one in-Hg lower than the reading recorded at idle, the exhaust system may be restricted (the catalytic converter could be plugged or an exhaust pipe or muffler could be restricted).

REPLACEMENT

☀☀ WARNING:

Do NOT service a catalytic converter until it has completely cooled down.

Upstream catalytic converters

Four-cylinder models

➡Note: On 2001 through 2003 four-cylinder engines, there are two upstream catalysts. Toyota calls them Warm-Up Three-Way Catalysts (WU-TWCs) and they're integral parts of the exhaust manifold. On 2004 and later four-cylinder models, there is a single upstream catalyst, or WU-TWC, and it too is an integral part of the exhaust manifold.

7   Refer to *Exhaust manifold/catalytic converter assembly - removal and installation* in Chapter 2A.

1999 through 2003 V6 models

▶ Refer to illustrations 18.9a, 18.9b and 18.10

➡Note: On 1999 through 2003 V6 engines, the Warm-Up Three-Way (WU-TWC) catalytic converters are located directly below the exhaust manifolds, to which they're bolted.

8   Raise the front of the vehicle and place it securely on jackstands.
9   To remove the front WU-TWC, remove the two nuts that secure the upper end of the WU-TWC to the exhaust manifold and the two nuts that secure the lower end of the WU-TWC to the connecting pipe (see illustrations).
10 To remove the rear WU-TWC, remove the nuts and bolts that secure it to the rear exhaust manifold, the connecting pipe and the downstream catalytic converter (see illustration).
11 Installation is the reverse of removal.

2004 and later V6 models

➡Note: On 2004 and later V6 engines, the Warm-Up Three-Way exhaust manifold and the WU-TWC are welded together, so they're not serviceable separately.

12 Refer to *Exhaust manifold/catalytic converter assemblies - removal and installation* in Chapter 2B.

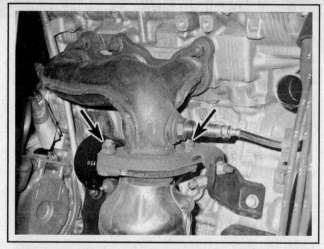

18.9a To disconnect the upper end of the front WU-TWC from the front exhaust manifold on a V6 engine, remove these two nuts

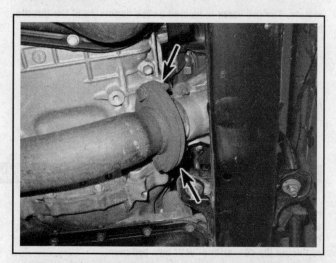

18.9b To disconnect the lower end of the front WU-TWC from the connecting pipe on a V6 engine, remove these two nuts and bolts

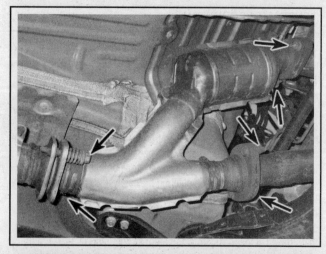

18.10 To disconnect the rear WU-TWC from the rear exhaust manifold, the connecting pipe and the downstream catalyst, remove these nuts and bolts

Downstream catalytic converter

➡**Note: The downstream catalytic converter, which Toyota and Lexus refer to as a Three-Way Catalyst (TWC), is located underneath the vehicle. It's located between the flange at the rear end of the front exhaust pipe and the muffler, to which it's con-** nected by a short pipe. The WU-TWC, short pipe and muffler are welded up into a single assembly.

13 Raise the front of the vehicle and place it securely on jackstands.
14 Refer to illustrations 16.1b and 16.1c in Chapter 4.

---

19 Evaporative emissions control (EVAP) system - description and component replacement

DESCRIPTION

1 The Evaporative Emissions Control (EVAP) system absorbs fuel vapors (unburned hydrocarbons) and, during engine operation, releases them into the intake manifold from which they're drawn into the intake ports where they mix with the incoming air-fuel mixture. The EVAP system consists of the fuel tank filler neck cap, the EVAP canister, the vapor pressure sensor, the canister closed valve, the pressure switching valve and the purge valve. Everything except the purge valve and the canister closed valve is located underneath the vehicle.

2 Modern EVAP systems are quite complex. But basically, here's how it works: When the gasoline inside the fuel tank warms up on a hot day, it evaporates and produces fuel vapors. These vapors, which are raw unburned hydrocarbons, elevate the pressure inside the sealed fuel tank. If there were no way to vent them somewhere, the fuel tank would eventually spring a leak somewhere. The vapor pressure sensor monitors the pressure inside the tank and keeps the Powertrain Control Module (PCM) informed. When the PCM senses that the pressure has exceeded the specified threshold, it energizes the pressure switching valve, which opens, allowing the vapors to migrate to the EVAP canister. The canister stores these vapors until the PCM energizes the purge valve, which opens and purges the EVAP system, i.e. allows intake manifold vacuum to pull the vapors from the canisters into the intake manifold.

3 The vapor pressure sensor is located on top of the fuel tank. The vapor pressure sensor monitors the pressure of fuel vapors inside the tank. When the vapor pressure exceeds the upper threshold, the vapor pressure sensor signals the PCM, which opens the Vacuum Switching Valve (VSV) for the pressure switching valve, allowing the fuel vapors to migrate to the EVAP canister, where they are stored until they're purged.

4 When the engine is cold or still warming up, no captive fuel vapors are allowed to escape from the EVAP canister. After the engine is warmed up (165 degrees F.) the PCM puts the system into closed loop operation. Then it energizes the canister purge valve, which regulates the flow of vapors from the canister to the intake manifold. The rate of vapor flow is regulated by the purge valve in response to commands from the PCM, which controls the duty cycle of the valve, which means that the valve's opening can be controlled to a fine degree (it's not just open or closed) in order to regulate the volume of the purged vapors in an appropriate way so that the air/fuel mixture doesn't become too rich. On four-cylinder models, the EVAP canister purge valve is located on the firewall, behind the throttle body. On V6 models, the canister purge valve is located in the engine compartment, at the right (timing belt) end of the engine.

➡**Note: On earlier models, the purge valve is referred to by Toyota and Lexus as the Vacuum Switching Valve for the Evaporative Emission Control system, or VSV for EVAP. However, in this book we refer to it as the canister purge valve, or simply, the purge valve.**

5 When the EVAP system is being purged and stored vapors are being drawn from the canister by intake manifold vacuum, a vacuum condition would quickly result inside the canister and the fuel tank if they were not vented to atmospheric pressure. So during purging, atmospheric air is drawn through the air filter housing, through the fresh air line, then into the canister. The canister closed valve opens and closes the EVAP system's fresh air line in response to signals from the Powertrain Control Module (PCM). When commanded by the PCM, the canister closed valve also closes the fresh air line to the EVAP canister for monitor testing. The canister closed valve is located in the engine compartment, on the backside of the air filter housing.

➡**Note: On earlier models, Lexus and Toyota refer to this device as the Vacuum Switching Valve (VSV) for Canister Closed Valve (CCV). However, in this book we refer to it simply as the canister closed valve.**

The EVAP system monitor

6 The EVAP system diagnostic monitor is an OBD-II test that the PCM runs to check the EVAP system and the fuel tank for leaks. Before the monitor sequence begins, several things must happen. First, you must start the engine. If the engine is cold, the engine coolant temperature and the intake air temperature are about the same. The PCM watches the progress of the warm-up sequence closely. Once the oxygen sensors and the catalysts have warmed up enough for the PCM to put the system into closed loop operation, the PCM initiates the EVAP purge sequence. The purge valve opens, the canister closed valve opens and the EVAP canister's contents are purged, i.e. drawn into the intake manifold. During a fast-idle warm-up intake vacuum is high and the extra-rich mixture caused by purging the vapors stored inside the EVAP canister actually helps to smooth out the idle.

7 During purging, the pressure inside the EVAP system is neutral because the opened canister closed valve allows atmospheric pressure to be drawn into the canister as vapors are drawn from the canister into the intake manifold. During this initial period of operation, the PCM is also monitoring the fuel tank pressure with the vapor pressure sensor. As soon as the purge sequence is complete, the PCM closes the canister closed valve. When the canister closed valve is first closed, the pressure switching valve and the purge valve are still open, so a (relative) vacuum develops inside the purge line from the air intake to the canister and to the EVAP line from the canister to the fuel tank. The PCM then closes the purge valve to produce a (relative) vacuum in the line between the tank and the purge valve. Then it monitors any change in pressure (through the vapor pressure sensor) to check for EVAP system leaks. If there's a leak, the Malfunction Indicator Light (MIL) or Check Engine light comes on and the PCM stores a Diagnostic Trouble Code (DTC) that indicates a malfunction in the EVAP system (see Section 2).

8 At a certain point in the monitoring sequence the PCM closes the canister closed valve and opens the pressure switching valve, which causes a pressure drop in the EVAP system. The PCM keeps the purge valve open until the pressure inside the EVAP system drops to a specified threshold, at which point the PCM closes the purge valve. If the

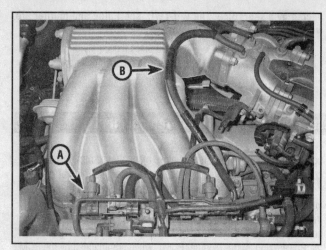

19.13a On V6 models, the purge valve (A) for the Evaporative Emission Control (EVAP) system is located on this mounting bracket on front of the intake manifold. The EVAP purge hose (B) connects a metal pipe on the bracket to the throttle body

pressure doesn't drop, or drops too much, the PCM turns on the MIL or Check Engine light and stores a DTC (see Section 2) that indicates an incorrect purge flow in the EVAP system.

9   Then the PCM monitors the operation of the canister closed valve and the venting (air inlet) function of the system. When vapor pressure rises to a specified threshold, the PCM opens the canister closed valve. The pressure inside the system goes up quickly because of the air drawn into the system. If the PCM detects no increase in pressure or if the pressure is below the specified increase, the PCM concludes that either the canister closed valve is malfunctioning or there's a restriction somewhere in the venting and, again, stores a DTC (see Section 2) and turns on the MIL or Check Engine light.

10   Finally, the PCM closes the pressure switching valve, which prevents atmospheric air from entering the fuel tank side of the system. When the pressure switching valve is operating correctly, this should produce a slight pressure rise inside the tank (because the fuel inside the tank is still slowly getting warmer). But if there's no change in pressure, the PCM concludes that the pressure switching valve isn't closing and, again, stores a DTC (see Section 2) and turns on the MIL or Check Engine light. The monitoring sequence is now completed. The PCM immediately repeats the entire sequence again, conditions permitting, and continues to do so as long as the engine is operating in closed loop.

REPLACEMENT

❊❊ WARNING:

Gasoline and gasoline vapor is extremely flammable, so take extra precautions when you work on any part of the fuel system or EVAP system. Don't smoke or allow open flames or bare light bulbs near the work area, and don't work in a garage where a gas-type appliance (such as a water heater or a clothes dryer) is present. Since gasoline is carcinogenic, wear fuel resistant gloves when there's a possibility of being exposed to fuel, and, if you spill any fuel on your skin, rinse it off immediately with soap and water. Mop up any spills immediately and do not store fuel-soaked rags where they could ignite. When you perform any kind of work on the fuel system, wear safety glasses and have a Class B type fire extinguisher on hand.

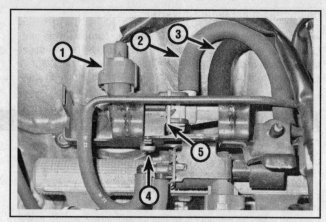

19.13b A typical EVAP purge valve assembly (this one is on a V6 model). To remove the purge valve:

1   *Disconnect the electrical connector*
2   *Disconnect the EVAP hose (coming from the EVAP canister)*
3   *Disconnect the EVAP purge hose (goes to a metal pipe on the mounting bracket, which is connected to the throttle body by the purge hose shown in illustration 19.13a)*
4   *Remove the purge valve mounting fastener*
5   *Remove the purge valve from the mounting bracket (on four-cylinder models, the bracket, which is part of the purge valve, is bolted to the firewall)*

Purge valve

▶ **Refer to illustrations 19.13a and 19.13b**

➡ Note: On four-cylinder models, the purge valve is mounted on the firewall, behind the throttle body. On V6 models, the purge valve is located on a mounting bracket bolted to the front of the intake manifold. The purge valve depicted in the accompanying illustration is on a V6 engine, but the purge valve used on four-cylinder models is similar except for its location.

11   Disconnect the cable from the negative battery terminal (see Chapter 5, Section 1).

12   Remove the engine cover, if applicable (see Intake manifold - removal and installation in Chapter 2).

13   Disconnect the electrical connector from the purge valve (see illustrations).

14   Disconnect the vacuum hoses from the purge valve.

15   Remove the purge valve mounting screw (V6 models) or mounting bolt (four-cylinder models) and remove the purge valve.

16   Installation is the reverse of removal.

All other components (under vehicle)

▶ **Refer to illustration 19.19**

17   Disconnect the cable from the negative battery terminal (see Chapter 5, Section 1).

18   Raise the rear of the vehicle and support it securely on jackstands.

19   Remove the protective cover from the EVAP canister assembly (see illustration).

Vacuum Switching Valve (VSV) for the pressure switching valve

▶ **Refer to illustration 19.20**

➡ Note: The VSV for the pressure switching valve is located at the EVAP canister.

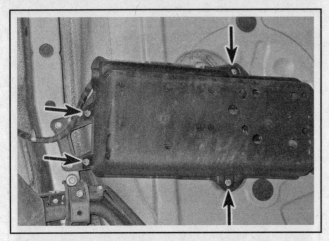

19.19 To detach the protective cover from the EVAP canister, remove these mounting nuts

20 Disconnect the electrical connector from the Vacuum Switching Valve (VSV) for the pressure switching valve (see illustration).

21 Clearly label the two EVAP hoses that connect the VSV to the pressure switching valve and to the EVAP canister, then disconnect both of them from the VSV.

22 Remove the VSV mounting screw and remove the VSV.

23 Installation is the reverse of removal.

Air drain valve

▶ Refer to illustration 19.24

➡Note: The air drain valve is located on the EVAP canister.

24 Disconnect the EVAP hoses from the air drain valve (see illustration).

25 Remove the air drain valve mounting bolt and remove the air drain valve.

26 Installation is the reverse of removal.

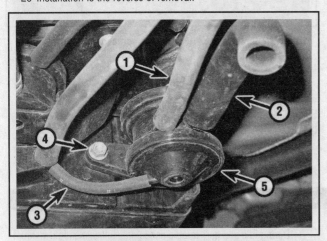

19.24 Air drain valve removal and installation details:

1 Disconnect the EVAP inlet hose (to the Closed Canister Valve on the air filter housing)
2 Air drain hose (you don't have to remove this hose to remove the air drain valve, but you should inspect it)
3 EVAP connecting hose (between EVAP canister and air drain valve)
4 Air drain valve mounting bolt
5 Air drain valve

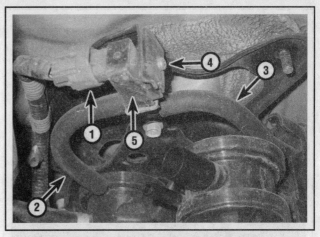

19.20 Removal and installation details for the Vacuum Switching Valve (VSV) for the pressure switching valve:

1 Disconnect the electrical connector
2 Disconnect the hose that connects the VSV to the pressure switching valve
3 Disconnect the hose that connects the VSV to the EVAP canister
4 Remove the VSV mounting screw
5 Remove the VSV

Pressure switching valve

▶ Refer to illustration 19.27

➡Note: The pressure switching valve is located on the EVAP canister.

27 Detach the VSV electrical harness from the pressure switching valve mounting bracket (see illustration).

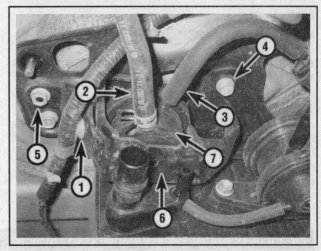

19.27 Pressure switching valve removal and installation details:

1 Using a small screwdriver, pry the VSV wiring harness clip loose from the pressure switching valve mounting bracket
2 EVAP hose (to fuel tank)
3 EVAP connecting hose (between VSV and pressure switching valve)
4 Mounting bolt
5 Mounting nut
6 Mounting bracket
7 Pressure switching valve

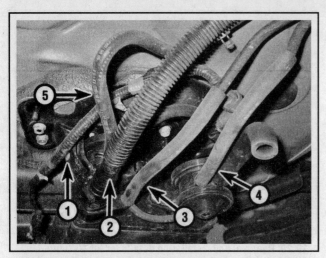

19.32a Before removing the EVAP canister:

1 *Detach the VSV wiring harness from the pressure switching valve mounting bracket*
2 *Disconnect the EVAP vent hose (goes to the fuel tank) (see illustration 19.32b for help with the fitting)*
3 *Disconnect the EVAP purge hose (goes to purge valve in engine compartment)*
4 *Disconnect the EVAP air inlet hose (goes to canister closed valve on air filter housing)*
5 *Disconnect the EVAP hose (goes to fuel tank)*

28 Disconnect the EVAP hoses from the pressure switching valve.
29 Remove the pressure switching valve mounting bracket bolt and nut and remove the pressure switching valve and mounting bracket.
30 Installation is the reverse of removal.

EVAP canister

♦ **Refer to illustrations 19.32a, 19.32b and 19.33**

➡**Note: The EVAP canister is located underneath the rear of the vehicle, behind the rear suspension crossmember.**

31 Disconnect the electrical connector from the VSV (see illustration 19.20).
32 Detach the VSV wiring harness from the pressure switching valve mounting bracket and disconnect the EVAP hoses from the EVAP canister assembly (see illustrations).
33 Remove the EVAP canister mounting bracket bolts (see illustration) and remove the EVAP canister and mounting bracket as a single assembly.

19.32b To disconnect the EVAP vent hose fitting, squeeze the release tabs, then pull off the connector from the pipe on the EVAP canister

19.33 To detach the EVAP canister, remove these three mounting bolts

34 If you are replacing the EVAP canister, remove the VSV, air drain valve and pressure switching valve from the canister assembly (see Steps 20 through 30), then detach the EVAP canister from its mounting bracket.
35 Installation is the reverse of removal.

20 Positive Crankcase Ventilation (PCV) system

♦ **Refer to illustrations 20.1a, 20.1b and 20.1c**

1 The Positive Crankcase Ventilation (PCV) system (see illustrations) reduces hydrocarbon emissions by directing blow-by gases and crankcase vapors into the intake manifold, where they're mixed with intake air before drawn into the combustion chambers where they're consumed along with the air/fuel mixture. The PCV system does this by circulating fresh air from the air filter housing through a series of hoses into the crankcase, where the fresh air mixes with blow-by gases before being drawn from the crankcase by intake vacuum, through the PCV valve and then into the intake manifold.

2 During idle and part-throttle conditions, intake manifold vacuum is high. Blow-by gases and crankcase vapors flow from the crankcase through the PCV valve and the crankcase ventilation hose (also known as the PCV hose) into the intake manifold. The strong intake manifold vacuum also pulls fresh air from the air intake duct or the air filter housing through the fresh air inlet hose into the crankcase.

3 There is no scheduled maintenance interval for the PCV valve or the PCV system hoses. But over time the PCV system might become less efficient as an oily residue of sludge builds up inside the PCV valve and the hoses. One symptom of a clogged PCV system is leaking

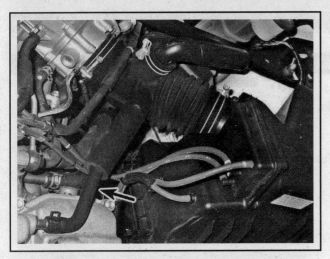

20.1a On all engines, the fresh air inlet hose connects the air intake duct to a pipe on the valve cover, from which the fresh air is pulled down into the crankcase (this is a V6 model, but four-cylinder models use a similar hose)

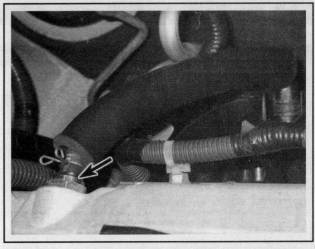

20.1b On four-cylinder engines, the PCV valve is located at the right rear corner of the valve cover, and the crankcase ventilation hose (or PCV hose) connects the PCV valve to the intake manifold on the backside of the engine

seals. When crankcase vapors can't escape, pressure builds inside the bottom end and eventually causes crankshaft seals to leak. So anytime that you're changing the oil filter, air filter, fuel filter, spark plugs, etc. it's a good idea to pull off the PCV hoses and inspect them. If the hoses are clogged, remove them and clean them out. If they're cracked, torn or deteriorated, replace them.

4   Checking and replacement of the PCV valve is covered in Chapter 1.

20.1c On V6 engines, the PCV valve is located at the right rear corner of the rear valve cover, and the crankcase ventilation hose (or PCV hose) connects the PCV valve to a pipe on the underside of the manifold (the manifold has been disconnected and raised up in this photo so that you can see the connection to the manifold)

Specifications

Torque specifications	Ft-lbs (unless otherwise indicated)	Nm
Knock sensor(s)		
Four-cylinder models (retaining nut)	168 in-lbs	20
V6 models		
1999 through 2003 (screw-in type)	29	39
2004 on (retaining nuts)	168 in-lbs	20
Oxygen sensors	35	44

Section

Reference to other Chapters

7

AUTOMATIC TRANSAXLE

1 General information

There are several models of automatic transaxles on the vehicles covered in this manual. Four-cylinder engines have been equipped with the U241-E 4-speed. 1999 through 2003 models with V6 engines are equipped with the U140-E (2WD models) or U140-F (4WD models). 2004 and later V6 models are equipped with either the U151-E (2WD models) or the U151-F (4WD models). The transaxles vary in oil capacity, bearing and gear size to compensate for the engine horsepower and torque.

1999 through 2003 models use the MF2AV or the MF2A transfer case with viscous coupling in the differential. 2004 and later models are equipped with the TRAC traction control system that eliminates wheel slip by applying brake to the appropriate wheels using the ABS system.

Due to the complexity of the automatic transaxles covered in this manual and to the specialized equipment necessary to perform most service operations, this Chapter contains only those procedures related to general diagnosis, routine maintenance, adjustment and removal and installation.

If the transaxle requires major repair work, it should be left to a dealer service department or an automotive or transmission shop. You can, however, remove and install the transaxle yourself and save the expense, even if a transmission shop does the repair work (provided that a proper diagnosis has been made prior to removal).

2 Diagnosis - general

Automatic transaxle malfunctions may be caused by five general conditions:

a) poor engine performance
b) improper adjustments
c) hydraulic malfunctions
d) mechanical malfunctions
e) malfunctions in the computer or its signal network

Diagnosis of these problems should always begin with a check of the easily repaired items: fluid level and condition (see Chapter 1), shift linkage adjustment and throttle linkage adjustment. Next, perform a road test to determine if the problem has been corrected or if more diagnosis is necessary. If the problem persists after the preliminary tests and corrections are completed, additional diagnosis should be done by a dealer service department or transmission shop. Refer to the *Troubleshooting* section at the front of this manual for information on symptoms of transaxle problems.

PRELIMINARY CHECKS

1 Drive the vehicle to warm the transaxle to normal operating temperature.
2 Check the fluid level as described in Chapter 1:
a) If the fluid level is unusually low, add enough fluid to bring the level within the designated area of the dipstick, then check for external leaks (see below).
b) If the fluid level is abnormally high, drain off the excess, then check the drained fluid for contamination by coolant. The presence of engine coolant in the automatic transaxle fluid indicates that a failure has occurred in the internal radiator walls that separate the coolant from the transaxle fluid (see Chapter 3).
c) If the fluid is foaming, drain it and refill the transaxle, then check for coolant in the fluid, or a high fluid level.
3 Check the engine idle speed.
➡Note: If the engine is malfunctioning, do not proceed with the preliminary checks until it has been repaired and runs normally.
4 On early models, check the accelerator cable for adjustment. Adjust it if necessary (see Chapter 4).
5 Inspect the shift cable (see Section 3). Make sure that it's properly adjusted and that the linkage operates smoothly.

FLUID LEAK DIAGNOSIS

6 Most fluid leaks are easy to locate visually. Repair usually consists of replacing a seal or gasket. If a leak is difficult to find, the following procedure may help.
7 Identify the fluid. Make sure it's automatic transaxle fluid and not engine oil or brake fluid (automatic transaxle fluid is a deep red color).
8 Try to pinpoint the source of the leak. Drive the vehicle several miles, then park it over a large sheet of cardboard. After a minute or two, you should be able to locate the leak by determining the source of the fluid dripping onto the cardboard.
9 Make a careful visual inspection of the suspected component and the area immediately around it. Pay particular attention to gasket mating surfaces. A mirror is often helpful for finding leaks in areas that are hard to see.
10 If the leak still cannot be found, clean the suspected area thoroughly with a degreaser or solvent, then dry it.
11 Drive the vehicle for several miles at normal operating temperature and varying speeds. After driving the vehicle, visually inspect the suspected component again.
12 Once the leak has been located, the cause must be determined before it can be properly repaired. If a gasket is replaced but the sealing flange is bent, the new gasket will not stop the leak. The bent flange must be straightened.
13 Before attempting to repair a leak, check to make sure that the following conditions are corrected or they may cause another leak.
➡Note: Some of the following conditions cannot be fixed without highly specialized tools and expertise. Such problems must be referred to a transmission shop or a dealer service department.

Gasket leaks

14 Check the pan periodically. Make sure the bolts are tight, no bolts are missing, the gasket is in good condition and the pan is flat (dents in the pan may indicate damage to the valve body inside).
15 If the pan gasket is leaking, the fluid level or the fluid pressure may be too high, the vent may be plugged, the pan bolts may be too tight, the pan sealing flange may be warped, the sealing surface of the transaxle housing may be damaged, the gasket may be damaged or the transaxle casting may be cracked or porous. If sealant instead of gasket material has been used to form a seal between the pan and the transaxle housing, it may be the wrong sealant.

Seal leaks

16 If a transaxle seal is leaking, the fluid level or pressure may be too high, the vent may be plugged, the seal bore may be damaged, the seal itself may be damaged or improperly fitted, the surface of the shaft protruding through the seal may be damaged or a loose bearing may be causing excessive shaft movement.

17 Make sure the dipstick tube seal is in good condition and the tube is properly seated. Periodically check the area around the speedometer gear or sensor for leakage. If fluid is evident, check the O-ring for damage.

Case leaks

18 If the case itself appears to be leaking, the casting is porous and will have to be repaired or replaced.

19 Make sure the oil cooler hose fittings are tight and in good condition.

Fluid comes out vent pipe or fill tube

20 If this condition occurs, the transaxle is overfilled, there is coolant in the fluid, the case is porous, the dipstick is incorrect, the vent is plugged or the drain-back holes are plugged.

---

3 Shift cable - replacement and adjustment

> ✳✳ **CAUTION:**
>
> **The shift cable replacement procedure for Lexus models will require the instrument panel and the heating/air conditioning unit to be removed for access to the shift cable (see Chapters 11 and 3).**

REPLACEMENT

▶ **Refer to illustrations 3.6, 3.7, 3.8, 3.10, 3.11 and 3.12**

1 Disconnect the cable from the negative terminal of the battery (see Chapter 5, Section 1).

2 Remove the center console trim panel (see Chapter 11).

3 Remove the floor carpet and the instrument panel center lower trim panel (see Chapter 11).

4 Remove the battery and the battery tray (see Chapter 5).

5 Remove the air filter housing and the air intake ducts (see Chapter 4).

6 Disconnect the shift cable from the manual lever on the transaxle (see illustration).

7 Remove the large C-clip cable retainer (see illustration) from the bracket located on the side of the transaxle.

3.6 Remove the nut and disconnect the shift cable from the manual lever on the side of the transaxle

8 Remove the shift cable from the wire hanger (see illustration).

9 On Lexus models, remove the air conditioning unit (see Chapter 3).

10 Disconnect the shift cable from the shift lever assembly on the center console (see illustration).

3.7 Remove the large C-clip cable retainer from the bracket on the side of the transaxle

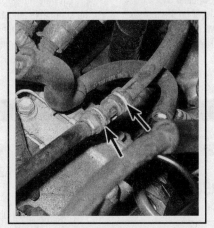

3.8 Remove the shift cable from the wire hanger on the top of the transaxle

3.10 Use a trim panel tool (A) to separate the shift cable from the shift lever (B) at the center console

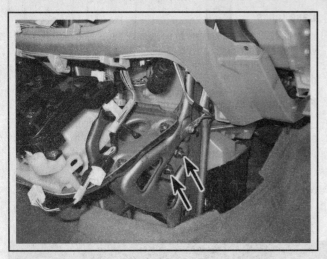

3.11 Remove the two mounting nuts and bolts at the center console

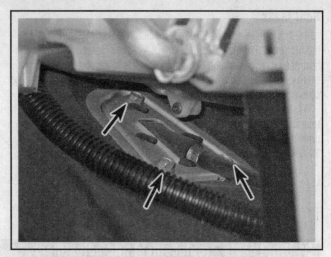

3.12 Remove the bolts from the shift cable bracket at the firewall

11 On Highlander models, working on the side of the center console, remove the two mounting bolts (see illustration).

➡**Note: Lexus models are equipped with a stopper cap that secures the cable to the shift lever assembly. Rotate the stopper cap counterclockwise, lift the cap and slide the cable from the shift lever bracket.**

12 Remove the bolts from the shift cable bracket at the firewall (see illustration).

13 Pull the cable through the floor.

14 Installation is the reverse of removal.

15 Be sure to adjust the cable when you're done (see Steps 16 through 21).

ADJUSTMENT

▶ **Refer to illustration 3.19**

16 When the shift lever inside the vehicle is moved from the NEUTRAL position to other positions, it should move smoothly and accurately to each position and the shift indicator should indicate the correct gear position. If the indicator isn't aligned with the correct position, adjust the shift cable as follows:

17 Remove the battery and the battery tray (see Chapter 5).

18 Disconnect the shift cable from the manual lever on the transaxle.

19 Move the manual lever on the transaxle down into PARK (fully downward), then return it two notches to the NEUTRAL position (see illustration).

3.19 To adjust the shift cable, push the manual lever all the way DOWN, return it two clicks to the Neutral position, place the shift lever inside the vehicle at the Neutral position, then tighten the swivel nut

20 While holding the manual lever in the NEUTRAL position, tighten the shift cable adjustment bolt.

21 Check the operation of the transaxle in each shift lever position (try to start the engine in each gear - the starter should operate in the Park and Neutral positions only).

4   Shift lever - replacement

1 Disconnect the cable from the negative terminal of the battery (see Chapter 5, Section 1).

HIGHLANDER MODELS AND 1999 THROUGH 2003 RX 300 MODELS

▶ **Refer to illustrations 4.5 and 4.6**

2 Remove the center console trim panel (see Chapter 11).

3 Remove the floor carpet and the lower trim panel (see Chapter 11).

4 Disconnect the shift cable from the shift lever at the console (see Section 3).

5 Disconnect the shift lever electrical connectors (see illustration).

6 Remove the mounting bolts and separate the shift lever from the console (see illustration).

7 The remainder of installation is the reverse of removal.

4.5 Disconnect the shift lever electrical connectors from the shift lever assembly at the center console

4.6 Location of the shift lever mounting bolts - 2001 Highlander shown, other models similar

2004 AND LATER RX 330 MODELS

8   Remove the shift lever knob by rotating it counterclockwise.

9   Remove the four clips and lift the console panel from the shift lever assembly.

10   Remove the two clips and lift the trim panel from below the radio.

11   Remove the front door cowl side trim panels and the scuff plates from the right and left door openings (see Chapter 11).

12   Remove the center console trim panel (see Chapter 11).

13   Remove the glove compartment door assembly (see Chapter 11).

14   Remove the floor carpet and the lower trim panel (see Chapter 11).

15   Disconnect the shift cable from the shift lever at the console (see Section 3).

16   Disconnect the shift lever electrical connectors.

17   Remove the mounting bolts and separate the shift lever from the console.

18   The remainder of installation is the reverse of removal.

5   Brake Transmission Shift Interlock (BTSI) system - description, replacement and adjustment

※※ WARNING:

The models covered by this manual are equipped with a Supplemental Restraint System (SRS), more commonly known as airbags. Always disable the airbag system before working in the vicinity of any airbag system component to avoid the possibility of accidental deployment of the airbag(s), which could cause personal injury (see Chapter 12). Do not use a memory saving device to preserve the PCM or radio memory when working on or near airbag system components.

DESCRIPTION

1   The Brake Transmission Shift Interlock (BTSI) system incorporates two solenoid-operated devices; one mounted next to the ignition key lock cylinder and the other mounted on the shift lever assembly under the center console. The solenoid mounted on the shift lever assembly is also equipped with a Shift Lock Module that receives information from the brake pedal switch and the transaxle range (TR) sensor for proper activation of the key lock solenoid (ignition lock cylinder) and the shift lock module (console shift lever). The shift lock module locks the shift lever in the PARK position when the ignition key is in the LOCK position. When the ignition key is in the RUN position, a magnetic holding device is energized. When the system is functioning correctly, the only way to unlock the shift lever and move it out of PARK is to depress the brake pedal. The BTSI system also prevents the ignition key from being turned to the LOCK position unless the shift lever is fully located in the PARK position.

CHECK

2   Verify that the ignition key can be removed only in the PARK position.

3   When the shift lever is in the PARK position, you should be able to rotate the ignition key from OFF to LOCK. But when the shift lever is in any gear position other than PARK (including NEUTRAL), you should not be able to rotate the ignition key to the LOCK position.

4   You should not be able to move the shift lever out of the PARK position when the ignition key is turned to the OFF position.

5   You should not be able to move the shift lever out of the PARK position when the ignition key is turned to the RUN or START position until you depress the brake pedal.

6   You should not be able to move the shift lever out of the PARK position when the ignition key is turned to the ACC or LOCK position.

7   Once in gear, with the ignition key in the RUN position, you should be able to move the shift lever between gears, or put it into NEUTRAL, without depressing the brake pedal.

8   If the BTSI system doesn't operate as described, try adjusting it as follows.

REPLACEMENT

9   Disconnect the cable from the negative terminal of the battery (see Chapter 5, Section 1).

Key lock solenoid

▶ **Refer to illustration 5.12**

10   Remove the steering column trim covers (see Chapter 11).
11   Access the ignition key lock cylinder and disconnect the key lock solenoid electrical connector.
12   Remove the key lock solenoid mounting screws (see illustration) and separate the solenoid from the ignition key lock cylinder.
13   Installation is the reverse of removal.

Shift lock module

Highlander models and 1999 through 2003 RX 300 models

▶ **Refer to illustrations 5.15, 5.16, 5.17, 5.18 and 5.19**

14   Remove the shift lever assembly (see Section 4).
15   Remove the console bulb connector (see illustration).
16   Separate the shift overdrive harness pins from the main connector (see illustration).

5.12 Location of the key lock solenoid mounting screws

17   Remove the two set screws and separate the shift lever knob button from the shaft (see illustration).
18   Remove the indicator cover base mounting screws (see illustration).

5.15 Twist the console bulb holder counterclockwise to remove it from the side of the indicator base

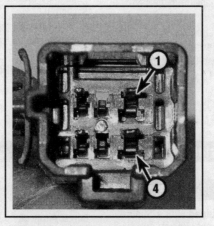

5.16 Use a small jeweler's screwdriver to separate pins 1 and 4 from the backside of the connector and remove the two-wire harness

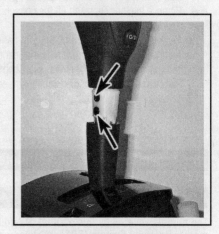

5.17 Remove the two set screws and lift the shift lever knob from the shaft

5.18 Remove the indicator cover base mounting screws

5.19 Location of the shift lock module mounting screws (2001 Highlander shown, other models similar)

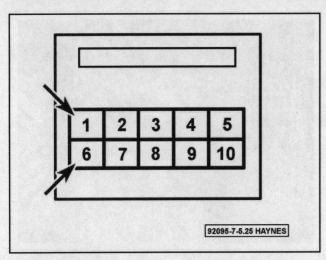

5.25 Use a small jeweler's screwdriver to separate pins 1 and 6 from the backside of the connector and remove the two-wire harness

5.30 Location of the shift lock override button on a 2001 Highlander - other models similar

19 Remove the shift lock module mounting screws (see illustration) and separate the shift lock module from the assembly.

20 Installation is the reverse of removal.

2004 and later RX 330 models

▶ Refer to illustration 5.25

21 Remove the shift lever assembly (see Section 4).

22 Remove the console bulb connector.

23 Remove the position indicator slide cover.

24 Remove the position indicator lower housing.

25 On multi-mode type shift lever systems, separate the harness pins from the main connector (see illustration).

26 Remove the shift lock module mounting screws and separate the shift lock module from the assembly.

27 Installation is the reverse of removal.

Transmission Range (TR) sensor

28 The transmission range sensor incorporates the Park/Neutral function as well as the backup light switch and transmission gear position information to the PCM. The Park/Neutral function of the switch

prevents the engine from starting in any gear other than Park or Neutral. If the engine starts with the shift lever in any position other than Park or Neutral, adjust the switch. The transmission position switch is also an information sensor for the electronic controlled transaxle. When the shift lever is placed in position (Park/Neutral), the transmission range sensor sends a voltage signal to the PCM. Refer to Chapter 6 for the replacement procedure.

SHIFT LOCK OVERRIDE FEATURE

▶ Refer to illustration 5.30

29 In the event the Brake Transmission Shift Interlock (BTSI) system fails and the shift lever cannot be moved out of gear, the system is equipped with an override feature. The BTSI system can be bypassed and the shift lever can be used in manual operation with the parking brake ON and the engine OFF.

30 Locate the override button alongside the shift lever on the center console (see illustration), carefully pry the protective cover off, turn the ignition switch to LOCK position, press the override button and shift the select lever into NEUTRAL.

6   Driveaxle oil seals - replacement

▶ Refer to illustrations 6.4 and 6.6

1 Fluid leaks occasionally occur due to wear of the driveaxle oil seals. Replacement of these seals is relatively easy, since the repairs can be performed without removing the transaxle from the vehicle.

2 The driveaxle oil seals are located in either sides of the transaxle, where the driveaxle shaft is splined into the differential. If leakage at the seal is suspected, raise the vehicle and support it securely on jackstands. If the seal is leaking, fluid will be found on the side of the transaxle.

3 Remove the driveaxle (see Chapter 8). If you're replacing the right side driveaxle seal, remove the intermediate shaft and the driveaxle

assembly as a single unit.

4 Using a screwdriver or prybar, carefully pry the oil seal out of the transaxle bore (see illustration).

5 If the oil seal cannot be removed with a screwdriver or prybar, a special oil seal removal tool (available at auto parts stores) will be required.

6 Using a seal driver or a large deep socket as a drift, install the new oil seal. Drive it into the bore squarely and make sure that it is completely seated (see illustration). Lubricate the lip of the new seal with multi-purpose grease.

7 Install the driveaxle assembly (see Chapter 8). Be careful not to damage the lip of the new seal.

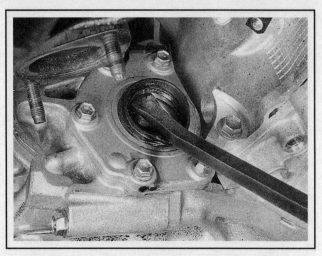

6.4 Carefully pry out the old driveaxle seal with a prybar, screwdriver or a special seal removal tool; make sure you don't gouge or nick the surface of the seal bore

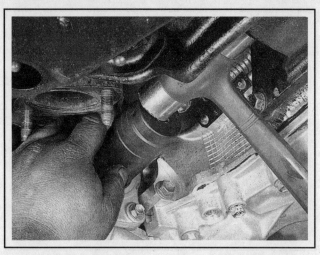

6.6 Drive in the new driveaxle seal with a large socket or a special seal installer

7   Transaxle fluid cooler - removal and installation

▶ Refer to illustrations 7.4 and 7.6

1   Remove the radiator grille (see Chapter 11).
2   Remove the left side inner fender splash shield (see Chapter 11).
3   Remove the front bumper (see Chapter 11).
4   Remove the cooling duct from the transaxle fluid cooler (see illustration).

5   Detach the transaxle fluid cooler lines from the transaxle fluid cooler.
6   Remove the mounting bolts (see illustration) and lift the transaxle fluid cooler from the front fender area.
7   Installation is the reverse of removal.

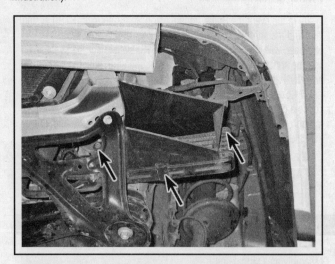

7.4 Remove the cooling duct from the transaxle fluid cooler

7.6 Location of the transaxle fluid cooler mounting bolts

8   Transfer case - removal and installation

1   Remove the engine/transaxle assembly with the transfer case attached (see Chapter 2C).
2   Remove the transverse engine mounting bracket from the transfer case and engine.

3   Remove the transfer case mounting bolts and nuts.
4   Separate the transfer case from the engine/transaxle assembly.
5   Installation is the reverse of removal.

9   Automatic transaxle - removal and installation

➡**Note: The following procedure requires the engine and transaxle to be removed as a unit, then separated once they are out of the vehicle (see Chapter 2, Part C for the engine/transaxle removal procedure).**

REMOVAL

1   Remove the engine/transaxle assembly (see Chapter 2C).
2   On 4WD models, remove the transfer case (see Section 8).
3   Remove the transaxle-to-engine bolts (see illustration 9.7) and separate the transaxle from the engine.
4   Check the engine and transaxle mounts (see Chapter 2A or 2B). If any of these components are worn or damaged, replace them.

INSTALLATION

▶ **Refer to illustration 9.7**

5   If removed, install the torque converter on the transaxle input shaft. Make sure the converter hub splines are properly engaged with the splines on the transaxle input shaft.

6   With an assistant holding the torque converter in place inside the transaxle, mate the transaxle to the engine and install a couple of the transaxle-to-engine bolts. Do not use excessive force to install the transaxle - if something binds and the transaxle won't mate with the engine, alter the angle of the transaxle slightly until it does mate.

❋ CAUTION:

Do NOT use transaxle-to-engine bolts to force the engine and transaxle into alignment. Doing so could crack or damage major components. If you experience difficulties, have an assistant help you line up the dowel pins on the block with the transaxle. Some wiggling of the engine and/or the transaxle will probably be necessary to secure proper alignment of the two.

Turn the converter to align the bolt holes in the converter with the bolt holes in the driveplate. Install the converter-to-driveplate bolts.

➡**Note: Install all of the bolts before tightening any of them.**

7   Install the remaining transaxle-to-engine bolts. Tighten all of the bolts to the torque listed in this Chapter's Specifications (see illustration).
8   On 4WD models, install the transfer case (see Section 8).
9   Install the engine/transaxle assembly (see Chapter 2C).
10 Reinstall the remaining components in the reverse order of removal.
11 Remove all jacks and hoists and lower the vehicle. Tighten the wheel nuts to the torque listed in the Chapter 1 Specifications. Tighten the driveaxle/hub nuts to the torque listed in the Chapter 8 Specifications.
12 Add the specified type and amount of engine oil and transaxle fluid (see Chapter 1).
13 Connect the negative battery cable. Run the engine, cycle the transaxle through the gears, then check the fluid level (see Chapter 1). Check for proper operation and leaks.
14 Road test the vehicle to check for proper transaxle operation and fluid leakage. Recheck the fluid level.

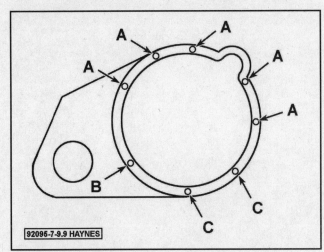

92095-7-9.9 HAYNES

9.7 Transaxle-to-engine bolt locations and identification

10   Automatic transaxle overhaul - general information

In the event of a problem occurring, it will be necessary to establish whether the fault is electrical, mechanical or hydraulic in nature, before repair work can be contemplated. Diagnosis requires detailed knowledge of the transaxle's operation and construction, as well as access to specialized test equipment, and so is deemed to be beyond the scope of this manual. It is therefore essential that problems with the automatic transaxle are referred to a dealer service department or other qualified repair facility for assessment.

Note that a faulty transaxle should not be removed before the vehicle has been diagnosed by a knowledgeable technician equipped with the proper tools, as troubleshooting must be performed with the transaxle installed in the vehicle.

Notes

Section

8

DRIVELINE

1 General information

The information in this Chapter deals with the components from the rear of the engine to the front wheels, except for the transaxle, which is dealt with in the previous Chapter. For the purposes of this Chapter, these components are grouped into two categories - driveshaft and driveaxles. Separate Sections within this Chapter offer general descriptions and checking procedures for components in each of the two groups.

Since nearly all the procedures covered in this Chapter involve working under the vehicle, make sure it's securely supported on sturdy jackstands or on a hoist where the vehicle can be easily raised and lowered.

2 Driveaxles - removal and installation

FRONT

Removal

▶ **Refer to illustrations 2.1, 2.2 and 2.7**

✳✳ WARNING:

If the vehicle is equipped with electronically modulated air suspension, make sure that the height control switch is turned off.

1 Remove the wheel cover or hub cap. Unstake the nut with a punch or chisel (see illustration).

2 Break the hub nut loose with a socket and large breaker bar (see illustration).

3 Loosen the wheel lug nuts, raise the vehicle and support it securely on jackstands. Remove the wheel. Drain the transaxle lubricant (see Chapter 1).

4 Remove the nuts and bolt securing the balljoint to the control arm, then pry the control arm down and separate the lower control arm from the balljoint (see Chapter 10). Now remove the driveaxle/hub nut.

5 Swing the knuckle/hub assembly out (away from the vehicle) until the end of the driveaxle is free of the hub.

➡**Note: If the driveaxle splines stick in the hub, tap on the end of the driveaxle with a plastic hammer. Support the outer end of the driveaxle with a piece of wire to avoid unnecessary strain on the inner CV joint.**

6 If you're working on the right-side axle on a four-cylinder model, remove the bolts from the support bearing bracket. If you're working

2.1 If the driveaxle nut is 'staked,' use a center punch to unstake it (wheel removed for clarity)

on a V6 model, remove the snap-ring from the support bearing of the intermediate shaft using a pair of pliers.

7 Carefully pry the inner end of the driveaxle from the transaxle - or, on models so equipped, the intermediate shaft - using a large screwdriver or prybar positioned between the transaxle or bearing support and the CV joint housing (see illustration). Support the CV joints and carefully remove the driveaxle from the vehicle.

2.2 Loosen the driveaxle/hub nut with a long breaker bar

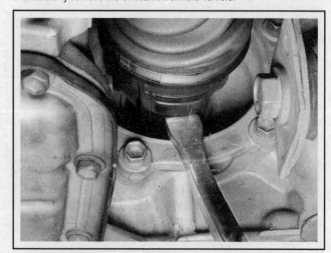

2.7 To separate the inner end of the driveaxle from the transaxle, pry on the CV joint housing like this with a large screwdriver or prybar - you may need to give the prybar a sharp rap with a brass hammer

2.8a Pry the old spring clip from the inner end of the driveaxle with a small screwdriver or awl

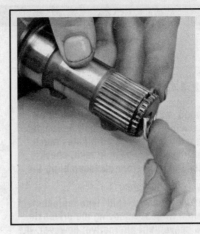

2.8b To install the new spring clip, start one end in the groove and work the clip over the shaft end, into the groove

Installation

▶ **Refer to illustrations 2.8a and 2.8b**

8 Pry the old spring clip from the inner end of the driveaxle and install a new one (see illustrations). Lubricate the differential or inter-mediate shaft seal with multi-purpose grease and raise the driveaxle into position while supporting the CV joints.

9 Insert the splined end of the inner CV joint or the intermediate shaft into the differential side gear and make sure the spring clip locks in its groove. If you're installing a driveaxle/intermediate shaft assem-bly, install the center support bearing bolts or snap-ring, as applicable.

10 Apply a light coat of multi-purpose grease to the outer CV joint splines, pull out on the strut/steering knuckle assembly and install the stub axle into the hub.

11 Reconnect the balljoint to the lower control arm and tighten the nuts (see the torque specifications in Chapter 10).

12 Install a new driveaxle/hub nut. Tighten the hub nut securely, but don't try to tighten it to the actual torque specification until you've low-ered the vehicle to the ground.

13 Grasp the inner CV joint housing (not the driveaxle) and pull out to make sure the driveaxle has seated securely in the transaxle. If the inner CV joint-to-flange bolts were removed, tighten them to the torque listed in this Chapter's Specifications.

14 Install the wheel and lug nuts, then lower the vehicle.

15 Tighten the lug nuts to the torque listed in the Chapter 1 Specifi-cations. Tighten the hub nut to the torque listed in this Chapter's Speci-fications. Stake the nut to the groove in the driveaxle, using a hammer and punch.

16 Refill the transaxle with the recommended type and amount of lubricant (see Chapter 1).

REAR (4WD MODELS)

Removal

✳✳ WARNING:

If the vehicle is equipped with electronically modulated air sus-pension, make sure that the height control switch is turned off.

17 Remove the wheel cover or hub cap. Unstake the nut with a punch or chisel (see illustration 2.1).

18 Break the hub nut loose with a socket and large breaker bar (see illustration 2.2).

19 Loosen the wheel lug nuts, raise the vehicle and support it

securely on jackstands. Remove the wheel.

20 Remove the ABS sensor lead clamps and the parking brake cable bracket from the trailing arm.

21 Remove the bolts that secure the control arms to the rear knuckle (see Chapter 10).

22 Remove the driveaxle nut, then pull the driveaxle assembly out of the rear knuckle. Make sure you don't damage the lip of the inner rear knuckle seal. If the splines on the outer CV joint spindle hang up on the splines in the hub, knock them loose with a hammer and punch.

✳✳ CAUTION:

Don't let the driveaxle hang by the inner CV joint.

23 Carefully pry the inner end of the driveaxle from the differential using a large screwdriver or prybar positioned between the differential and the CV joint housing. Support the CV joints and carefully remove the driveaxle from the vehicle.

Installation

24 Pry the old spring clip from the inner end of the driveaxle and install a new one (see illustrations 2.8a and 2.8b). Lubricate the differ-ential or intermediate shaft seal with multi-purpose grease and raise the driveaxle into position while supporting the CV joints.

25 Insert the splined end of the inner CV joint or the intermediate shaft into the differential and make sure the spring clip locks in its groove.

26 Apply a light coat of multi-purpose grease to the outer CV joint splines, pull out on the rear knuckle assembly and install the stub axle into the hub.

27 Reconnect the control arms to the rear knuckle (see the torque specifications in Chapter 10).

28 Install a new driveaxle/hub nut. Tighten the hub nut securely, but don't try to tighten it to the actual torque specification until you've low-ered the vehicle to the ground.

29 Grasp the inner CV joint housing (not the driveaxle) and pull out to make sure the driveaxle has seated securely in the transaxle.

30 Install the wheel and lug nuts, then lower the vehicle.

31 Tighten the lug nuts to the torque listed in the Chapter 1 Specifi-cations. Tighten the hub nut to the torque listed in this Chapter's Speci-fications. Stake the nut to the groove in the driveaxle, using a hammer and punch.

32 Refill the transaxle with the recommended type and amount of lubricant (see Chapter 1).

3 Driveaxle boot - replacement

DISASSEMBLY

▶ Refer to illustrations 3.3, 3.4, 3.6 and 3.7

➡ Note 1: If the CV joint boots must be replaced, explore all options before beginning the job. Complete rebuilt driveaxles are available on an exchange basis, which eliminates much time and work. Whichever route you choose to take, check on the cost and availability of parts before disassembling the vehicle.

➡ Note 2: Some auto parts stores carry 'split' type replacement boots, which can be installed without removing the driveaxle from the vehicle. This is a convenient alternative; however, the driveaxle should be removed and the CV joint disassembled and cleaned to ensure the joint is free from contaminants such as moisture and dirt which will accelerate CV joint wear. Do NOT disassemble the outboard CV joint.

1   Remove the driveaxle (see Section 2).

2   Mount the driveaxle in a vise with wood lined jaws (to prevent damage to the axleshaft). Check the CV joint for excessive play in the radial direction, which indicates worn parts. Check for smooth operation throughout the full range of motion for each CV joint. If a boot is torn, disassemble the joint, clean the components and inspect for damage due to loss of lubrication and possible contamination by foreign matter.

3   Using a small screwdriver, pry the retaining tabs of the clamps up to loosen them and slide them off (see illustration).

4   Using a screwdriver, carefully pry up on the edge of the outer boot and push it away from the CV joint. Old and worn boots can be cut off. Pull the inner CV joint boot back from the housing and slide the housing off the tripod (see illustration).

5   Mark the tripod and axleshaft to ensure that they are reassembled properly.

6   Remove the tripod joint snap-ring with a pair of snap-ring pliers (see illustration).

7   Use a hammer and a brass punch to drive the tripod joint from

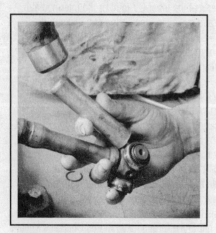

3.3 Lift the tabs on the boot clamps with a small screwdriver, then open the clamps

3.4 Remove the boot from the inner CV joint and slide the joint housing from the tripod

3.6 Remove the snap-ring with a pair of snap-ring pliers

3.7 Drive the tripod joint from the driveaxle with a brass punch and hammer; be careful not to damage the bearing surfaces or the splines on the shaft

3.10a Wrap the splined area of the axleshaft with tape to prevent damage to the boots when removing or installing them

3.10b Install the tripod with the recessed portion of the splines facing the axle shaft

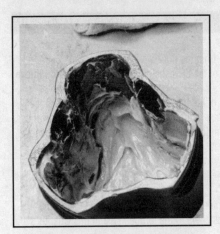

3.10c Place grease at the bottom of the CV joint housing

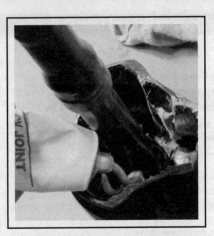

3.10d Install the boot clamps onto the axleshaft, then insert the tripod into the housing, followed by the rest of the grease

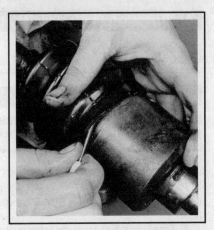

3.12a Equalize the pressure inside the boot by inserting a small, dull screwdriver between the boot and the outer race

the driveaxle (see illustration).

8   If you haven't already cut them off, remove both boots.

➡**Note: Do NOT disassemble the outboard CV joint.**

If you're working on a right side driveaxle, you'll also have to cut off the clamp for the dynamic damper and slide the damper off.

➡**Note: The damper may have to be pressed off with a hydraulic press. Also, before removing the damper, measure its position from the end of the driveaxle - when reassembling, it must be returned to the same spot.**

CHECK

9   Thoroughly clean all components, including the outer CV joint assembly, with solvent until the old CV joint grease is completely removed. Inspect the bearing surfaces of the inner tripods and housings for cracks, pitting, scoring and other signs of wear. It's not possible to inspect the bearing surfaces of the inner and outer races of the outer CV joint, but you can at least check the surfaces of the ball bearings themselves. If they're in good shape, so are the races; if they're not, neither are the races. If the inner CV joint is worn, you can buy a new inner

CV joint and install it on the old axleshaft; if the outer CV joint is worn, you'll have to purchase a new outer CV joint and axleshaft (they're sold pre-assembled).

REASSEMBLY

▶ **Refer to illustrations 3.10a, 3.10b, 3.10c, 3.10d, 3.12a, 3.12b, 3.12c and 3.12d**

10   Wrap the splines on the inner end of the axleshaft with electrical or duct tape to protect the boots from the sharp edges of the splines. Slide the clamps and boot(s) onto the axleshaft, then place the tripod on the shaft. Apply grease to the tripod assembly and inside the housing. Insert the tripod into the housing and pack the remainder of the grease around the tripod (see illustrations).

11   Slide the boot into place, making sure both ends seat in their grooves. Adjust the length of the driveaxle, positioning it midway through its travel.

12   Equalize the pressure in the boot, then tighten and secure the boot clamps (see illustrations).

13   Install the driveaxle assembly (see Section 2).

3.12b To install the new clamps, bend the tang down . . .

3.12c . . . then tap the tabs over to hold it in place

3.12d If your replacement boot came with crimp-type clamps, a special tool such as this one (available at most auto parts stores) will be required to tighten them properly

4 Driveshaft (4WD models) - check, removal and installation

CHECK

1   Raise the rear of the vehicle and support it securely on jackstands. Block the front wheels to keep the vehicle from rolling off the stands. Release the parking brake and place the transmission in Neutral.

2   Crawl under the vehicle and visually inspect the driveshaft. Look for any dents or cracks in the tubing. If any are found, the driveshaft must be replaced.

3   Check for oil leakage at the front and rear of the driveshaft. Leakage where the driveshaft connects to the transfer case indicates a defective transfer case seal. Leakage where the driveshaft connects to the differential indicates a defective pinion seal.

4   While under the vehicle, have an assistant rotate a rear wheel so the driveshaft will rotate. As it does, make sure the universal joints are operating properly without binding, noise or looseness. Listen for any noise from the center bearing, indicating it's worn or damaged. Also check the rubber portion of the center bearing for cracking or separation.

5   The universal joints can also be checked with the driveshaft motionless, by gripping your hands on either side of the joint and attempting to twist the joint. Any movement at all in the joint is a sign of considerable wear. Lifting up on the shaft will also indicate movement in the universal joints. If the joints are worn, front or rear portion of the driveshaft must be replaced as an assembly.

6   Finally, check the driveshaft mounting bolts at the ends to make sure they're tight.

REMOVAL AND INSTALLATION

▶ **Refer to illustrations 4.8 and 4.10**

7   Raise the rear of the vehicle and support it securely on jackstands. Block the front wheels to prevent the vehicle from rolling. Place the transmission in Neutral with the parking brake off.

8   Make reference marks on the driveshaft flange and the differential pinion flange in line with each other (see illustration). This is to make sure the driveshaft is reinstalled in the same position to preserve the balance.

9   Remove the rear universal joint bolts. Turn the driveshaft (or wheels) as necessary to bring the bolts into the most accessible position. Remove the four bolts, nuts and washers.

10   Mark the position of the center support bearings, then unbolt the center support bearings from the floorpan (see illustration).

11   Pull out the front of the driveshaft from the transaxle and remove the driveshaft assembly.

12   Lubricate the lips of the transfer case seal with multi-purpose grease. Carefully guide the intermediate shaft yoke into the transfer case and then install the mounting bolts through the center support bearing, but don't tighten them yet.

13   Reconnect the driveshaft to the pinion flange. Be sure to align the marks and tighten the fasteners to the torque listed in this Chapter's Specifications.

14   Reconnect the center support bearings. Be sure to align the marks and tighten the fasteners to the torque listed in this Chapter's Specifications.

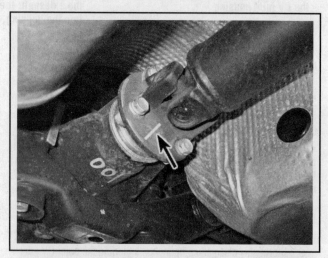

4.8 Mark the relationship of the driveshaft to the differential pinion flange, then using a back-up wrench to hold each bolt, break loose all four bolts

4.10 Remove the fasteners securing the driveshaft center support bearings

5   Differential (4WD models) - removal and installation

▶ Refer to illustration 5.6

※ **WARNING:**

If the vehicle is equipped with electronically modulated air suspension, make sure that the height control switch is turned off.

1   Raise the rear of the vehicle and support it securely on jackstands. Block the front wheels to prevent the vehicle from rolling. Place the transmission in Neutral with the parking brake off.
2   Drain the differential lubricant (see Chapter 1).
3   Remove the driveaxles from the from the differential (see Section 2).
4   Mark the relationship of the driveshaft to the pinion flange, then unbolt the driveshaft from the flange (see Section 4). Suspend the driveshaft with a piece of wire (don't let it hang by the center support bearing).
5   Remove the rear exhaust pipe assembly (see Chapter 4).
6   Support the differential with a floor jack. Remove the mounting bracket and differential mounting bolts (see illustration). Slowly lower the jack and remove the differential out from under the vehicle.

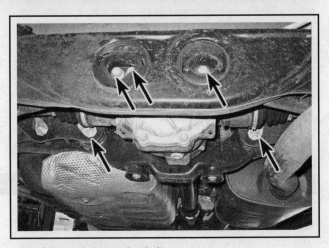

5.6 Differential mounting bolts

7   Installation is the reverse of the removal procedure. Tighten all fasteners to the torque values listed in this Chapter's Specifications. Fill the differential with the proper lubricant (see Chapter 1).

6   Differential oil seals (4WD models) - removal and installation

PINION OIL SEAL

Removal

※ **WARNING:**

If the vehicle is equipped with electronically modulated air suspension, make sure that the height control switch is turned off.

1   Raise the rear of the vehicle and support it securely on jackstands. Block the front wheels to prevent the vehicle from rolling. Place the transmission in Neutral with the parking brake off.
2   Mark the relationship of the driveshaft to the pinion flange, then unbolt the driveshaft from the flange (see Section 4). Suspend the driveshaft with a piece of wire (don't let it hang by the center support bearing).
3   Using a hammer and a punch, unstake the pinion flange nut.
4   A flange holding tool will be required to keep the companion flange from moving while the self-locking pinion nut is loosened. A chain wrench will also work.
5   Remove the pinion nut.
6   Withdraw the flange. It may be necessary to use a two-jaw puller engaged behind the flange to draw it off. Do not attempt to pry or hammer behind the flange or hammer on the end of the pinion shaft.
7   Pry out the old seal and discard it.

Installation

8   Lubricate the lips of the new seal and fill the space between the seal lips with wheel bearing grease, then tap it evenly into position with a seal installation tool or a large socket. Make sure it enters the housing squarely and is tapped in to its full depth.
9   Install the pinion flange; if necessary, tighten the pinion nut to draw the flange into place. Do not try to hammer the flange into position. Tighten the nut to the initial torque listed in this Chapter's Specifications.
10   Using an inch-pound torque wrench (dial or beam-type), measure the torque required to rotate the pinion and tighten the nut in small increments (no more than 108 in-lbs) until it matches the pinion shaft bearing preload listed in this Chapter's Specifications. If the maximum torque listed in this Chapter's Specifications is reached before the specified preload is obtained, the bearing spacer in the differential must be replaced.
11   Once the proper preload is reached, stake the collar of the nut into the slot in the pinion shaft.
12   Reconnect the driveshaft to the pinion flange (see Section 4). Check the differential lubricant level and add some, if necessary, to bring it to the appropriate level (see Chapter 1).

DRIVEAXLE OIL SEALS

※ **WARNING:**

If the vehicle is equipped with electronically modulated air suspension, make sure that the height control switch is turned off.

13   Raise the rear of the vehicle and support it securely on jackstands. Block the front wheels to prevent the vehicle from rolling. Place the transmission in Neutral with the parking brake off.

14 Remove the driveaxles (see Section 2).

15 Carefully pry out the side gear shaft oil seal with a seal removal tool or a large screwdriver; make sure you don't scratch the seal bore.

16 Using a seal installer or a large deep socket as a drift, install the new oil seal. Drive it into the bore squarely and make sure it's completely seated.

17 Lubricate the lip of the new seal with multi-purpose grease, then install the driveaxles (see Section 2). Be careful not to damage the lip of the new seal.

18 Check the differential lubricant level and add some, if necessary, to bring it to the appropriate level (See Chapter 1).

Specifications

General

Differential pinion shaft bearing preload
Rear, 4WD models, with used bearing   5.2 to 7.8 in-lbs (0.6 to 0.9 Nm)

Torque specifications

	Ft-lbs	Nm
CV joint-to-flange bolts	48	65
Differential front mounting bracket bolts (4WD models)		
Differential pinion flange nut (4WD models)		
Initial	80	108
Maximum	174	235
Differential rear mounting bolts-to-subframe		
(4WD models)	70	95
Differential mounting bracket-to-suspension member -		
mounting bolts (4WD models)	101	137
Driveaxle/hub nut	217	294
Driveshaft flange-to-rear differential bolts		
(4WD models)	54	74
Driveshaft center support bearing bolts		
(4WD models)	27	37
Right-side driveaxle support bearing bolt(s)		
Four-cylinder models	47	64
V6 models	24	32.4

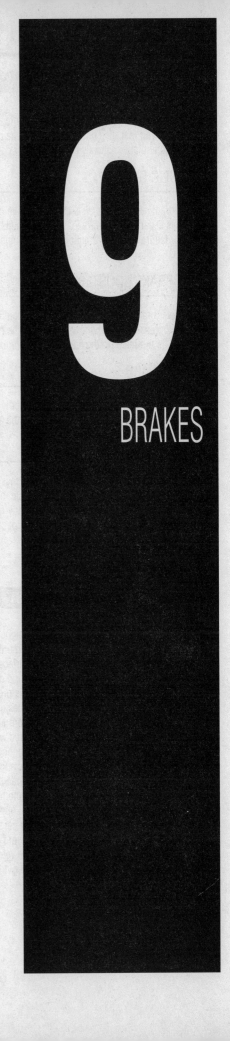

9

BRAKES

Section

1 General information

The vehicles covered by this manual are equipped with hydraulically operated front and rear disc brake systems. These disc type brakes are self-adjusting and automatically compensate for pad wear.

HYDRAULIC SYSTEM

The hydraulic system consists of two separate circuits. The master cylinder has separate reservoirs for the two circuits, and, in the event of a leak or failure in one hydraulic circuit, the other circuit will remain operative.

POWER BRAKE BOOSTER

The power brake booster, utilizing engine manifold vacuum and atmospheric pressure to provide assistance to the hydraulically operated brakes, is mounted on the firewall in the engine compartment.

PARKING BRAKE

The parking brake operates the rear brakes only, through cable actuation. It's activated by a pedal mounted on the left-side kick panel.

SERVICE

After completing any operation involving disassembly of any part of the brake system, always test-drive the vehicle to check for proper braking performance before resuming normal driving. When testing the brakes, perform the tests on a clean, dry, flat surface. Conditions other than these can lead to inaccurate test results.

Test the brakes at various speeds with both light and heavy pedal pressure. The vehicle should stop evenly without pulling to one side or the other. Avoid locking the brakes, because this slides the tires and diminishes braking efficiency and control of the vehicle.

Tires, vehicle load and wheel alignment are factors which also affect braking performance.

PRECAUTIONS

There are some general cautions and warnings involving the brake system on this vehicle:

a) *Use only brake fluid conforming to DOT 3 specifications.*

b) *The brake pads and linings contain fibers which are hazardous to your health if inhaled. Whenever you work on brake system components, clean all parts with brake system cleaner. Do not allow the fine dust to become airborne. Also, wear an approved filtering mask.*

c) *Safety should be paramount whenever any servicing of the brake components is performed. Do not use parts or fasteners which are not in perfect condition, and be sure that all clearances and torque specifications are adhered to. If you are at all unsure about a certain procedure, seek professional advice. Upon completion of any brake system work, test the brakes carefully in a controlled area before putting the vehicle into normal service. If a problem is suspected in the brake system, don't drive the vehicle until it's fixed.*

d) *Clean up any spilled brake fluid immediately and then wash the area with large amounts of water. This is especially true for any finished or painted surfaces.*

2 Anti-lock Brake System (ABS) - general information

1   The Anti-lock Brake System (ABS) is designed to maintain vehicle steerabilty, directional stability and optimum deceleration under severe braking conditions and on most road surfaces. It does so by monitoring the rotational speed of each wheel and controlling the brake line pressure to each wheel during braking. This prevents the wheel from locking up. The ABS system is primarily designed to prevent wheel lockup during heavy braking, but the information provided by the wheel speed sensors of the ABS system is shared with several optional systems that use the data to control vehicle handling. EBD (Electronic Brakeforce Distribution), varies the front-to-rear and side-to-side braking balance under different vehicle loads. The Trac system controls only the front (driving) wheels, and is designed to automatically adjust front wheel speed when starting or accelerating on slippery surfaces. The VSC system (Vehicle Skid Control) affects your car's handling during cornering, using information from the ABS sensors and the yaw-rate sensor (which senses the side-to-side tilt of the vehicle). When the skid-control ECU senses oversteer or understeer, it reduces engine power and selectively applies the brakes.

COMPONENTS

Actuator assembly

▶ **Refer to illustration 2.2**

2   The actuator assembly is mounted in the engine compartment and consists of an electric hydraulic pump and solenoid valves (see illustration).

a) *The electric pump provides hydraulic pressure to charge the reservoirs in the actuator, which supplies pressure to the braking system. The pump and reservoirs are housed in the actuator assembly.*

b) *The solenoid valves modulate brake line pressure during ABS operation.*

Speed sensors

3   These sensors are located at each wheel and generate small elec-

2.2 The ABS actuator assembly (mounted in the right front portion of the engine compartment or beneath the power brake booster, depending on model)

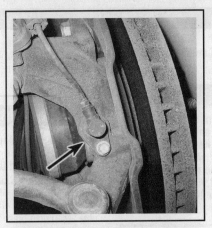

2.12a Front wheel speed sensor (rear wheel speed sensor on 4WD models similar)

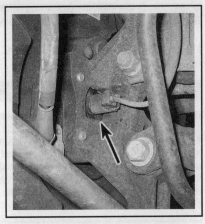

2.12b The rear wheel speed sensor connector on a 2WD model (the sensor itself is incorporated into the rear hub and bearing assembly)

trical pulsations when the toothed sensor rings are turning, sending a signal to the electronic controller indicating wheel rotational speed.

4   The front speed sensors are mounted to the front steering knuckle in close relationship to the toothed sensor rings, which are integral with the front driveaxle outer CV joints.

5   The rear wheel sensors are bolted to the rear suspension knuckles. The sensor rings are integrated with the rear hub assemblies.

ABS computer

6   The ABS computer is mounted with the actuator and is the brain for the ABS system. The function of the computer is to accept and process information received from the wheel speed sensors to control the hydraulic line pressure, avoiding wheel lock up. The computer also constantly monitors the system, even under normal driving conditions, to find faults within the system.

DIAGNOSIS AND REPAIR

7   If a dashboard warning light comes on and stays on while the vehicle is in operation, the ABS system requires attention. Although special electronic ABS diagnostic testing tools are necessary to properly diagnose the system, you can perform a few preliminary checks before taking the vehicle to a dealer service department.

   a) Check the brake fluid level in the reservoir.
   b) Verify that the computer electrical connectors are securely connected.
   c) Check the electrical connectors at the hydraulic control unit.
   d) Check the fuses.
   e) Follow the wiring harness to each wheel and verify that all connections are secure and that the wiring is undamaged.

8   If the above preliminary checks do not rectify the problem, the vehicle should be diagnosed by a dealer service department or other qualified repair shop. Due to the complexity of this system, all actual repair work must be done by a qualified automotive technician.

✳ WARNING:

Do NOT try to repair an ABS wiring harness. The ABS system is sensitive to even the smallest changes in resistance. Repairing

the harness could alter resistance values and cause the system to malfunction. If the ABS wiring harness is damaged in any way, it must be replaced.

✳ CAUTION:

Make sure the ignition is turned off before unplugging or reattaching any electrical connections.

WHEEL SPEED SENSOR - REMOVAL AND INSTALLATION

▶ **Refer to illustrations 2.12a and 2.12b**

9   Loosen the wheel lug nuts, raise the vehicle and support it securely on jackstands. Remove the wheel.

✳ WARNING:

If the vehicle is equipped with an electronically modulated air suspension, make sure that the height control switch is turned off before raising the vehicle.

10 Make sure the ignition key is turned to the Off position.

11 Trace the wiring back from the sensor, detaching all brackets and clips while noting its correct routing, then disconnect the electrical connector.

12 For front wheel speed sensors on all models and rear wheel speed sensors on 4WD models, remove the mounting bolt and carefully pull the sensor out from the knuckle (see illustration). The rear wheel speed sensor on 2WD models is integrated with the rear hub and cannot be removed without removing the hub (see illustration). See Chapter 10 for the rear hub removal procedure.

13 Installation is the reverse of the removal procedure. Tighten the mounting fastener securely.

14 Install the wheel and lug nuts, tightening them securely. Lower the vehicle and tighten the lug nuts to the torque listed in the Chapter 1 Specifications.

3 Disc brake pads - replacement

▶ Refer to illustrations 3.5, 3.6a through 3.6p and 3.7a through 3.7n

✳ WARNING:

Disc brake pads must be replaced on both front or rear wheels at the same time - never replace the pads on only one wheel. Also, the dust created by the brake system is harmful to your health. Never blow it out with compressed air and don't inhale any of it. An approved filtering mask should be worn when working on the brakes. Do not, under any circumstances, use petroleum-based solvents to clean brake parts. Use brake system cleaner only!

➡Note 1: This procedure applies to both the front and rear disc brakes.

➡Note 2: The manufacturer recommends replacing the pad shims and wear indicators whenever the pads are replaced.

1 Remove the cap from the brake fluid reservoir.

2 Loosen the wheel lug nuts, raise the front or rear of the vehicle and support it securely on jackstands. Block the wheels at the opposite end.

✳ WARNING:

If the vehicle is equipped with an electronically modulated air suspension, make sure that the height control switch is turned off before raising the vehicle.

3 Remove the wheels. Work on one brake assembly at a time, using the assembled brake for reference if necessary.

4 Inspect the brake disc carefully as outlined in Section 5. If machining is necessary, follow the information in that Section to remove the disc, at which time the pads can be removed as well.

5 Push the piston back into its bore to provide room for the new brake pads. A C-clamp can be used to accomplish this (see illustration). As the piston is depressed to the bottom of the caliper bore, the fluid in the master cylinder will rise. Make sure that it doesn't overflow.

3.5 Before removing the caliper, slowly depress the piston in the caliper bore by using a large C-clamp between the outer brake pad and the back of the caliper

3.6a Always wash the brakes with brake cleaner before disassembling anything

3.6b Remove the lower caliper bolt. To detach the caliper completely, remove both caliper bolts but do not let the caliper hang by the brake hose

3.6c Pivot the caliper up (be careful not to damage the boot for the upper slide pin). . .

3.6d . . . and secure it to the strut bracket with a piece of wire; do not allow the caliper to hang by the flexible brake hose

3.6e Remove inner pad . . .

3.6f . . . and the outer pad

If necessary, siphon off some of the fluid.

6 If you're replacing the front brake pads, follow the accompanying photos, beginning with illustration 3.6a. Be sure to stay in order and

read the caption under each illustration.

7 If you're replacing the rear brake pads, wash the brake assembly (see illustration 3.6a), then follow the accompanying photos beginning

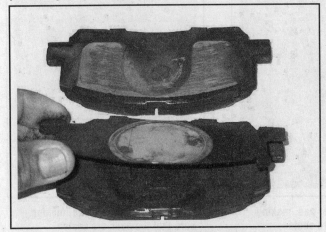

3.6g Remove the shim(s) from each brake pad, noting the order and position (some pads may only have one shim). The manufacturer recommends shim replacement when the pads are replaced

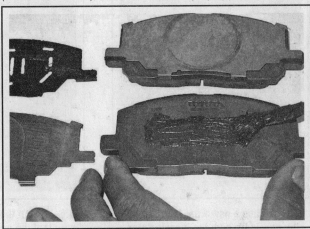

3.6h Apply a thin film of disc brake grease in between the shims and the back of the brake pads

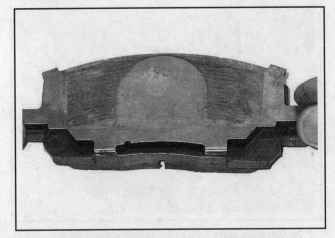

3.6i Place the shims on the pads in the correct order and position as the original ones. If necessary, refer to the installed pads on the other side of the vehicle if there is any question about shim placement

3.6j Remove the upper and lower pad support plates; make sure they are a tight fit and aren't worn. If necessary, replace them

3.6k Install the upper and lower pad support plates

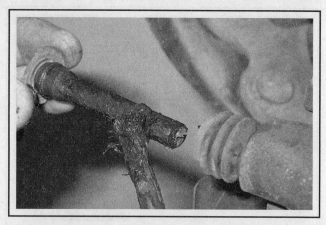

3.6l Pull out the upper and lower sliding pins and clean them. Apply a coat of high-temperature grease to the pins and reinstall them. Be careful not to damage the pin boots, replace any boots that are worn or damaged

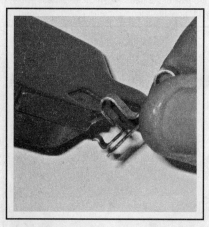

3.6m Place a new wear indicator on the new inner pad; the manufacturer recommends wear indicator replacement when the pads are replaced

3.6n Install the inner pad, making sure that the ends are seated correctly on the pad support plates

3.6o Install the outer pad, making sure that the ends are seated correctly on the pad support plates

3.6p Install the caliper and tighten the caliper bolts to the torque listed in this Chapter's Specifications

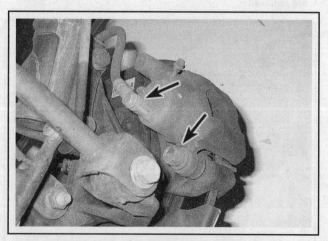

3.7a Wash the brake with brake system cleaner (see illustration 3.6a), then remove the caliper retaining bolt (lower arrow); the upper arrow points to the brake hose banjo fitting bolt, which shouldn't be unscrewed unless the caliper is being removed from the vehicle

3.7b Pivot the caliper up and support it in that position. Secure it to the strut coil spring with a piece of wire and do not let it hang by the brake hose.

3.7c Remove the inner brake pad . . .

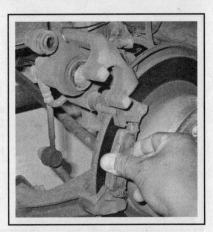

3.7d . . . and the outer brake pad

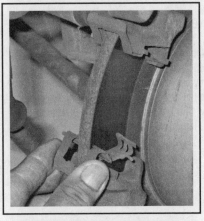

3.7e Remove the upper and lower pad support plates; make sure they are a tight fit and aren't worn. If necessary, replace them

3.7f Remove the shim(s) and pad wear indicator from each brake pad, noting the order and position (some pads may only have one shim). The manufacturer recommends shim and indicator replacement when the pads are replaced

3.7g Replace the wear indicator on each brake pad

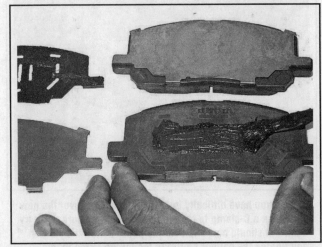

3.7h Apply a thin film of disc brake grease between the shims and the back of the brake pads

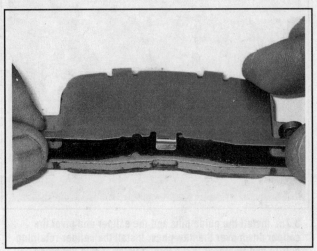

3.7i Replace the shim(s) on each pad in the original position as the old ones. If necessary, refer to the installed pads on the other side of the vehicle if there is any question about shim placement

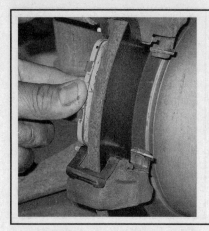

3.7j Install the pad support plates onto the caliper bracket, then install the inner brake pad

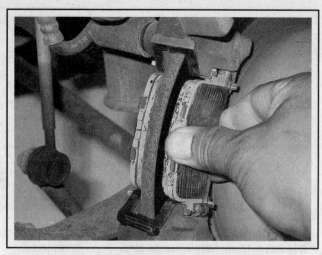

3.7k Install the outer brake pad

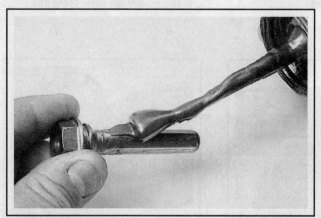

3.7l Pull out the upper and lower sliding pins and clean them. Apply a coat of high-temperature grease to the pins and then reinstall them. Be careful not to damage the pin boots (replace any boots that are worn or damaged). On some models, the pin will be exposed by simply removing the caliper from the bracket. Be careful not to damage the pin boot when removing the caliper

with illustration 3.7a. Be sure to stay in order and read the caption under each illustration.

8   When reinstalling the caliper, be sure to tighten the mounting bolts to the torque listed in this Chapter's Specifications. After the job has been completed, firmly depress the brake pedal a few times to bring the pads into contact with the disc. Check the level of the brake fluid, adding some if necessary. Check the operation of the brakes carefully before placing the vehicle into normal service.

3.7m Install the guide pins and the caliper and pivot the caliper down over the new pads. Install the caliper-retaining bolt(s) and tighten it to the torque listed in this Chapter's Specifications

3.6n If you have difficulty installing the caliper over the new pads, use a C-clamp to bottom the piston in its bore, then try again - it should now slip over the pads. Install the caliper and tighten the caliper bolt(s) to the torque listed in this Chapter's Specifications

4 Disc brake caliper - removal and installation

✷✷ WARNING:

Dust created by the brake system is harmful to your health. Never blow it out with compressed air and don't inhale any of it. An approved filtering mask should be worn when working on the brakes. Do not, under any circumstances, use petroleum-based solvents to clean brake parts. Use brake system cleaner only!

➡Note: Always replace the calipers in pairs - never replace just one of them.

REMOVAL

▶ Refer to illustrations 4.2 and 4.3

1 Loosen the wheel lug nuts, raise the vehicle and support it securely on jackstands. Remove the wheels.

✷✷ WARNING:

If the vehicle is equipped with an electronically modulated air suspension, make sure that the height control switch is turned off before raising the vehicle.

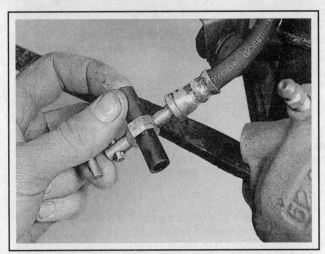

4.2 Using a piece of rubber hose of the appropriate size, plug the brake line banjo fitting to prevent brake fluid from leaking out and to prevent dirt and moisture from contaminating the fluid in the hose

2 Remove the brake hose banjo bolt and disconnect the hose from the caliper. Plug the hose to keep contaminants out of the brake system and to prevent losing any more brake fluid than is necessary (see illustration).

➡Note: If you're just removing the caliper for access to other components, don't detach the hose.

3 Remove the caliper mounting bolts (see illustration).

4 Remove the caliper. If necessary, remove the caliper bracket from the steering knuckle or rear knuckle (see illustrations 5.2a and 5.2b).

INSTALLATION

5 Install the caliper by reversing the removal procedure. Tighten the caliper mounting bolts (and bracket bolts, if removed) to the torque listed in this Chapter's Specifications. Install new sealing washers on either side of the brake hose banjo fitting, then tighten the banjo bolt to the torque listed in this Chapter's Specifications.

6 Bleed the brake system (see Section 8).

7 Install the wheels and lug nuts. Lower the vehicle and tighten the lug nuts to the torque listed in the Chapter 1 Specifications.

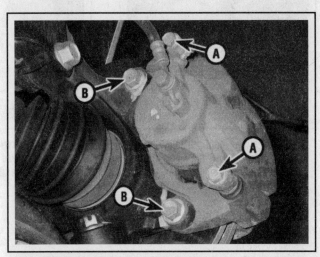

4.3 Front brake caliper mounting details (rear caliper similar)

A   Caliper mounting bolts
B   Caliper bracket mounting bolts

5 Brake disc - inspection, removal and installation

INSPECTION

▶ Refer to illustrations 5.2a, 5.2b, 5.3, 5.4a, 5.4b, 5.5a and 5.5b

1 Loosen the wheel lug nuts, raise the vehicle and support it securely on jackstands. Remove the wheel and install the lug nuts to hold the disc in place. If the rear brake disc is being worked on, release the parking brake.

✷✷ WARNING:

If the vehicle is equipped with an electronically modulated air suspension, make sure that the height control switch is turned off before raising the vehicle.

2 Remove the brake caliper as outlined in Section 4. It isn't necessary to disconnect the brake hose. After removing the caliper bolts, suspend the caliper out of the way with a piece of wire (see illustra-

5.2a To remove the front caliper bracket from the steering knuckle, remove these two bolts

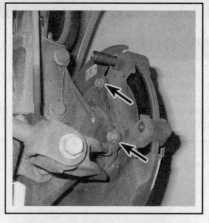

5.2b To remove the rear caliper bracket from the knuckle, remove these two bolts

5.3 The brake pads on this vehicle were obviously neglected, as they wore down completely and cut deep grooves into the disc - wear this severe means the disc must be replaced

5.4a Use a dial indicator to check disc runout; if the reading exceeds the maximum allowable runout limit, the disc will have to be machined or replaced

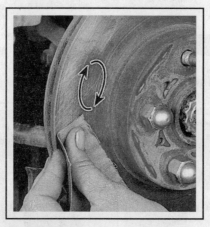

5.4b Using a swirling motion, remove the glaze from the disc surface with sandpaper or emery cloth

5.5a The minimum wear dimension is cast into the back side of the disc (typical)

tion 3.6d). Remove the caliper bracket mounting bolts and then remove the bracket (see illustrations).

3   Visually inspect the disc surface for score marks and other damage. Light scratches and shallow grooves are normal after use and may not always be detrimental to brake operation, but deep scoring - over 0.039-inch (1.0 mm) - requires disc removal and refinishing by an automotive machine shop. Be sure to check both sides of the disc (see illustration). If pulsating has been noticed during application of the brakes, suspect disc runout.

4   To check disc runout, place a dial indicator at a point about 1/2-inch from the outer edge of the disc (see illustration). Set the indicator to zero and turn the disc. The indicator reading should not exceed the specified allowable runout limit. If it does, the disc should be refinished by an automotive machine shop.

➡Note: Professionals recommend resurfacing the discs whenever the pads are replaced regardless of the dial indicator reading, as this will impart a smooth finish and ensure a perfectly flat surface, eliminating any brake pedal pulsation or other undesirable symptoms. At the very least, if you elect not to have the discs resurfaced, remove the glaze from the surface with sandpaper or emery cloth using a swirling motion (see illustration).

5   It's absolutely critical that the disc not be machined to a thickness under the specified minimum allowable refinish thickness. The minimum wear (or discard) thickness is cast into the inside of the disc (see illustration). The disc thickness can be checked with a micrometer (see illustration).

REMOVAL

▶ Refer to illustrations 5.6a and 5.6b

6   Remove the lug nuts that were put on to hold the disc in place and slide the disc off the hub. If the rear disc won't come off, it may be interfering with the parking brake shoes; remove the plug (see illustration) and rotate the adjuster to back the parking brake shoes away from the drum surface within the disc (see illustration).

INSTALLATION

7   Place the disc in position over the threaded studs.
8   Install the caliper bracket, tightening the bolts to the torque listed in this Chapter's Specifications.
9   Install the caliper, tightening the bolts to the torque listed in this

5.5b Use a micrometer to measure disc thickness

5.6a If the rear disc is difficult to remove, remove this plug . . .

5.6b . . . insert a screwdriver through the hole (the hole must be at the 6 o'clock position, because that's where the adjuster is located) and rotate the adjuster to back the parking brake shoes away from the drum surface in the disc

Chapter's Specifications. Bleeding won't be necessary unless the brake hose was disconnected from the caliper.

10 If you're installing a rear disc, adjust the parking brake shoes as described in Section 10.

11 Install the wheel and lug nuts. Lower the vehicle and tighten the lug nuts to the torque listed in the Chapter 1 Specifications. Depress the brake pedal a few times to bring the brake pads into contact with the disc. Check the operation of the brakes carefully before driving the vehicle.

6 Master cylinder - removal and installation

REMOVAL

▶ **Refer to illustration 6.5**

✳✳ CAUTION:

Brake fluid will damage paint or finished surfaces. Cover all body parts and be careful not to spill fluid during this procedure. Clean up any spilled brake fluid immediately and then wash the area with large amounts of water.

1 Remove top engine cover, if equipped (see Chapter 2).

2 Remove the strut brace from between the strut towers, if equipped (see Chapter 10). On 2004 and later Lexus models, remove the cowl tray from the engine compartment (see Chapter 11).

3 Remove the air filter housing (see Chapter 4).

➡**Note: On late model vehicles, remove the air duct between the throttle body and the air filter housing for access if necessary (see Chapter 4).**

4 Remove as much fluid as possible from the reservoir with a syringe.

➡**Note: On late model Lexus vehicles, the reservoir is not directly attached to the master cylinder. Drain the reservoir and detach any hoses and then remove it for access.**

5 Unplug the electrical connector for the brake fluid level warning switch (see illustration).

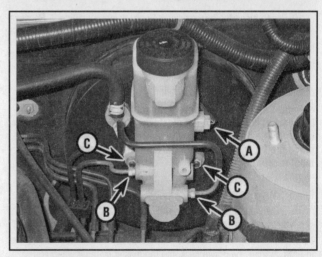

6.5 Master cylinder installation details

A	Electrical connector	C	Mounting nuts
B	Brake line fittings		

6 Place rags under the fittings and prepare caps or plastic bags to cover the ends of the lines once they're disconnected.

7 Loosen the fittings at the ends of the brake lines where they enter the master cylinder. To prevent rounding off the flats, use a flare-nut wrench, which wraps around the fitting hex (see illustration 6.5).

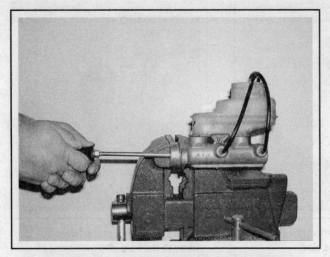

6.11 The best way to bleed air from the master cylinder before installing it on the vehicle is with a pair of bleed tubes

6.19 Loosen the fittings on the master cylinder to bleed it when it is installed on the vehicle (typical shown)

8 Carefully move the brake lines away from the master cylinder and plug the ends to prevent contamination.

9 Remove the nuts attaching the master cylinder to the power booster. Pull the master cylinder off the studs to remove it. Again, be careful not to spill fluid or bend the brake lines as this is done (see illustration 6.5).

→**Note: If necessary, remove brake lines going to the ABS actuator from the master cylinder if they cannot be moved aside without damaging them.**

INSTALLATION

▶ **Refer to illustrations 6.11 and 6.19**

10 Bench bleed the new master cylinder before installing it. Because it will be necessary to depress the master cylinder piston and, at the same time, control flow from the brake line outlets, it is recommended that the master cylinder be mounted in a vise.

11 Attach a pair of master cylinder bleeder tubes to the outlet ports of the master cylinder (see illustration).

→**Note: On models that have a detached reservoir, feed the bleeder tubes back into the master cylinder where the reservoir hoses connect. On models with an auxiliary type reservoir, place the bleeder tubes into the small tank mounted on the master cylinder.**

12 Fill the reservoir with brake fluid of the recommended type (see Chapter 1).

13 Slowly push the pistons into the master cylinder (a large Phillips screwdriver can be used for this) - air will be expelled from the pressure chambers and into the reservoir. Because the tubes are submerged in fluid, air can't be drawn back into the master cylinder when you release the pistons.

14 Repeat the procedure until no more air bubbles are present.

15 Remove the bleed tubes, one at a time, and install plugs in the open ports to prevent fluid leakage and air from entering. If installing a new master cylinder, adjust the booster pushrod length (see Section 9).

16 Install the reservoir cover, then install the master cylinder over the studs on the power brake booster and tighten the attaching nuts only finger tight at this time.

17 Thread the brake line fittings into the master cylinder. Since the master cylinder is still a bit loose, it can be moved slightly in order for the fittings to thread in easily. Do not strip the threads as the fittings are tightened.

18 Tighten the mounting nuts to the torque listed in this Chapter's Specification and then the brake line fittings securely.

19 Fill the master cylinder reservoir with fluid, then bleed the master cylinder and the brake system as described in Section 8. To bleed the cylinder on the vehicle, have an assistant pump the brake pedal several times slowly and then hold the pedal to the floor. Loosen the fitting nut to allow air and fluid to escape. Repeat this procedure on both fittings until the fluid is clear of air bubbles (see illustration).

⁂ **CAUTION:**

Have plenty of rags on hand to catch the fluid - brake fluid will ruin painted surfaces.

20 If it was necessary to remove the master cylinder-to-ABS actuator brake lines, also bleed the lines at the ABS actuator.

21 The remainder of installation is the reverse of removal. Test the operation of the brake system carefully before placing the vehicle into normal service.

⁂ **WARNING:**

Do not operate the vehicle if you are in doubt about the effectiveness of the brake system. On models equipped with ABS, it is possible for air to become trapped in the anti-lock brake system hydraulic control unit, so, if the pedal continues to feel spongy after repeated bleedings or the BRAKE or ANTI-LOCK light stays on, have the vehicle towed to a dealer service department or other qualified shop to be bled with the aid of a scan tool.

7 Brake hoses and lines - inspection and replacement

INSPECTION

1 About every six months, with the vehicle raised and supported securely on jackstands, the rubber hoses which connect the steel brake lines with the front and rear brake assemblies should be inspected for cracks, chafing of the outer cover, leaks, blisters and other damage. These are important and vulnerable parts of the brake system and inspection should be complete. A light and mirror will be helpful for a thorough check. If a hose exhibits any of the above conditions, replace it with a new one.

REPLACEMENT

Front brake hose

▶ **Refer to illustrations 7.3 and 7.4**

2 Loosen the wheel lug nuts, raise the vehicle and support it securely on jackstands. Remove the wheel.

❇❇ WARNING:

If the vehicle is equipped with an electronically modulated air suspension, make sure that the height control switch is turned off before raising the vehicle.

3 At the frame bracket, unscrew the brake line fitting from the hose (see illustration). Use a flare-nut wrench to prevent rounding off the corners.

4 Remove the U-clip from the female fitting at the bracket with a pair of pliers (see illustration), then pass the hose through the bracket.

5 At the caliper end of the hose, remove the banjo fitting bolt, then separate the hose from the caliper. Note that there are two copper sealing washers on either side of the fitting - they should be replaced with new ones during installation.

6 Remove the fastener from the strut bracket and then detach the hose from it.

7 To install the hose, pass the caliper fitting end through the strut bracket, then connect the fitting to the caliper with the banjo bolt and new sealing washers. Make sure the locating lug on the fitting is engaged with the hole in the caliper, then tighten the bolt to the torque listed in this Chapter's Specifications.

8 Push the metal support into the strut bracket and install the U-clip. Make sure the hose isn't twisted between the caliper and the strut bracket.

9 Route the hose into the frame bracket, again making sure it isn't twisted, then connect the brake line fitting, starting the threads by hand. Install the clip and E-ring, if equipped, then tighten the fitting securely.

10 Bleed the caliper (see Section 8).

11 Install the wheel and lug nuts, lower the vehicle and tighten the lug nuts to the torque listed in the Chapter 1 Specifications.

Rear brake hose

12 Perform Steps 2, 3 and 4 above, then repeat Steps 3 and 4 at the other end of the hose. Be sure to bleed the caliper (see Section 8).

Metal brake lines

13 When replacing brake lines, be sure to use the correct parts. Don't use copper tubing for any brake system components. Purchase genuine steel brake lines from a dealer or auto parts store.

14 Prefabricated brake line, with the tube ends already flared and fittings installed, is available at auto parts stores and dealer parts departments.

15 When installing the new line, make sure it's securely supported in the brackets and has plenty of clearance between moving or hot components.

16 After installation, check the master cylinder fluid level and add fluid as necessary. Bleed the brake system (see Section 8) and test the brakes carefully before driving the vehicle in traffic.

7.3 Unscrew the brake line threaded fitting with a flare-nut wrench to protect the fitting corners from being rounded off

7.4 Remove the brake hose-to-brake line U-clip with a pair of pliers

8 Brake hydraulic system - bleeding

▶ Refer to illustration 8.8

➡Note: Bleeding the hydraulic system is necessary to remove any air that manages to find its way into the system when it's been opened during removal and installation of a hose, line, caliper or master cylinder.

1   You'll probably have to bleed the system at all four brakes if air has entered it due to low fluid level, or if the brake lines have been disconnected at the master cylinder.

2   If a brake line was disconnected only at a wheel, then only that caliper must be bled.

3   If a brake line is disconnected at a fitting located between the master cylinder and any of the brakes, that part of the system served by the disconnected line must be bled.

4   Remove any residual vacuum from the brake power booster by applying the brake several times with the engine off.

5   Remove the master cylinder reservoir cover and fill the reservoir with brake fluid. Reinstall the cover.

➡Note: Check the fluid level often during the bleeding operation and add fluid as necessary to prevent the fluid level from falling low enough to allow air bubbles into the master cylinder.

6   Have an assistant on hand, as well as a supply of new brake fluid, a clear container partially filled with clean brake fluid, a length of tubing to fit over the bleeder valve and a wrench to open and close the bleeder valve.

7   Beginning at the right rear wheel, loosen the bleeder valve slightly, then tighten it to a point where it's snug but can still be loosened quickly and easily.

8   Place one end of the tubing over the bleeder valve and submerge the other end in brake fluid in the container (see illustration).

9   Have the assistant depress the brake pedal slowly, then hold the pedal down firmly.

10 While the pedal is held down, open the bleeder valve just enough to allow a flow of fluid to leave the valve. Watch for air bubbles to exit the submerged end of the tube. When the fluid flow slows after a couple of seconds, close the valve and have your assistant release the pedal.

8.8 When bleeding the brakes, a hose is connected to the bleeder valve at the caliper or wheel cylinder and then submerged in brake fluid. Air will be seen as bubbles in the tube and container. All air must be expelled before moving to the next wheel

11 Repeat Steps 9 and 10 until no more air is seen leaving the tube, then tighten the bleeder valve and proceed to the left rear wheel, the right front wheel and the left front wheel, in that order, and perform the same procedure. Be sure to check the fluid in the master cylinder reservoir frequently.

12 Never use old brake fluid. It contains moisture that will deteriorate the brake system components.

13 Refill the master cylinder with fluid at the end of the operation.

14 Check the operation of the brakes. The pedal should feel solid when depressed, with no sponginess. If necessary, repeat the entire process.

9 Power brake booster - check, removal and installation

OPERATING CHECK

1   Depress the brake pedal several times with the engine off and make sure there's no change in the pedal reserve distance.

2   Depress the pedal and start the engine. If the pedal goes down slightly, operation is normal.

AIRTIGHTNESS CHECK

3   Start the engine and turn it off after one or two minutes. Depress the brake pedal slowly several times. If the pedal depresses less each time, the booster is airtight.

4   Depress the brake pedal while the engine is running, then stop

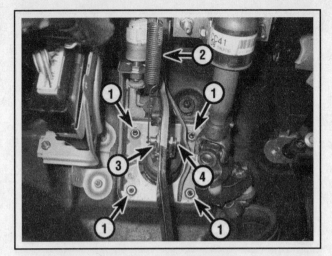

9.11 Power brake booster mounting details:

1 Mounting fasteners	*3 Clevis retaining clip*
2 Return spring	*4 Clevis pin*

9.14a Measure the distance that the pushrod protrudes from the brake booster at the master cylinder mounting surface (including the gasket)

the engine with the pedal depressed. If there's no change in the pedal reserve travel after holding the pedal for 30 seconds, the booster is airtight.

REMOVAL

▶ **Refer to illustration 9.11**

➡**Note: On 2003 and earlier Highlander models with ABS and Traction Control, power brake booster removal and installation requires special tools and procedures. The ABS and Traction Control actuator assembly must be removed and then pressure bled with special tools during installation. It is recommended that removal and installation of the power brake booster be referred to a dealership or qualified repair facility.**

5   Power brake booster units shouldn't be disassembled. They require special tools not normally found in most automotive repair stations or shops. They're fairly complex and, because of their critical relationship to brake performance, should be replaced with a new or rebuilt one.

6   On 2004 and later Lexus models, remove the cowl tray from the engine compartment (see Chapter 11).

7   Remove the air filter housing (see Chapter 4).

8   Remove the brake master cylinder (see Section 6). To provide room for booster removal, some models may require that the brake lines that cross in front of the master cylinder and be removed entirely (at both ends) and not just disconnected from the master cylinder. Other brake lines that cross in front of the brake booster can be separated from the firewall and carefully moved aside.

9   Carefully disconnect the vacuum hose from the brake booster.

10 Inside the vehicle, remove the knee bolster and the brace behind it (see Chapter 11).

11 Remove the brake pedal return spring near the top of the brake pedal. Remove the clevis pin-retaining clip with pliers and then pull out the pin (see illustration).

12 Remove the four fasteners holding the brake booster to the firewall and the brake pedal return spring. Then slide the booster straight out from the firewall until the studs clear the holes (see illustration 9.11).

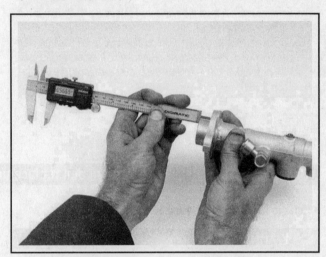

9.14b Measure the distance from the mounting flange to the end of the master cylinder

INSTALLATION

▶ **Refer to illustrations 9.14a, 9.14b, 9.14c and 9.10d**

13 Installation procedures are basically the reverse of removal. Tighten the booster mounting nuts to the torque listed in this Chapter's Specifications.

14 If a new power brake booster unit is being installed, check the pushrod clearance (see illustration) as follows:

a) *Measure the distance that the pushrod protrudes from the master cylinder mounting surface on the front of the power brake booster, including the gasket (if equipped) (see illustration). Write down this measurement. This is dimension A.*

b) *Measure the distance from the mounting flange to the end of the master cylinder (see illustration). Write down this measurement. This is dimension B.*

c) *Measure the distance from the end of the master cylinder to the bottom of the pocket in the piston (see illustration). Write down this measurement. This is dimension C.*

d) *Subtract measurement B from measurement C, then subtract*

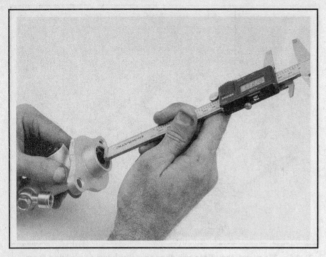

9.14c Measure the distance from the piston pocket to the end of the master cylinder

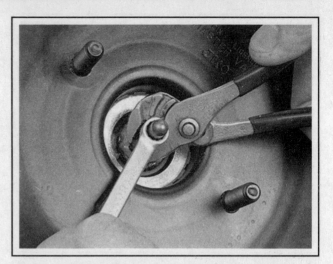

9.14d To adjust the length of the booster pushrod, hold the serrated portion of the rod with a pair of pliers and turn the adjusting screw in or out, as necessary, to achieve the desired setting

measurement A from the difference between B and C. This is the pushrod clearance.

e) Compare your calculated pushrod clearance to the pushrod clearance listed in this Chapter's Specifications. If necessary, adjust the pushrod length to achieve the correct clearance (see illustration 9.10d).

15 After the final installation of the master cylinder and brake hoses and lines, the brake pedal height and freeplay must be adjusted and the system must be bled. See the appropriate Sections of this Chapter for those procedures.

10 Parking brake shoes - inspection and replacement

▶ Refer to illustrations 10.4 and 10.5a through 10.5u

✳✳ WARNING 1:

Dust created by the brake system is hazardous to your health. Never blow it out with compressed air and don't inhale any of it. An approved filtering mask should be worn when working on the brakes. Do not, under any circumstances, use petroleum-based solvents to clean brake parts. Use brake system cleaner only!

✳✳ WARNING 2:

Parking brake shoes must be replaced on both wheels at the same time - never replace the shoes on only one wheel.

1 Remove the brake disc (see Section 5).
2 Inspect the thickness of the lining material on the shoes. If the lining has worn down to 1/32-inch or less, the shoes must be replaced.
3 Remove the hub and bearing assembly (see Chapter 10).

10.4 Before disassembling it, be sure to wash the parking brake assembly with brake cleaner

10.5a Remove the rear parking brake shoe return spring from the anchor pin . . .

10.5b . . . and unhook it from the rear shoe

10.5c Remove the front parking brake shoe return spring from the anchor pin . . .

10.5d . . . and unhook it from the front shoe

10.5e Remove the rear shoe hold-down spring and pull out the pin

10.5f Remove the shoe strut from between the shoes

10.5g Remove the front shoe hold-down spring and pull out the pin

10.5h Remove the adjuster and the tension spring (the tension spring, which is not visible in this photo, is behind the adjuster and is attached to both shoes)

➡**Note: It is possible to perform the shoe replacement procedure without removing the hub and bearing assembly, although working room is limited.**

4   Wash off the brake parts with brake system cleaner (see illustration).

5   Follow the accompanying illustrations for the brake shoe replacement procedure (see illustrations 10.5a through 10.5u). Be sure to stay in order and read the caption under each illustration.

6   Install the brake disc. Temporarily thread three of the wheel lug nuts onto the studs to hold the disc in place.

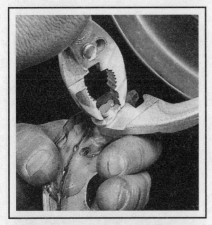

10.5i Pop the C-washer off the pivot pin on the back of the rear shoe . . .

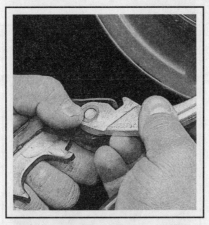

10.5j . . . and pull the parking brake lever off the pivot pin

10.5k Apply a thin coat of high-temperature grease to the contact surfaces of the backing plate

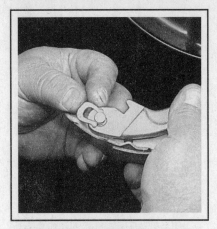

10.5l Slide the parking brake lever onto the pivot pin and install a new C-washer

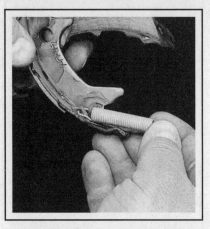

10.5m Attach the tension spring to the back side of the rear shoe . . .

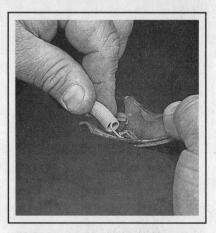

10.5n . . . and to the back side of the front shoe

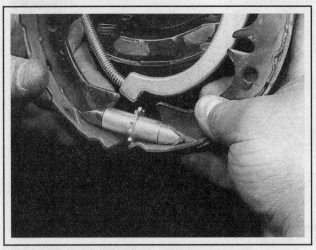

10.5o Flip the shoes around and install the adjuster; make sure both ends of the adjuster are properly engaged with the shoes as shown

10.5p Place the shoes in position and install the strut and spring as shown; make sure the ends of the strut are properly engaged with the shoes as shown

7   Remove the hole plug from the brake disc. Adjust the parking brake shoe clearance by turning the adjuster star wheel with a brake adjusting tool or screwdriver until the shoes contact the disc and the disc can't be turned (see illustrations 5.6a and 5.6b). Back off the

adjuster eight notches, then install the hole plug.
8   Install the caliper bracket (see illustration 5.2b) and brake caliper (see Section 4). Be sure to tighten the bolts to the torque listed in this Chapter's Specifications.

10.5q Install the front shoe return spring . . .

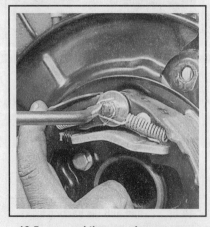

10.5r . . . and the rear shoe return spring

10.5s Install the rear shoe hold-down spring

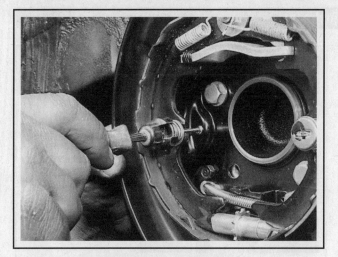

10.5t . . . and the front shoe hold-down spring

10.5u This is how the parking brake assembly should look when you're done!

9   Install the wheel and tighten the lug nuts to the torque specified in Chapter 1.

10 Set the parking brake and count the number of clicks that it trav-

els. It should be between about five to seven clicks - if it's not, adjust the parking brake as described in the next Section.

11 Parking brake - adjustment

▶ **Refer to illustrations 11.3 and 11.4**

1   Slowly depress the parking brake pedal all the way and count the number of clicks. It should take about five to seven clicks to apply the parking brake. If it travels less than five clicks, there's a chance the parking brake might not be releasing completely. If it travels more than seven clicks, the parking brake may not hold adequately on an incline, allowing the car to roll.

2   Release the pedal.

3   On 2004 and later models, the parking brake adjustment is performed at the pedal assembly (see illustration).

4   On 2003 and earlier models, the parking brake adjustment is performed at the equalizer (see illustration).

5   In both designs, loosen the locknut(s) and tighten or loosen the adjuster nut. Tighten the locknut(s) after the desired travel is attained.

11.3 The location of the parking brake cable adjuster and locknut on 2004 and later models

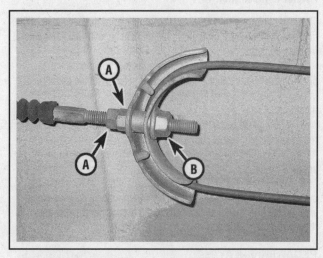

11.4 Parking brake cable locknuts (A) and adjuster nut (B) (2003 and earlier models)

12 Brake light switch - removal, installation and adjustment

REMOVAL AND INSTALLATION

▶ **Refer to illustration 12.1**

1   The brake light switch is located on a bracket at the top of the brake pedal (see illustration).

2   Disconnect the wiring harness at the brake light switch.

3   Loosen the locknut and unscrew the switch from the pedal bracket.

4   Installation is the reverse of removal.

ADJUSTMENT

▶ **Refer to illustration 12.6**

5   Check and, if necessary, adjust brake pedal height (see Chapter 1).

6   Loosen the switch locknut, adjust the switch so that the distance the plunger protrudes is within the range listed in this Chapter's Specifications (see illustration). (If you're unable to measure this distance, adjust the plunger so that it lightly contacts the pedal stop.) Tighten the locknut.

7   Plug the electrical connector into the switch. Make sure the brake lights come on when the brake pedal is depressed and go off when the pedal is released. If not, repeat the adjustment procedure until the brake lights function properly.

8   Check and, if necessary, adjust brake pedal freeplay (see Section 13).

12.1 The brake light switch (A) is located at the top of the brake pedal. Loosen the locknut (B) and then unscrew the switch from its bracket

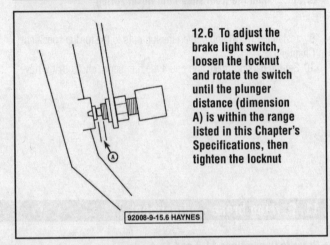

12.6 To adjust the brake light switch, loosen the locknut and rotate the switch until the plunger distance (dimension A) is within the range listed in this Chapter's Specifications, then tighten the locknut

92008-9-15.6 HAYNES

13 Brake pedal - adjustment

PEDAL HEIGHT

▶ **Refer to illustration 13.1**

1   The height of the brake pedal is the distance the pedal sits off the floor (see illustration). If the pedal height is not within Specifications, it must be adjusted.

2   To adjust the brake pedal, loosen the locknut and back the pushrod out for clearance. Turn the pushrod to adjust the pedal height in the middle of the specified range, then retighten the locknut (see illustration 13.1).

3   At the brake pedal, loosen the locknut on the brake switch and retract the switch. Before measuring the brake pedal height, make sure the pedal is in the fully-returned position. Measure the pedal height and adjust if necessary (see Step 2).

4   Adjust the brake pedal switch by turning it clockwise until the switch body just contacts the pedal arm, then rotate it counter-clockwise to gain the specified clearance at the end of this Chapter and tighten the switch locknut.

PEDAL FREEPLAY

5   The freeplay is the pedal slack, or the distance the pedal can be

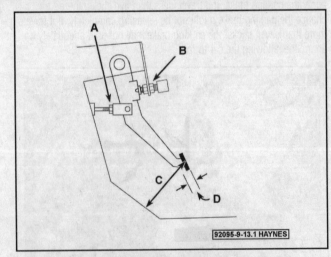

92095-9-13.1 HAYNES

13.1 Brake pedal height and freeplay measuring and adjustment points

A   Clevis locknut
B   Brake light switch adjusting nut/locknut
C   Pedal height measurement point
D   Freeplay measurement point

depressed before it begins to have any effect on the brake system (see illustration 13.1). If the pedal freeplay is not within the specified range, it must be adjusted.

6   To adjust the pedal freeplay, loosen the locknut on the pushrod. Then back out the pushrod to adjust the freeplay to the specified range, then retighten the locknut.

7   Before adjusting brake pedal freeplay, depress the brake pedal several times (with the engine off). Measure the freeplay and adjust if necessary. Loosen the locknut on the pushrod, then back off the pushrod to adjust the pedal freeplay to the specified range and retighten the locknut.

BRAKE PEDAL RESERVE DISTANCE

8   With the parking brake released and the engine running, depress the pedal with normal braking effort and have an assistant measure the distance from the center of the pedal pad to the floor. Compare the measurement to the reserve distance listed in this Chapter's Specifications. If the distance is less than specified, refer to troubleshooting section of this book.

Specifications

General

Brake fluid type	DOT 3
Brake pedal	
Height	5 5/8 to 6 5/16 inches (150 to 160 mm)
Freeplay	5/64 to 1/8 inch (2 to 3 mm)
Reserve	3.1 inches (80 mm) or above
Power brake booster pushrod-to-master	
cylinder piston clearance	0.0 inch (0 mm)
Brake light switch plunger (dimension A)	5/64 to 1/8-inch (2 to 3 mm)

Disc brakes

Minimum brake pad thickness	See Chapter 1
Disc minimum thickness	Cast into disc
Disc runout limit	
Front	0.0020 inch (0.05 mm)
Rear	0.0059 inch (0.15 mm)
Parking brake shoe minimum thickness	1/32-inch (0.8 mm)

Torque specifications

	Ft-lbs (unless otherwise indicated)	Nm
Caliper mounting bolts		
Highlander		
Front caliper	25	34
Rear caliper		
2003 and earlier (upper)	32	43
2003 and earlier (lower)	25	34
2004 and later	32	43
Lexus		
Front caliper	25	34
Rear caliper		
2003 and earlier (upper)	20	26
2003 and earlier (lower)	14	20
2004 and later	32	43
Caliper bracket mounting bolts		
Highlander		
Front	79	107
Rear	58	78
2003 and earlier (2WD)	44	62
2003 and earlier (4WD)	43	58
2004 and later	58	78
Lexus		
Front		
2003 and earlier	77	104
2004 and later	79	107
Rear		
2003 and earlier	34	47
2004 and later	58	78
Brake hose-to-caliper banjo bolt	21	29
Master cylinder-to-brake booster nuts	108 in-lbs	13
Power brake booster mounting nuts	108 in-lbs	13
Wheel lug nuts	See Chapter 1	

Section

Reference to other Chapters

10

SUSPENSION AND STEERING SYSTEMS

1 General information

▶ **Refer to illustrations 1.1 and 1.2**

The front suspension is a MacPherson strut design. The upper end of each strut is attached to the vehicle's body strut support. The lower end of the strut is connected to the upper end of the steering knuckle. The steering knuckle is attached to a balljoint mounted on the outer end of the suspension control arm (see illustration).

The rear suspension also utilizes strut/coil spring assemblies. The upper end of each strut is attached to the vehicle body by a strut support. The lower end of the strut is attached to an axle carrier. The carrier is located by a pair of suspension arms on each side, and a longitudinally mounted strut rod between the body and each carrier (see illustration).

Lexus models are available with an optional electronically modulated air suspension. This suspension allows the driver to control the ride height depending on road conditions.

The power-assisted rack-and-pinion steering gear, which is located behind the engine/transaxle assembly, is mounted on the engine cradle. The steering gear actuates the tie-rods, which are attached to the steering knuckles. The steering column is designed to collapse in the event of an accident.

Frequently, when working on the suspension or steering system components, you may come across fasteners that seem impossible to loosen. These fasteners on the underside of the vehicle are continually subjected to water, road grime, mud, etc., and can become rusted or "frozen," making them extremely difficult to remove. In order to unscrew these stubborn fasteners without damaging them (or other components), be sure to use lots of penetrating oil and allow it to soak in for a while. Using a wire brush to clean exposed threads will also ease removal of the nut or bolt and prevent damage to the threads. Sometimes a sharp blow with a hammer and punch will break the bond between nut and bolt threads, but care must be taken to prevent the punch from slipping off the fastener and ruining the threads. Heating the stuck fastener and surrounding area with a torch sometimes helps too, but isn't recommended because of the obvious dangers associated with fire. Long breaker bars and extension, or "cheater," pipes will increase leverage, but never use an extension pipe on a ratchet - the ratcheting mechanism could be damaged. Sometimes tightening the nut or bolt first will help to break it loose. Fasteners that require drastic measures to remove should always be replaced with new ones.

Since most of the procedures dealt with in this Chapter involve jacking up the vehicle and working underneath it, a good pair of jackstands will be needed. A hydraulic floor jack is the preferred type of jack to lift the vehicle, and it can also be used to support certain components during various operations.

✳✳ WARNING:

Never, under any circumstances, rely on a jack to support the vehicle while working on it. Whenever any of the suspension or steering fasteners are loosened or removed they must be inspected and, if necessary, replaced with new ones of the same part number or of original equipment quality and design. Torque specifications must be followed for proper reassembly and component retention. Never attempt to heat or straighten any suspension or steering components. Instead, replace any bent or damaged part with a new one.

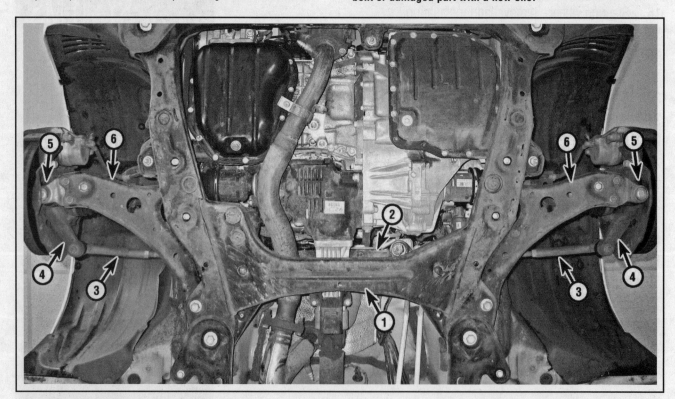

1.1 Front suspension and steering components

1	Subframe	3	Tie-rod end	5	Balljoint
2	Steering gear	4	Steering knuckle	6	Control arm

1.2 Rear suspension components

1 Rear suspension crossmember	3 Rear axle carrier	5 Suspension arm (no. 1)
2 Suspension arm (no. 2)	4 Strut rod	

2 Stabilizer bar bushings and links (front) - removal and installation

▶ **Refer to illustrations 2.3 and 2.4**

✳✳ WARNING:

If the vehicle is equipped with an electronically modulated air suspension, make sure that the height control switch is turned off before raising the vehicle.

➡Note: Stabilizer bar removal involves removing the steering gear and then removing the stabilizer bar out the left wheel-well. If one becomes damaged, it is most likely the result of an accident that was severe enough to damage other major components (such as the subframe itself). Damage this severe will require the services of an auto body shop.

1   Loosen the front wheel lug nuts, then raise the front of the vehicle and support it securely on jackstands.

2   Remove the front wheels.

3   Disconnect the stabilizer bar links from the bar (see illustration). If the ballstud turns with the nut, use an Allen wrench to hold the stud.

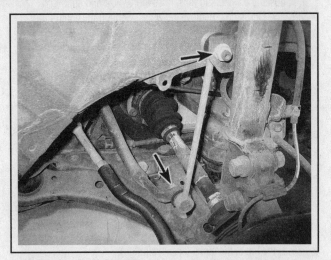

2.3 To detach the stabilizer bar link from the bar, remove the lower nut; if you're removing the strut (or replacing the link), remove the upper nut

4   Detach both stabilizer bar bushing retainers from the subframe (see illustration).

5   While the stabilizer bar is detached, slide off the retainer bushings and inspect them. If they're cracked, worn or deteriorated, replace them. It's also a good idea to inspect the stabilizer bar link. To check it, flip the balljoint stud side-to-side five or six times and then install the nut. Using an inch-pound torque wrench, turn the nut continuously one turn every two to four seconds and note the torque reading on the fifth turn. It should be about 0.4 to 17 in-lbs. If it isn't, replace the link assembly.

6   Clean the bushing area of the stabilizer bar with a stiff wire brush to remove any rust or dirt.

7   Lubricate the inside and outside of the new bushings with vegetable oil (used in cooking) to simplify reassembly.

✳✳ CAUTION:

Don't use petroleum or mineral-based lubricants or brake fluid - they will lead to deterioration of the bushings.

➡**Note: These bushings are split so that you can install them without having to slide them onto the ends of the stabilizer bar. Install the bushings with the slit in each bushing facing towards the rear of the vehicle.**

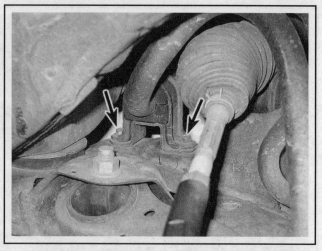

2.4 Bushing retainer fasteners

8   Install the links, tightening the link nuts to the torque listed in this Chapter's Specifications.

9   Install the retainers and bolts, tightening the bolts to the torque listed in this Chapter's Specifications.

3   Strut assembly (front) - removal, inspection and installation

✳✳ WARNING:

If the vehicle is equipped with an electronically modulated air suspension, make sure that the height control switch is turned off before raising the vehicle.

REMOVAL

▶ **Refer to illustrations 3.3, 3.7 and 3.9**

1   Loosen the wheel lug nuts, raise the vehicle and support it securely on jackstands. Remove the wheel. Support the control arm with a floor jack.

2   Disconnect the stabilizer bar link from the strut (see illustration 2.4).

3   Remove the brake hose bracket and the wheel speed sensor wiring harness from the strut (see illustration).

4   Mark the position of the steering knuckle to the strut to help preserve the camber alignment adjustment. Remove the strut-to-knuckle nuts and knock the bolts out with a hammer and punch (see illustration 3.3).

5   Separate the strut from the steering knuckle. Be careful not to overextend the inner CV joint. Also, don't let the steering knuckle fall outward, as the brake hose could be damaged.

6   If the strut is to be disassembled, loosen, but do not remove, the damper shaft nut (in the center).

7   On 2004 and later Lexus models, remove the panels on the cowl to access the top of the struts (see illustration).

8   On 2004 and later Lexus models equipped with electronic modulated air suspension, remove the shock absorber cap and detach air-line fitting from the top of the strut.

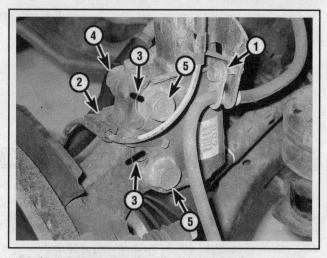

3.3 Strut mounting details at the lower bracket:

1   *Brake hose bracket*
2   *ABS harness bracket*
3   *Marks to preserve camber alignment*
4   *Mounting nuts (one shown)*
5   *Mounting bolts*

9   Support the strut and spring assembly with one hand (or have an assistant hold it) and remove the three strut-to-shock tower nuts (see illustration). Remove the assembly out from the fenderwell.

➡**Note: Some models have a strut brace between the tops of the two strut towers that is retained by the same nuts that secure the strut to the body. The brace can be removed without removing the struts if necessary.**

3.7 On 2004 and later Lexus models, remove the screen(s) from the cowl panel

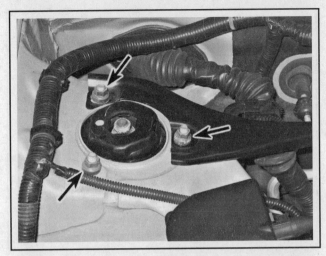

3.9 Mounting fasteners for the upper part of the strut and the strut brace

INSPECTION

10 Check the strut body for leaking fluid, dents, cracks and other obvious damage that would warrant repair or replacement.

11 Check the coil spring for chips or cracks in the spring coating (this can cause premature spring failure due to corrosion). Inspect the spring seat for cuts, hardness and general deterioration.

12 If any undesirable conditions exist, proceed to the strut disassembly procedure (see Section 4).

➡**Note: On 2004 and later Lexus models with modulated air suspension, the strut assembly must be replaced if it is found to be faulty.**

INSTALLATION

13 Guide the strut assembly up into the fenderwell and insert the three upper mounting studs through the holes in the shock tower. Once the three studs protrude from the shock tower, install the nuts so the strut won't fall back through. This is most easily accomplished with the

✹✹ WARNING:

Don't remove the large nut in the center

help of an assistant, as the strut is quite heavy and awkward.

➡**Note: Don't forget to install the strut brace, on models so equipped.**

14 Slide the steering knuckle into the strut flange and insert the two bolts. Install the nuts, match the position of the strut to the steering knuckle that was previously marked and tighten them to the torque listed in this Chapter's Specifications.

15 Reattach the brake hose bracket to the strut and the wheel speed sensor wiring harness bracket.

16 Install the wheel and lug nuts, then lower the vehicle and tighten the lug nuts to the torque listed in the Chapter 1 Specifications.

17 Tighten the three upper mounting nuts to the torque listed in this Chapter's Specifications.

18 If you're working on a model equipped with the electronic modulated air suspension, be sure to attached the air-line to the top of the strut and then install the cap.

4 Strut/coil spring assembly - replacement

➡**Note: The following procedure applies to strut assemblies that are not used in the electronically modulated air suspension.**

1 If the struts or coil springs exhibit the telltale signs of wear (leaking fluid, loss of damping capability, chipped, sagging or cracked coil springs) explore all options before beginning any work. The strut/shock absorber assemblies are not serviceable and must be replaced if a problem develops. However, strut assemblies complete with springs may be available on an exchange basis, which eliminates much time and work. Whichever route you choose to take, check on the cost and availability of parts before disassembling your vehicle.

✹✹ WARNING:

Disassembling a strut is potentially dangerous and utmost attention must be directed to the job, or serious injury may result. Use only a high-quality spring compressor and carefully

follow the manufacturer's instructions furnished with the tool. After removing the coil spring from the strut assembly, set it aside in a safe, isolated area.

DISASSEMBLY

⬧ **Refer to illustrations 4.3, 4.4, 4.5, 4.6 and 4.7**

2 Remove the strut and spring assembly (see Section 3 [front] or Section 10 [rear]). Mount the strut assembly in a vise. Line the vise jaws with wood or rags to prevent damage to the unit and don't tighten the vise excessively.

3 Following the tool manufacturer's instructions, install the spring compressor (which can be obtained at most auto parts stores or equipment yards on a daily rental basis) on the spring and compress it

4.3 Install the spring compressor in accordance with the tool manufacturer's instructions and compress the spring until all pressure is relieved from the upper spring seat

sufficiently to relieve all pressure from the upper spring seat (see illustration). This can be verified by wiggling the spring.

4 Loosen the damper shaft nut (see illustration).

5 Remove the nut and suspension support (see illustration). Inspect the bearing in the suspension support for smooth operation. If it doesn't turn smoothly, replace the suspension support. Check the rubber portion of the suspension support for cracking and general deterioration. If there is any separation of the rubber, replace it.

6 Remove the upper spring seat from the damper shaft (see illustration). Check the spring seat for cracking and hardness; replace it if necessary. Remove the upper insulator.

7 Carefully lift the compressed spring from the assembly (see illustration) and set it in a safe place.

✳✳ WARNING:

Never place your head near the end of the spring!

8 Slide the rubber bumper off the damper shaft.

9 Check the lower insulator for wear, cracking and hardness and replace it if necessary.

REASSEMBLY

▶ **Refer to illustrations 4.11, 4.12 and 4.14**

10 If the lower insulator is being replaced, set it into position with

4.4 Remove the damper shaft nut

4.5 Lift the suspension support off the damper shaft

4.6 Remove the spring seat from the damper shaft

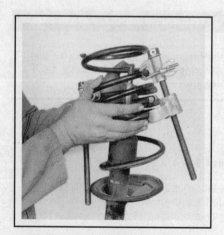

4.7 Remove the compressed spring assembly - keep the ends of the spring pointed away from your body

4.11 When installing the spring, make sure the end fits into the recessed portion of the lower seat

4.12 The flats on the damper shaft must match up with the flats in the spring seat

4.14 Make sure the arrows on the upper spring seat face toward the outside of the vehicle

the dropped portion seated in the lowest part of the seat. Extend the damper rod to its full length and install the rubber bumper.

11 Carefully place the coil spring onto the lower insulator, with the end of the spring resting in the lowest part of the insulator (see illustration).

12 Install the upper insulator on the spring, with the mark on the top of the insulator pointing the same direction as the strut bracket (for the steering knuckle). Install the spring seat, making sure that the flats in the hole in the seat match up with the flats on the damper shaft (see illustration).

13 Align the OUT mark of the spring upper seat with the mark of the upper insulator.

14 If you're working on a front strut, make sure the arrow on the spring seat faces toward the lower bracket, where the steering knuckle fits (see illustration).

15 Install the dust seal and suspension support to the damper shaft.

16 Install the damper nut and tighten it to the torque listed in this Chapter's Specifications. Remove the spring compressor tool.

17 Install the strut/spring assembly (see Section 3 [front] or 10 [rear]).

5   Control arm - removal, inspection and installation

✳✳ WARNING:

If the vehicle is equipped with an electronically modulated air suspension, make sure that the height control switch is turned off before raising the vehicle.

REMOVAL

▶ **Refer to illustrations 5.3a, 5.3b, 5.4 and 5.5**

1 Loosen the wheel lug nuts on the side to be dismantled, raise the front of the vehicle, support it securely on jackstands and remove the wheel.

2 Remove the transverse engine mount at the subframe that covers the front mounting bolts for the control arms.

➡**Note: The procedure is difficult and requires the use of an engine lifting hoist or support fixture (refer to Chapter 2 for engine mount removal/installation).**

3 Remove the balljoint retaining bolt and nuts (see illustration). Use a prybar to disconnect the balljoint from the control arm (see illustration).

4 Remove the two bolts that attach the front of the control arm to the engine cradle (see illustration).

➡**Note: On Lexus vehicles with the electronically modulated air suspension, detach the height sensor linkage from the control arm before removing the mounting bolts.**

5 Remove the bolt and nut that attach the rear of the control arm to the engine cradle (see illustration).

6 Remove the control arm.

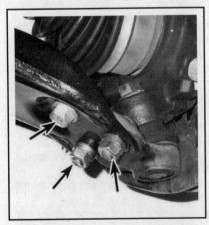

5.3a Remove these nuts and this bolt to disconnect the control arm from the balljoint

5.3b Separate the control arm from the balljoint with a prybar

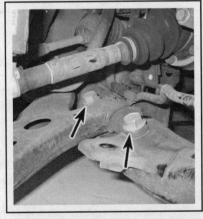

5.4 To detach the front end of the control arm from the engine cradle, remove these two bolts

INSPECTION

7   Make sure the control arm is straight. If it's bent, replace it. Do not attempt to straighten a bent control arm.

8   Inspect the bushings. If they're cracked, torn or worn out, replace the control arm.

INSTALLATION

9   Installation is the reverse of removal. Be sure to tighten all fasteners to the torque listed in this Chapter's Specifications.

10   Install the wheel and lug nuts, lower the vehicle and tighten the lug nuts to the torque listed in the Chapter 1 Specifications.

11   It's a good idea to have the front wheel alignment checked, and if necessary, adjusted after this job has been performed.

5.5 To detach the rear end of the control arm from the engine cradle, remove this nut and bolt

---

6   Balljoints - replacement

✳✳ WARNING:

If the vehicle is equipped with an electronically modulated air suspension, make sure that the height control switch is turned off before raising the vehicle.

1   Loosen the wheel lug nuts, raise the vehicle and support it securely on jackstands. Remove the wheel.

➡Note: If you're going to remove the balljoint using a puller (as described in Steps 9 through 13), loosen the driveaxle/hub nut before raising the vehicle (see Chapter 8).

PICKLEFORK METHOD

✳✳ CAUTION:

The following procedure is the quickest way to detach a balljoint from the steering knuckle, but it will very likely damage the balljoint boot. If you want to save the boot, proceed to Step 9.

2   Remove the cotter pin from the balljoint stud and loosen the nut a few turns (but don't remove it yet).

3   Separate the balljoint from the steering knuckle with a picklefork-type balljoint separator. Remove the balljoint stud nut.

4   Remove the bolt and nuts securing the balljoint to the control arm (see illustration 5.3a). Separate the balljoint from the control arm with a prybar (see illustration 5.3b).

5   To install the balljoint, position it on the steering knuckle and install the nut, but don't tighten it yet.

6   Attach the balljoint to the control arm and install the bolt and nuts, tightening them to the torque listed in this Chapter's Specifications.

7   Tighten the balljoint stud nut to the torque listed in this Chapter's Specifications and install a new cotter pin. If the cotter pin hole doesn't line up with the slots on the nut, tighten the nut additionally until it does line up - don't loosen the nut to insert the cotter pin.

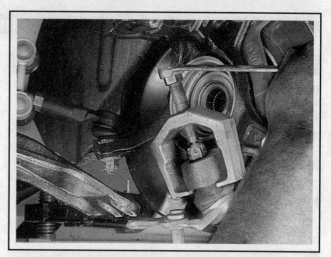

6.13 To separate the balljoint from the steering knuckle, install a small puller and pop the balljoint stud loose

8   Install the wheel and lug nuts. Lower the vehicle and tighten the lug nuts to the torque listed in the Chapter 1 Specifications.

PULLER METHOD

▶ **Refer to illustration 6.13**

9   Remove the wheel speed sensor (see Chapter 9).

10   Separate the control arm from the balljoint (see Section 5).

11   Pull the outer end of the driveaxle from the steering knuckle (see Chapter 8) and suspend the driveaxle with a piece of wire.

12   Remove the cotter pin from the balljoint stud and loosen the nut a few turns (but don't remove it yet).

13   Install a small puller (see illustration) and pop the balljoint stud from the steering knuckle.

14   Remove the nut and remove the balljoint.

15   Install the new balljoint into the steering knuckle and tighten the

SUSPENSION AND STEERING SYSTEMS 10-9

nut to the torque listed in this Chapter's Specifications. Install a new cotter pin. If the cotter pin hole doesn't line up with the slots on the nut, tighten the nut additionally until it does line up - don't loosen the nut to insert the cotter pin.

16 Insert the outer end of the driveaxle through the steering knuckle and install the nut. Tighten it securely, but don't attempt to tighten it completely yet.

7 Steering knuckle and hub - removal and installation

✳✳ WARNING 1:

Dust created by the brake system is harmful to your health. Never blow it out with compressed air and don't inhale any of it. Do not, under any circumstances, use petroleum-based solvents to clean brake parts. Use brake system cleaner only.

✳✳ WARNING 2:

If the vehicle is equipped with an electronically modulated air suspension, make sure that the height control switch is turned off before raising the vehicle.

REMOVAL

1 Loosen the driveaxle/hub nut (see Chapter 8). Loosen the wheel lug nuts, raise the vehicle and support it securely on jackstands. Remove the wheel.

2 Remove the wheel speed sensor from the knuckle and remove the brake disc from the hub (see Chapter 9).

3 Loosen, but do not remove, the strut-to-steering knuckle bolts (see illustration 3.3).

4 Separate the tie-rod end from the steering knuckle arm (see Section 18).

5 Remove the balljoint-to-lower arm bolt and nuts (see illustration 5.3a and 5.3b). The strut-to-knuckle bolts can now be removed.

17 Connect the balljoint to the lower arm and tighten the fasteners to the torque listed in this Chapter's Specifications.

18 Install the wheel speed sensor.

19 Install the wheel and lug nuts, lower the vehicle and tighten the lug nuts to the torque listed in the Chapter 1 Specifications.

20 Tighten the driveaxle/hub nut to the torque listed in the Chapter 8 Specifications, then stake it in place (see Chapter 8).

6 Push the driveaxle from the hub as described in Chapter 8. Support the end of the driveaxle with a piece of wire.

7 Separate the steering knuckle from the strut. If necessary, detach the balljoint from the steering knuckle.

INSTALLATION

8 Guide the knuckle and hub assembly into position, inserting the driveaxle into the hub.

9 Push the knuckle into the strut flange and install the bolts and nuts, but don't tighten them yet.

10 Connect the balljoint to the control arm and install the bolt and nuts (don't tighten them yet).

11 Attach the tie-rod to the steering knuckle arm (see Section 18). Tighten the strut bolt nuts, the balljoint-to-control arm bolt and nuts and the tie-rod nut to the torque values listed in this Chapter's Specifications. Replace all cotter pins with new ones.

12 Place the brake disc on the hub and install the caliper and wheel speed sensor as outlined in Chapter 9.

13 Install the driveaxle/hub nut and tighten it securely (final tightening will be carried out when the vehicle is lowered).

14 Install the wheel and lug nuts, lower the vehicle and tighten the lug nuts to the torque listed in the Chapter 1 Specifications.

15 Install the driveaxle/hub nut and tighten it to the torque listed in the Chapter 8 Specifications, then stake it in place (see Chapter 8).

8 Hub and bearing assembly (front) - removal and installation

Due to the special tools and expertise required to press the hub and bearing from the steering knuckle, this job should be left to a professional shop. However, the steering knuckle and hub may be removed and the assembly taken to a dealer service department or other qualified repair shop. See Section 7 for the steering knuckle and hub removal procedure.

9 Stabilizer bar and bushings (rear) - removal and installation

▶ Refer to illustration 9.3

✳✳ WARNING:

If the vehicle is equipped with an electronically modulated air suspension, make sure that the height control switch is turned off before raising the vehicle.

1 Loosen the rear wheel lug nuts. Raise the rear of the vehicle and support it securely on jackstands. Remove the rear wheels.

2 Remove the heat insulator from the exhaust system.

3 Disconnect the stabilizer bar links from the bar (see illustration). If the ballstud turns with the nut, use an Allen wrench to hold the stud.

4 Unbolt the stabilizer bar bushing retainers. Remove the two retainer brackets from the body, if necessary.

5 The stabilizer bar can now be removed from the vehicle. Remove the bushings from the stabilizer bar, noting their positions.

6 Check the bushings for wear, hardness, distortion, cracking and other signs of deterioration, replacing them if necessary. Check the stabilizer bar links as described in Section 2. When installing the bushings, the slit should be positioned towards the top.

7 Using a wire brush, clean the areas of the bar where the bushings ride. Lubricate the inside and outside of the new bushings with vegetable oil (used in cooking).

✳✳ CAUTION:

Don't use petroleum or mineral-based lubricants or brake fluid - they will lead to deterioration of the bushings.

8 Installation is the reverse of removal.

9.3 To detach the stabilizer bar link from the bar, remove the lower nut

10 Strut assembly (rear) - removal, inspection and installation

REMOVAL

▶ **Refer to illustrations 10.1 and 10.3**

✳✳ WARNING:

If the vehicle is equipped with an electronically modulated air suspension, make sure that the height control switch is turned off before raising the vehicle.

➡Note: When removing/replacing any rear suspension arms, loosely tighten all the bolts, move the suspension to its normal ride-height angle and position, then fully tighten the bolts.

1 Remove the interior trim covering the upper mounting fasteners for the strut (see illustration).

2 Loosen the rear wheel lug nuts, raise the rear of the vehicle and support it securely on jackstands. Remove the wheel.

3 Detach the brake hose and the wheel speed sensor harness from the strut (see illustration).

4 Disconnect the stabilizer bar link from the strut (see Section 9).

5 Support the axle carrier with a floor jack.

6 Loosen the strut-to-axle carrier bolt nuts (see illustration 10.3).

7 On 2004 and later Lexus models equipped with electronic modulated air suspension, disconnect the air hose going to the strut assembly.

➡Note: A line fitting tool is required for separating the air hose from the strut assembly. Check with your local auto parts store or tool dealer.

8 If the strut is to be disassembled, loosen, but do not remove, the damper shaft nut (in the center).

9 Remove the three upper strut-to-body mounting nuts.

10 Lower the axle carrier with the jack, remove the two strut-to-axle carrier bolts and then remove the strut assembly.

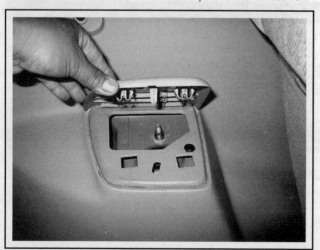

10.1 Pull the plastic trim cover up to remove it

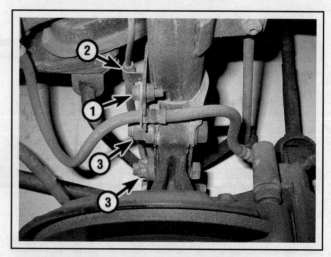

10.3 Strut mounting details at the lower bracket:

| 1 | Brake hose bracket bolt | 3 | Strut mounting fasteners |
| 2 | ABS harness bracket | | |

INSPECTION

11 Follow the inspection procedures described in Section 3. If you determine that the strut assembly must be disassembled for replacement of the strut or the coil spring, refer to Section 4.

➡Note: On 2004 and later Lexus models with modulated air suspension, the strut assembly must be replaced if it is found to be faulty.

INSTALLATION

12 Have an assistant maneuver the assembly up into the fenderwell and insert the mounting studs through the holes in the body. Install the nuts, but don't tighten them yet.

13 Push the axle carrier into the strut lower bracket and install the bolts and nuts, tightening them to the torque listed in this Chapter's Specifications.

14 Connect the stabilizer bar link to the strut bracket.

15 Attach the brake hose bracket and wheel speed sensor harness to the strut.

16 Install the wheel and lug nuts, lower the vehicle and tighten the lug nuts to the torque listed in the Chapter 1 Specifications.

17 Tighten the three strut upper mounting nuts to the torque listed in this Chapter's Specifications and replace the interior trim that covers the mounting fasteners.

11 Strut rod - removal and installation

▶ Refer to illustration 11.2

✴✴ WARNING:

If the vehicle is equipped with an electronically modulated air suspension, make sure that the height control switch is turned off before raising the vehicle.

➡Note: When removing/replacing any rear suspension arms, loosely tighten all the bolts, move the suspension to its normal ride-height angle and position, then fully tighten the bolts.

1 Loosen the wheel lug nuts, raise the vehicle and support it securely on jackstands. Remove the wheel.

2 Remove the strut rod-to-axle carrier bolt (see illustration).

3 Remove the strut rod-to-body bracket bolt and detach the rod from the vehicle.

➡Note: The parking brake cable bracket may need to be removed to allow removal of the strut bolt.

4 Installation is the reverse of the removal procedure. Be sure to tighten the bolts to the torque listed in this Chapter's Specifications.

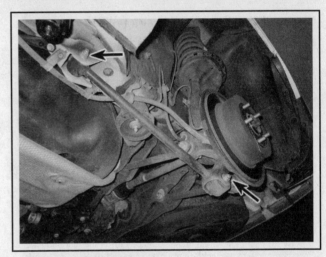

11.2 Strut rod fasteners

12 Suspension arms (rear) - removal and installation

✴✴ WARNING:

If the vehicle is equipped with an electronically modulated air suspension, make sure that the height control switch is turned off before raising the vehicle.

➡Note: Note that the paint marks on all of the suspension arms face to the rear of the vehicle. If no paint marks are visible, mark the suspension arms so that they can be installed back to their original positions.

REMOVAL

1 Loosen the rear wheel lug nuts, raise the rear of the vehicle and support it securely on jackstands. Remove the rear wheel.

Number 1 (front) suspension arm (2WD)

▶ Refer to illustration 12.3

2 Remove the rear stabilizer bar (see Section 9).

3 Remove the arm fasteners from the axle carrier and the rear suspension crossmember and then remove the arm (see illustration).

12.3 The number 1 suspension arm fasteners on 2WD models

Number 1 (front) suspension arm (4WD)

▶ Refer to illustrations 12.10a and 12.10b

4   On 4WD models, it is necessary to lower the rear suspension crossmember to access the bolt on the inner end of the number 1 suspension arm.

5   Disconnect the strut rods from the axle carrier on both sides of the vehicle (see Section 11).

6   Remove the exhaust center section and tailpipe (see Chapter 4).

7   Remove the driveshaft and center bearing assembly (see Chapter 8).

8   Remove both rear driveaxles (see Chapter 8).

9   Detach the suspension arms from both axle carriers.

10 Position a floor jack under the center of the crossmember (beneath the differential), then loosen and remove the crossmember-to-body fasteners (see illustration). Carefully lower the rear suspension crossmember with the jack (do not place any part of your body under the suspension while it is supported only by the jack) until the suspension arm bolts are accessible. Remove the bolts (see illustration).

✳✳ CAUTION:

A floor jack with a transmission adapter head is recommended for this procedure in order to securely support the crossmember.

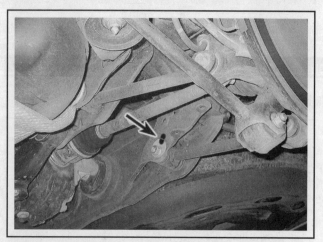

12.12 The cam bolt marked in relation to the rear suspension crossmember

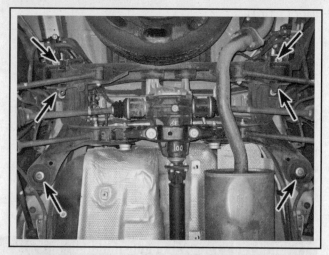

12.10a Rear suspension crossmember mounting fasteners on 4WD models

12.10b The number 1 suspension arm fasteners on 4WD models.

Number 2 (rear) suspension arm

▶ Refer to illustrations 12.12 and 12.13

11 Remove the rear stabilizer bar (see Section 9).

12 On 2003 and earlier models, mark the cam bolt (on both sides)

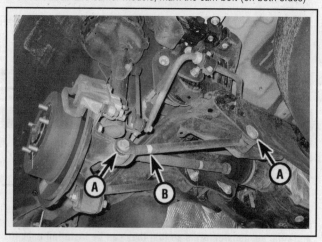

12.13 The number 2 suspension arm fasteners (A) on 4WD models. Note the paint mark (B), which faces the rear of the vehicle

to preserve the toe-in alignment angle (see illustration).

13 Remove the arm fasteners from the axle carrier and the suspension crossmember, then remove the arm (see illustration).

➡Note: On Lexus vehicles with the electronically modulated air suspension, detach the height sensor linkage from the arm before removing the mounting fasteners.

INSTALLATION

➡Note: When installing any rear suspension arms, loosely tighten all the bolts, move the suspension to its normal ride-height angle and position, then fully tighten the bolts.

14 Installation is the reverse of removal. Be sure to tighten all fasteners to the torque listed in this Chapter's Specifications.

➡Note: When reinstalling the suspension arms, the factory paint marks on the arms should face the rear of the vehicle.

15 Install the wheel and lug nuts, then lower the vehicle to the ground. Tighten the wheel lug nuts to the torque listed in the Chapter 1 Specifications. If you're working on a 4WD model, tighten the driveaxle/hub nut to the torque listed in the Chapter 8 Specifications.

16 Have the rear wheel alignment checked by a dealer service department or an alignment shop.

13 Hub and bearing assembly (rear) - removal and installation

✳✳ WARNING 1:

Dust created by the brake system is harmful to your health. Never blow it out with compressed air and don't inhale any of it. Do not, under any circumstances, use petroleum-based solvents to clean brake parts. Use brake system cleaner only.

✳✳ WARNING 2:

If the vehicle is equipped with an electronically modulated air suspension, make sure that the height control switch is turned off before raising the vehicle.

➡Note 1: The following procedure applies to 2WD models only. On 4WD models, special tools and expertise are required to press the hub and bearing from the rear axle carrier, so this job should be left to a professional shop. However, the axle carrier and hub may be removed and the assembly taken to a dealer service department or other repair shop. See Section 14 for the rear axle carrier and hub removal procedure.

➡Note 2: The rear hub and bearing assembly is not serviceable. If found to be defective, it must be replaced as a unit.

REMOVAL

▶ Refer to illustration 13.3

1 Loosen the wheel lug nuts, raise the vehicle and support it securely on jackstands. Remove the wheel.

2 Remove the disc from the hub and disconnect the wheel speed sensor electrical connector (see Chapter 9).

3 Remove the four hub-to-axle carrier bolts, accessible by turning the hub flange so that the large circular cutout exposes each bolt (see illustration).

13.3 To remove the four bolts that attach the hub and bearing assembly to the rear axle carrier, rotate the hub flange and align one of the holes in the flange with each of the bolts

4 Remove the hub and bearing assembly from its seat, maneuvering it out through the parking brake assembly.

INSTALLATION

5 Position the hub and bearing assembly on the axle carrier and align the holes in the backing plate. Install the bolts. A magnet is useful in guiding the bolts through the hub flange and into position. After all four bolts have been installed, tighten them to the torque listed in this Chapter's Specifications.

6 Install the disc and caliper, and the wheel. Lower the vehicle and tighten the lug nuts to the torque listed in the Chapter 1 Specifications.

14 Rear axle carrier - removal and installation

REMOVAL

1 Loosen the wheel lug nuts. On 4WD models, also loosen the driveaxle nut (see Chapter 8). Raise the vehicle and support it on jackstands. Block the front wheels and remove the rear wheel.

2 Remove the rear brake disc (see Chapter 9).

3 Remove the rear hub and bearing assembly (see Section 13). On 4WD models, remove the driveaxle nut in order to remove the driveaxle from the rear hub (see Chapter 8).

4 Detach the backing plate and parking brake assembly from the axle carrier. Suspend the backing plate and brake assembly from the coil spring with a piece of wire.

➡Note: It isn't necessary to disassemble the parking brake shoes or disconnect the parking brake cable from the backing plate.

5 Remove the wheel speed sensor or sensor harness from the axle carrier (see Chapter 9).

6 Loosen (but don't remove) the strut-to-axle carrier bolts (see illustration 10.3).

7 Detach the strut rod and suspension arms from the axle carrier (see Sections 11 and 12).

8 Remove the strut-to-axle carrier bolts while supporting the carrier so it doesn't fall, then detach the axle carrier from the strut bracket.

INSTALLATION

9 Inspect the carrier bushing for cracks, deformation and signs of wear. If it is worn out, take the carrier to a dealer service department or other repair shop to have the old one pressed out and a new one pressed in.

10 Push the axle carrier into the strut bracket, aligning the two bolt holes. Insert the two strut-to-carrier bolts and tighten them to the torque listed in this Chapter's Specifications. On 4WD models, place the driveaxle in the hub before installing the carrier into the strut bracket, then loosely install the driveaxle nut.

11 Connect the suspension arms to the axle carrier, but don't tighten the nut(s) yet.

12 Connect the strut rod to the axle carrier, but don't tighten the nut yet.

13 Place a jack under the carrier and raise it to simulate normal ride height.

14 Tighten the suspension arm bolt(s)/nut(s) and the strut rod bolt/nut to the torque listed in this Chapter's Specifications.

15 Reattach the wheel speed sensor or sensor harness to the axle carrier.

16 Attach the brake backing plate to the axle carrier, install the hub and tighten the four bolts to the torque listed in this Chapter's Specifications.

17 Install the rear brake disc and caliper and then reattach the wheel speed sensor or sensor harness (see Chapter 9).

18 Install the wheel and lug nuts.

19 Lower the vehicle and tighten the lug nuts to the torque listed in the Chapter 1 Specifications. On 4WD models, tighten the driveaxle nut to the torque listed in Chapter 8 Specifications.

15 Steering system - general information

All models are equipped with rack-and-pinion steering. The steering gear is bolted to the subframe and operates the steering knuckles via tie-rods. The inner ends of the tie-rods are protected by rubber boots that should be inspected periodically for secure attachment, tears and leaking lubricant.

The power assist system consists of a belt-driven pump and the associated lines and hoses. The fluid level in the power steering pump reservoir should be checked periodically (see Chapter 1).

The steering wheel operates the steering shaft, which actuates the steering gear through universal joints. Looseness in the steering can be caused by wear in the steering shaft universal joints, the steering gear, the tie-rod ends and loose retaining bolts.

16 Steering wheel - removal and installation

REMOVAL

⬧ Refer to illustrations 16.2, 16.3, 16.4 and 16.5

1 Turn the steering wheel so that the wheels are pointing straight ahead. Turn the ignition key to Off, then disconnect the cable from the negative terminal of the battery (see Chapter 5, Section 1).

2 Pry off the small covers on either side of the steering wheel and

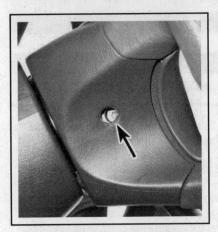

16.2 Pry off the small covers on either side of the steering wheel, then loosen the airbag module Torx screws

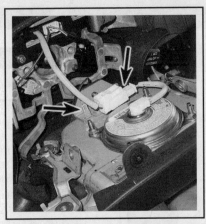

16.3 Disconnect the electrical connector for the airbag module

16.4 Mark the steering wheel hub in relation to the steering shaft

loosen the Torx screws that attach the airbag module to the steering wheel (see illustration). Loosen each screw until the groove in the circumference of the screw catches on the screw case.

➡Note: The screws do not need to be removed completely.

16.5 Use a steering wheel puller to remove the steering wheel

16.6 To center the spiral cable, turn the cable counterclockwise until it's harder to turn, rotate the cable clockwise two and a half turns and align the two small arrows (the cable should be able to rotate about two and a half turns in either direction when it's properly centered)

3   Pull the airbag module off the steering wheel and disconnect the electrical connector for the module (see illustration).

※ **WARNING:**

Carry the airbag module with the trim side facing away from you and set it down in an isolated area with the trim side facing up.

4   Remove the steering wheel retaining nut, then mark the relationship of the steering shaft to the hub (if marks don't already exist or don't line up) to simplify installation and ensure steering wheel alignment (see illustration).
5   Use a puller to disconnect the steering wheel from the shaft (see illustration).

※ **CAUTION:**

Don't hammer on the shaft in an attempt to remove the wheel. If it's necessary to remove the spiral cable, remove the steering column covers (see Chapter 11), unplug the electrical connector and disengage the three claws, then remove it from the column.

➡Note: On 2003 and earlier Lexus models, remove the mounting fasteners for the spiral cable from the body of the combination switch to remove it.

INSTALLATION

♦ **Refer to illustration 16.6**

6   Make sure that the front wheels are facing straight ahead. Turn the spiral cable counterclockwise by hand until it becomes harder to turn the cable. Rotate the cable clockwise about two and a half turns and align the two pointers (see illustration).
7   To install the wheel, align the mark on the steering wheel hub with the mark on the shaft and slip the wheel onto the shaft. Install the nut and tighten it to the torque listed in this Chapter's Specifications.
8   Plug in the electrical connectors for the airbag module and any other connectors. Make sure the connector locks are pushed back into position for the module.
9   Install the airbag module and tighten the Torx retaining screws to the torque listed in this Chapter's Specifications.
10   Connect the negative battery cable (see Chapter 5, Section 1).

17   Steering column - removal and installation

REMOVAL

▶ **Refer to illustrations 17.5, 17.6 and 17.7**

1   Park the vehicle with the wheels pointing straight ahead. Disconnect the cable from the negative terminal of the battery.

2   Remove the steering wheel (see Section 16), then turn the ignition key to the LOCK position to prevent the steering shaft from turning.

17.5 Location of the steering column harness connector

17.6 Mark the relationship of the U-joint to the intermediate shaft (A) and then remove the pinch bolt (B)

✳✳ CAUTION:

If this is not done, the airbag clockspring could be damaged.

3   Remove the lower finish panel under the column and the knee bolster behind it (see Chapter 11).

4   Remove the steering column covers (see Chapter 11).

5   Disconnect the electrical connector(s) for the steering column harness (see illustration).

6   Mark the relationship of the steering shaft U-joint to the intermediate shaft, then remove the pinch bolt (see illustration).

7   Remove the steering column mounting fasteners (see illustration), lower the column and pull it to the rear, making sure nothing is still connected. Separate the intermediate shaft from the steering shaft and remove the column.

INSTALLATION

8   Guide the steering column into position, connect the intermediate shaft, then install the mounting fasteners, but don't tighten them yet.

9   Tighten the column mounting fasteners to the torque listed in this Chapter's Specifications.

10 Install the pinch bolt, tightening it to the torque listed in this Chapter's Specifications.

11 The remainder of installation is the reverse of removal.

17.7 Mounting fasteners for the steering column (one hidden from view - vicinity given)

18   Tie-rod ends - removal and installation

✳✳ WARNING:

If the vehicle is equipped with an electronically modulated air suspension, make sure that the height control switch is turned off before raising the vehicle.

REMOVAL

▶ **Refer to illustrations 18.2a, 18.2b and 18.4**

1   Loosen the wheel lug nuts. Raise the front of the vehicle, support it securely on jackstands, block the rear wheels and set the parking

18.2a Hold the tie-rod end with a wrench and break the jam nut loose with another wrench

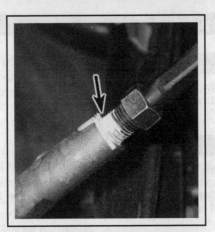

18.2b Back off the jam nut and mark the exposed threads to ensure that the new tie-rod end is threaded on the same number of turns

18.4 Install a small puller to separate the tie-rod end from the steering knuckle

brake. Remove the front wheel.

2 Hold the tie-rod with a pair of locking pliers or wrench and loosen the jam nut enough to mark the position of the tie-rod end in relation to the threads (see illustrations).

3 Remove the cotter pin and loosen the nut on the tie-rod end stud.

4 Disconnect the tie-rod from the steering knuckle arm with a puller (see illustration). Remove the nut and separate the tie-rod.

5 Unscrew the tie-rod end from the tie-rod.

INSTALLATION

6 Thread the tie-rod end on to the marked position and insert the tie-rod stud into the steering knuckle arm. Tighten the jam nut securely.

7 Install the castle nut on the stud and tighten it to the torque listed in this Chapter's Specifications. Install a new cotter pin.

8 Install the wheel and lug nuts. Lower the vehicle and tighten the lug nuts to the torque listed in the Chapter 1 Specifications.

9 Have the alignment checked by a dealer service department or an alignment shop.

19 Steering gear boots - replacement

◗ Refer to illustration 19.3

※※ WARNING:

If the vehicle is equipped with an electronically modulated air suspension, make sure that the height control switch is turned off before raising the vehicle.

1 Loosen the lug nuts, raise the vehicle and support it securely on jackstands. Remove the wheel.

2 Remove the tie-rod end and jam nut (see Section 18).

3 Remove the steering gear boot clamps (see illustration) and slide off the boot.

4 Before installing the new boot, wrap the threads and serrations on the end of the steering rod with a layer of tape so the small end of the new boot isn't damaged.

5 Slide the new boot into position on the steering gear until it seats in the groove in the steering rod and install new clamps.

6 Remove the tape and install the tie-rod end (see Section 18).

7 Install the wheel and lug nuts. Lower the vehicle and tighten the lug nuts to the torque listed in the Chapter 1 Specifications.

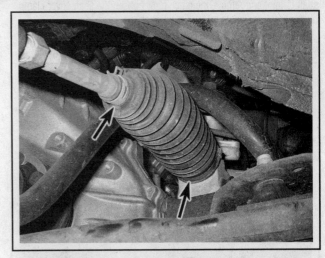

19.3 Remove the outer clamp from the steering gear boot with a pair of pliers; the inner clamp must be cut or pried off

20 Steering gear - removal and installation

✳ WARNING 1:

If the vehicle is equipped with an electronically modulated air suspension, make sure that the height control switch is turned off before raising the vehicle.

✳ WARNING 2:

Make sure the steering shaft is not turned while the steering gear is removed or you could damage the spiral cable for the airbag system. To prevent the shaft from turning, place the ignition key in the lock position or thread the seat belt through the steering wheel and clip it into place.

REMOVAL

▸ Refer to illustrations 20.2a, 20.02b, 20.3 and 20.8

1   Park the vehicle with the front wheels pointing straight ahead. Loosen the front wheel lug nuts, raise the front of the vehicle and support it securely on jackstands. Apply the parking brake and remove the wheels. Remove the engine under-covers on models so equipped.

2   Place a drain pan under the steering gear. Detach the power steering pressure and return lines (see illustration) and cap the ends to prevent excessive fluid loss and contamination. Detach the tube-support bracket from the top of the steering gear assembly.

✳ CAUTION:

Use a line fitting wrench for detaching the lines from the steering gear or the fittings could be severely damaged (see illustration).

3   Mark the relationship of the intermediate shaft to the steering shaft U-joint and remove the pinch bolt. Loosen the shaft cover clamp bolt and pull the cover up (see illustration).

4   On models equipped with the electronically modulated air suspension, detach the height control sensor linkage from the left control arm, then remove the sensor.

5   Mark the relationship of the lower intermediate shaft U-joint to the steering gear input shaft and then remove the U-joint pinch bolt (see illustration 20.2a).

6   Separate the tie-rod ends from the steering knuckle arms (see Section 18).

7   Detach the links from the ends of the stabilizer bar and remove the bushing retainer bolts (see Section 2).

➡Note: The steering gear mounting bolts cannot be removed until the stabilizer bar is lifted up out of the way.

8   Remove the steering gear mounting nuts (see illustration), lift up the stabilizer bar and knock out the steering gear mounting bolts.

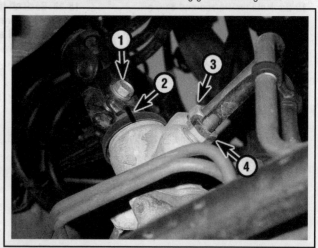

20.2a Steering gear connection details:

1   Lower intermediate shaft U-joint pinch bolt
2   Mark on steering input shaft in relation to U-joint
3   Return line fitting
4   Pressure line fitting

20.2b Disconnect the steering gear line fittings using a crow's-foot wrench (designed for line fittings) and a short extension

20.3 Upper intermediate shaft details:

1   Mark on upper intermediate shaft in relation to steering shaft U-joint
2   Shaft cover clamp bolt
3   Steering shaft U-joint pinch bolt

9 Separate the intermediate shaft U-joint from the steering gear input shaft and then remove the steering gear assembly out the left side of the vehicle.

10 Check the steering gear mounting grommets for excessive wear or deterioration, replacing them if necessary.

INSTALLATION

11 Raise the steering gear into position and connect the intermediate shaft U-joint to the steering input shaft, aligning the marks.

12 Install the mounting bolts and nuts and tighten them to the torque listed in this Chapter's Specifications.

13 Connect the tie-rod ends to the steering knuckle arms (see Section 18).

14 Install the steering shaft U-joint pinch bolt (in the passenger cab) and tighten it to the torque listed in this Chapter's Specifications. Make certain that the alignment marks match if the intermediate shaft was separated from the steering shaft U-joint.

15 Connect the power steering pressure and return hoses to the steering gear and fill the power steering pump reservoir with the recommended fluid (see Chapter 1). Reattach the bracket to the top of the steering gear assembly.

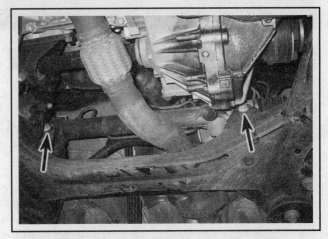

20.8 Steering gear mounting fasteners

16 Install the stabilizer bar bushing retainer bolts and bar end links and then tighten them to the torque listed in this Chapter's Specifications.

17 Lower the vehicle and bleed the steering system (see Section 22).

21 Power steering pump - removal and installation

✳ WARNING:

If the vehicle is equipped with an electronically modulated air suspension, make sure that the height control switch is turned off before raising the vehicle.

REMOVAL

▶ **Refer to illustration 21.6**

1 Disconnect the cable from the negative battery terminal (see Chapter 5, Section 1).

2 Using a large syringe or suction gun, suck as much fluid out of the power steering fluid reservoir as possible. Place a drain pan under the vehicle to catch any fluid that spills out when the hoses are disconnected.

3 Loosen the right front wheel lug nuts, raise the vehicle and support it securely on jackstands. Remove the right front wheel.

4 Remove the right front fender apron seal and liner, if necessary (see Chapter 11).

5 Remove the drivebelt (see Chapter 1).

6 Detach the fluid feed hose from the pump (see illustration). Disconnect the electrical connector from the power steering pressure switch.

7 Remove the pressure line-to-pump union bolt and separate the line from the pump (see illustration 21.6). Use two wrenches, one to hold the pressure port fitting and one to remove the union bolt. Remove the sealing washers on each side of the fitting - these should be replaced when installing the pump.

8 Remove the mounting bolts and then remove the pump (see illustration 21.6).

➡**Note: The upper bolt does not come out entirely; loosen it enough to release it from the engine, and the bolt will come out with the pump.**

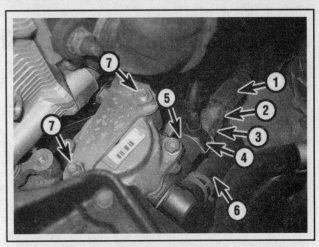

21.6 Installation details of the power steering pump (V6 engine shown, four-cylinder models similar):

1 *Power steering pressure switch electrical connector*
2 *Power steering pressure switch*
3 *Union bolt*
4 *Pressure feed line*
5 *Pressure port fitting*
6 *Fluid feed hose*
7 *Mounting bolts (viewed from behind - access can be made through a hole in the pulley and at the top of the pump)*

INSTALLATION

9 Installation is the reverse of removal. Be sure to tighten the pressure line union bolt and the mounting bolts to the torque listed in this Chapter's Specifications. Adjust the drivebelt tension (see Chapter 1).

10 Top up the fluid level in the reservoir (see Chapter 1) and bleed the system (see Section 22).

22 Power steering system - bleeding

※ WARNING:

If the vehicle is equipped with an electronically modulated air suspension, make sure that the height control switch is turned off before raising the vehicle.

1   Following any operation in which the power steering fluid lines have been disconnected, the power steering system must be bled to remove all air and obtain proper steering performance.
2   With the front wheels in the straight ahead position, check the power steering fluid level and, if low, add fluid until it reaches the Cold mark on the dipstick.
3   Start the engine and allow it to run at fast idle. Recheck the fluid level and add more if necessary to reach the Cold mark on the dipstick.
4   Bleed the system by turning the wheels from side to side, without hitting the stops. This will work the air out of the system. Keep the reservoir full of fluid as this is done.
→Note: **This procedure can be done with the front of the vehicle raised with a jack and supported on jackstands. This makes it easier to turn the wheels back and forth during the bleeding process.**

5   When the air is worked out of the system, return the wheels to the straight-ahead position and leave the vehicle running for several more minutes before shutting it off.
6   Road test the vehicle to be sure the steering system is functioning normally and noise free.
7   Recheck the fluid level to be sure it is up to the Hot mark on the dipstick while the engine is at normal operating temperature. Add fluid if necessary (see Chapter 1).

23 Subframe - removal and installation

▶ **Refer to illustration 23.11**

※ WARNING:

If the vehicle is equipped with an electronically modulated air suspension, make sure that the height control switch is turned off before raising the vehicle.

1   Disconnect the cable from the negative battery terminal (see Chapter 5, Section 1).
2   Loosen the front wheel lug nuts, raise the front of the vehicle and support it securely on jackstands. Remove both front wheels.
→Note: **The jackstands must be behind the front suspension subframe, not supporting the vehicle by the subframe.**

3   Remove the front bumper cover (see Chapter 11).
4   Disconnect the stabilizer bar bushing retainers from the subframe (see Section 2).
5   Disconnect the control arms from the steering knuckles (see Section 5). Also, detach height sensor linkage, if equipped.
6   Remove the steering gear mounting bolts from the subframe (see Section 20).
→Note: **Support the steering gear from above with a rope.**

7   Inspect the subframe for any hose, line or harness brackets that may be attached and detach them.
8   Support the engine from above using an engine hoist or equivalent (see Chapter 2C).

※ WARNING:

DO NOT place any part of your body under the engine when it's supported only by a hoist or other lifting device.

9   Detach all engine and transaxle mounts from the subframe (see Chapter 2)

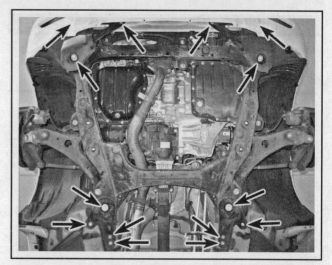

23.11 Subframe and bracket mounting bolts

10 Using two floor jacks, support the subframe. Position one jack on each side of the subframe, midway between the front and rear mounting points.
11 With the jacks sufficiently supporting the subframe, remove the fasteners securing the subframe brackets and the subframe (see illustration).
12 With the use of an assistant to steady the subframe, carefully lower the jacks until the subframe is sufficiently resting on the ground.

INSTALLATION

13 Installation is the reverse of removal. Tighten all suspension and subframe fasteners to the torque listed in this Chapter's Specifications. Tighten all other fasteners to the torque listed in the Chapter Specifications for Chapter's 2 and 11.

24 Wheels and tires - general information

▶ **Refer to illustration 24.1**

1   All vehicles covered by this manual are equipped with metric-sized fiberglass or steel belted radial tires (see illustration). Use of other size or type of tires may affect the ride and handling of the vehicle. Don't mix different types of tires, such as radials and bias belted, on the same vehicle as handling may be seriously affected. It's recommended that tires be replaced in pairs on the same axle, but if only one tire is being replaced, be sure it's the same size, structure and tread design as the other.

2   Because tire pressure has a substantial effect on handling and wear, the pressure on all tires should be checked at least once a month or before any extended trips (see Chapter 1).

3   Wheels must be replaced if they are bent, dented, leak air, have elongated bolt holes, are heavily rusted, out of vertical symmetry or if the lug nuts won't stay tight. Wheel repairs that use welding or peening are not recommended.

4   Tire and wheel balance is important in the overall handling, braking and performance of the vehicle. Unbalanced wheels can adversely affect handling and ride characteristics as well as tire life. Whenever a tire is installed on a wheel, the tire and wheel should be balanced by a shop with the proper equipment.

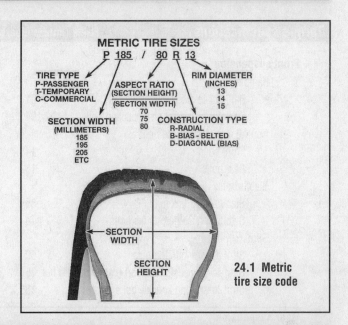

24.1 Metric tire size code

25 Wheel alignment - general information

▶ **Refer to illustration 25.1**

A wheel alignment refers to the adjustments made to the wheels so they are in proper angular relationship to the suspension and the ground. Wheels that are out of proper alignment not only affect vehicle control, but also increase tire wear. The alignment angles normally measured are camber, caster and toe-in (see illustration).

Getting the proper wheel alignment is a very exacting process, one in which complicated and expensive machines are necessary to perform the job properly. Because of this, you should have a technician with the proper equipment perform these tasks. We will, however, use this space to give you a basic idea of what is involved with a wheel alignment so you can better understand the process and deal intelligently with the shop that does the work.

Toe-in is the turning in of the wheels. The purpose of a toe specification is to ensure parallel rolling of the wheels. In a vehicle with zero toe-in, the distance between the front edges of the wheels will be the same as the distance between the rear edges of the wheels. The actual amount of toe-in is normally only a fraction of an inch. On the front end, toe-in is controlled by the tie-rod end position on the tie-rod. On the rear end, it's controlled either by a threaded adjuster on the rear (number two) suspension arm, or a cam at the inner end of the arm, depending on model year. Incorrect toe-in will cause the tires to wear improperly by making them scrub against the road surface.

Camber is the tilting of the wheels from vertical when viewed from one end of the vehicle. When the wheels tilt out at the top, the camber is said to be positive (+). When the wheels tilt in at the top the camber is negative (-). The amount of tilt is measured in degrees from vertical and this measurement is called the camber angle. This angle affects the amount of tire tread which contacts the road and compensates for changes in the suspension geometry when the vehicle is cornering or traveling over an undulating surface.

Caster is the tilting of the front steering axis from the vertical. A tilt toward the rear is positive caster and a tilt toward the front is negative caster. Too little caster will make the front end wander, while too much caster can make the steering effort higher.

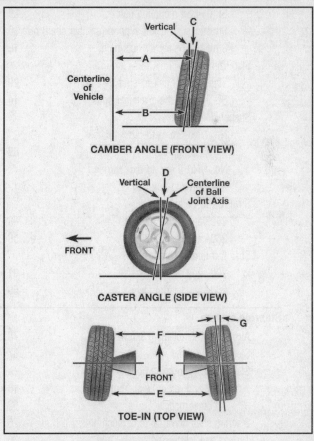

25.1 Camber, caster and toe-in angles

A minus B = C (degrees camber)
D = caster (expressed in degrees)
E minus F = toe-in (measured in inches)
G = toe-in (expressed in degrees)

Specifications

Torque specifications	Ft-lbs (unless otherwise indicated)	Nm
Front suspension		
Balljoint		
Balljoint-to-steering knuckle nut	91	123
Balljoint-to-control arm nuts and bolts	94	127
Control arm-to-subframe bolts		
Front bolts	148	200
Rear bolts	152	206
Stabilizer bar		
Stabilizer bar link nuts	55	74
Stabilizer bushing/retainer bolts	14	19
Strut		
Strut upper mounting nuts	59	80
Strut-to-suspension support (damper shaft) nut	36	49
Strut-to-steering knuckle bolts/nuts	155	210
Rear suspension		
Hub and bearing assembly-to-axle carrier bolts	59	80
Strut		
Strut upper mounting nuts	29	39
Strut-to-suspension support (damper shaft) nut	36	49
Strut-to-axle carrier nuts/bolts	188	255
Stabilizer bar		
Stabilizer bar link-to-strut assembly	29	39
Stabilizer bushing/retainer bolts	14	19
Suspension arms		
No. 1/No. 2 arm-to-crossmember		
through-bolt nut	83	113
No. 1/No. 2 arm-to-axle carrier		
through-bolt nut	131	177
Strut rod-to-body bolt		
2003 and earlier models	83	113
2004 and later models	59	80
Strut rod-to-axle carrier bolt		
2003 and earlier models	91	123
2004 and later models	59	80
Steering system		
Airbag module retaining screws	78 in-lbs	9
Steering wheel nut	37	50
Steering gear mounting bolts/nuts		
2003 and earlier Lexus models	134	81
All other models	52	70
Steering shaft universal joint pinch bolt	26	
Tie-rod end-to-steering knuckle nut	36	49
Power steering pump		
Mounting bolts	32	43
Pressure line banjo bolt	38	51
Wheel lug nuts	See Chapter 1	

Section

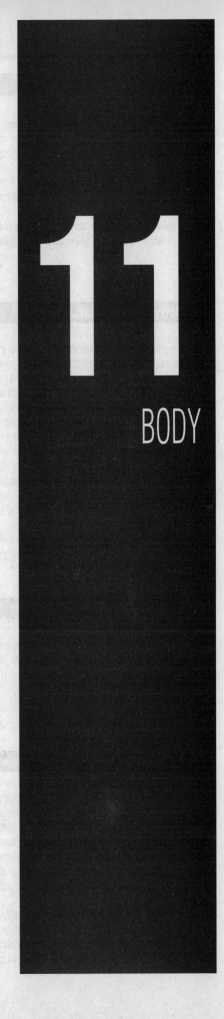

11

BODY

1 General information

These models feature a "unibody" layout, using a floor pan with integral side frame rails which support the body components, front and rear suspension systems and other mechanical components.

Certain components are particularly vulnerable to accident damage and can be unbolted and repaired or replaced. Among these parts are the body moldings, bumpers, front fenders, the hood and trunk lid, doors and all glass.

Only general body maintenance practices and body panel repair procedures within the scope of the do-it-yourselfer are included in this Chapter.

✳✳ WARNING:

The front seat belts on some models are equipped with pre-tensioners, which are pyrotechnic (explosive) devices designed to retract the seat belts in the event of a collision. On models equipped with pre-tensioners, do not remove the front seat belt retractor assemblies, and do not disconnect the electrical connectors leading to the assemblies. Problems with the pre-tensioners will turn on the SRS (airbag) warning light on the dash. If any pre-tensioner problems are suspected, take the vehicle to a dealer service department.

2 Body - maintenance

1   The condition of your vehicle's body is very important, because the resale value depends a great deal on it. It's much more difficult to repair a neglected or damaged body than it is to repair mechanical components. The hidden areas of the body, such as the wheel wells, the frame and the engine compartment, are equally important, although they don't require as frequent attention as the rest of the body.

2   Once a year, or every 12,000 miles, it's a good idea to have the underside of the body steam-cleaned. All traces of dirt and oil will be removed and the area can then be inspected carefully for rust, damaged brake lines, frayed electrical wires, damaged cables and other problems.

3   At the same time, clean the engine and the engine compartment with a steam cleaner or water-soluble degreaser.

4   The wheel wells should be given close attention, since undercoating can peel away and stones and dirt thrown up by the tires can cause the paint to chip and flake, allowing rust to set in. If rust is found, clean down to the bare metal and apply an anti-rust paint.

5   The body should be washed about once a week. Wet the vehicle thoroughly to soften the dirt, then wash it down with a soft sponge and plenty of clean soapy water. If the surplus dirt is not washed off very carefully, it can wear down the paint.

6   Spots of tar or asphalt thrown up from the road should be removed with a cloth soaked in kerosene. Scented lamp oil is available in most hardware stores and the smell is easier to work with than straight kerosene.

7   Once every six months, wax the body and chrome trim. If a chrome cleaner is used to remove rust from any of the vehicle's plated parts, remember that the cleaner also removes part of the chrome, so use it sparingly. On any plated parts where chrome cleaner is used, use a good paste wax over the plating for extra protection.

3 Vinyl trim - maintenance

Don't clean vinyl trim with detergents, caustic soap or petroleum-based cleaners. Plain soap and water works just fine, with a soft brush to clean dirt that may be ingrained. Wash the vinyl as frequently as the rest of the vehicle.

After cleaning, application of a high quality rubber and vinyl protectant will help prevent oxidation and cracks. The protectant can also be applied to weather stripping, vacuum lines and rubber hoses, which often fail as a result of chemical degradation, and to the tires.

4 Upholstery and carpets - maintenance

1   Every three months remove the floormats and clean the interior of the vehicle (more frequently if necessary). Use a stiff whisk broom to brush the carpeting and loosen dirt and dust, then vacuum the upholstery and carpets thoroughly, especially along seams and crevices.

2   Dirt and stains can be removed from carpeting with basic household or automotive carpet shampoos available in spray cans. Follow the directions and vacuum again, then use a stiff brush to bring back the "nap" of the carpet.

3   Most interiors have cloth or vinyl upholstery, either of which can be cleaned and maintained with a number of material-specific cleaners or shampoos available in auto supply stores. Follow the directions on the product for usage, and always spot-test any upholstery cleaner on an inconspicuous area (bottom edge of a backseat cushion) to ensure that it doesn't cause a color shift in the material.

4   After cleaning, vinyl upholstery should be treated with a protectant.

➡**Note: Make sure the protectant container indicates the product can be used on seats - some products may make a seat too slippery.**

✳✳ CAUTION:

Do not use protectant on steering wheels.

5 Leather upholstery requires special care. It should be cleaned regularly with saddlesoap or leather cleaner. Never use alcohol, gasoline, nail polish remover or thinner to clean leather upholstery.

6 After cleaning, regularly treat leather upholstery with a leather conditioner, rubbed in with a soft cotton cloth. Never use car wax on

leather upholstery.

7 In areas where the interior of the vehicle is subject to bright sunlight, cover leather seating areas of the seats with a sheet if the vehicle is to be left out for any length of time.

5 Body repair - minor damage

FLEXIBLE PLASTIC BODY PANELS (FRONT AND REAR BUMPER FASCIA)

The following repair procedures are for minor scratches and gouges. Repair of more serious damage should be left to a dealer service department or qualified auto body shop. Below is a list of the equipment and materials necessary to perform the following repair procedures on plastic body panels. Although a specific brand of material may be mentioned, it should be noted that equivalent products from other manufacturers may be used instead.

Wax, grease and silicone removing solvent

Cloth-backed body tape
Sanding discs
Drill motor with three-inch disc holder
Hand sanding block
Rubber squeegees
Sandpaper
Non-porous mixing palette
Wood paddle or putty knife
Curved-tooth body file
Flexible parts repair material

1 Remove the damaged panel, if necessary or desirable. In most cases, repairs can be carried out with the panel installed.

2 Clean the area(s) to be repaired with a wax, grease and silicone removing solvent applied with a water-dampened cloth.

3 If the damage is structural, that is, if it extends through the panel, clean the backside of the panel area to be repaired as well. Wipe dry.

4 Sand the rear surface about 1-1/2 inches beyond the break.

5 Cut two pieces of fiberglass cloth large enough to overlap the break by about 1-1/2 inches. Cut only to the required length.

6 Mix the adhesive from the repair kit according to the instructions included with the kit, and apply a layer of the mixture approximately 1/8-inch thick on the backside of the panel. Overlap the break by at least 1-1/2 inches.

7 Apply one piece of fiberglass cloth to the adhesive and cover the cloth with additional adhesive. Apply a second piece of fiberglass cloth to the adhesive and immediately cover the cloth with additional adhesive in sufficient quantity to fill the weave.

8 Allow the repair to cure for 20 to 30 minutes at 60-degrees to 80-degrees F.

9 If necessary, trim the excess repair material at the edge.

10 Remove all of the paint film over and around the area(s) to be repaired. The repair material should not overlap the painted surface.

11 With a drill motor and a sanding disc (or a rotary file), cut a "V" along the break line approximately 1/2-inch wide. Remove all dust and loose particles from the repair area.

12 Mix and apply the repair material. Apply a light coat first over the damaged area; then continue applying material until it reaches a level slightly higher than the surrounding finish.

13 Cure the mixture for 20 to 30 minutes at 60-degrees to 80-degrees F.

14 Roughly establish the contour of the area being repaired with a body file. If low areas or pits remain, mix and apply additional adhesive.

15 Block sand the damaged area with sandpaper to establish the actual contour of the surrounding surface.

16 If desired, the repaired area can be temporarily protected with several light coats of primer. Because of the special paints and techniques required for flexible body panels, it is recommended that the vehicle be taken to a paint shop for completion of the body repair.

STEEL BODY PANELS

▶ **See photo sequence**

Repair of minor scratches

17 If the scratch is superficial and does not penetrate to the metal of the body, repair is very simple. Lightly rub the scratched area with a fine rubbing compound to remove loose paint and built up wax. Rinse the area with clean water.

18 Apply touch-up paint to the scratch, using a small brush. Continue to apply thin layers of paint until the surface of the paint in the scratch is level with the surrounding paint. Allow the new paint at least two weeks to harden, then blend it into the surrounding paint by rubbing with a very fine rubbing compound. Finally, apply a coat of wax to the scratch area.

19 If the scratch has penetrated the paint and exposed the metal of the body, causing the metal to rust, a different repair technique is required. Remove all loose rust from the bottom of the scratch with a pocket knife, then apply rust inhibiting paint to prevent the formation of rust in the future. Using a rubber or nylon applicator, coat the scratched area with glaze-type filler. If required, the filler can be mixed with thinner to provide a very thin paste, which is ideal for filling narrow scratches. Before the glaze filler in the scratch hardens, wrap a piece of smooth cotton cloth around the tip of a finger. Dip the cloth in thinner and then quickly wipe it along the surface of the scratch. This will ensure that the surface of the filler is slightly hollow. The scratch can now be painted over as described earlier in this Section.

REPAIR OF DENTS

20 When repairing dents, the first job is to pull the dent out until the affected area is as close as possible to its original shape. There is no point in trying to restore the original shape completely as the metal in the damaged area will have stretched on impact and cannot be restored to its original contours. It is better to bring the level of the dent up to a point which is about 1/8-inch below the level of the surrounding metal. In cases where the dent is very shallow, it is not worth trying to pull it out at all.

These photos illustrate a method of repairing simple dents. They are intended to supplement Body repair - minor damage in this Chapter and should not be used as the sole instructions for body repair on these vehicles.

1 If you can't access the backside of the body panel to hammer out the dent, pull it out with a slide-hammer-type dent puller. In the deepest portion of the dent or along the crease line, drill or punch hole(s) at least one inch apart . . .

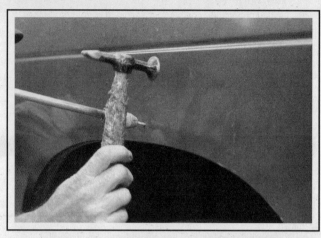

2 . . . then screw the slide-hammer into the hole and operate it. Tap with a hammer near the edge of the dent to help 'pop' the metal back to its original shape. When you're finished, the dent area should be close to its original contour and about 1/8-inch below the surface of the surrounding metal

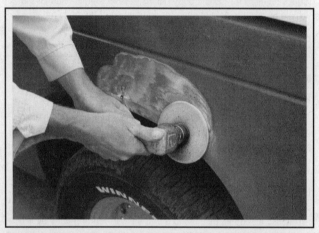

3 Using coarse-grit sandpaper, remove the paint down to the bare metal. Hand sanding works fine, but the disc sander shown here makes the job faster. Use finer (about 320-grit) sandpaper to feather-edge the paint at least one inch around the dent area

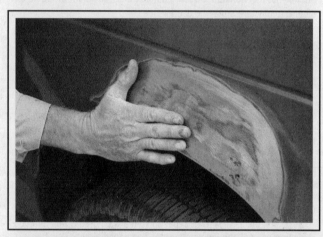

4 When the paint is removed, touch will probably be more helpful than sight for telling if the metal is straight. Hammer down the high spots or raise the low spots as necessary. Clean the repair area with wax/silicone remover

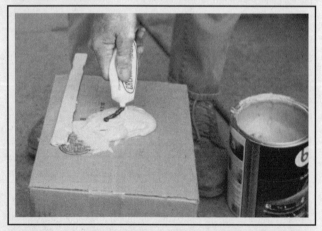

5 Following label instructions, mix up a batch of plastic filler and hardener. The ratio of filler to hardener is critical, and, if you mix it incorrectly, it will either not cure properly or cure too quickly (you won't have time to file and sand it into shape)

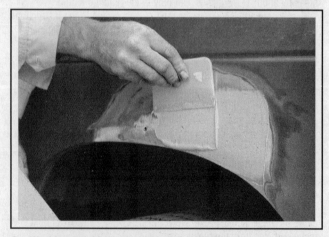

6 Working quickly so the filler doesn't harden, use a plastic applicator to press the body filler firmly into the metal, assuring it bonds completely. Work the filler until it matches the original contour and is slightly above the surrounding metal

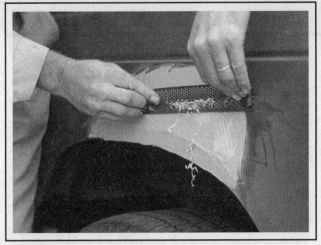

7 Let the filler harden until you can just dent it with your fingernail. Use a body file or Surform tool (shown here) to rough-shape the filler

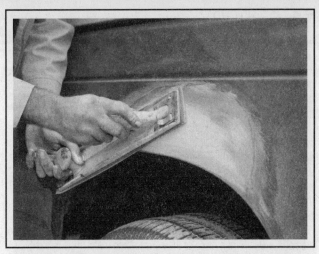

8 Use coarse-grit sandpaper and a sanding board or block to work the filler down until it's smooth and even. Work down to finer grits of sandpaper - always using a board or block - ending up with 360 or 400 grit

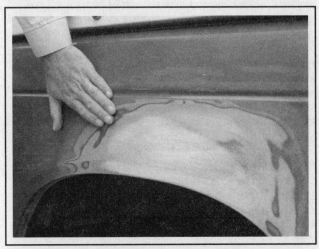

9 You shouldn't be able to feel any ridge at the transition from the filler to the bare metal or from the bare metal to the old paint. As soon as the repair is flat and uniform, remove the dust and mask off the adjacent panels or trim pieces

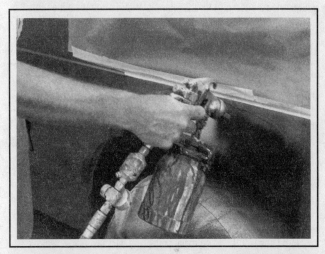

10 Apply several layers of primer to the area. Don't spray the primer on too heavy, so it sags or runs, and make sure each coat is dry before you spray on the next one. A professional-type spray gun is being used here, but aerosol spray primer is available inexpensively from auto parts stores

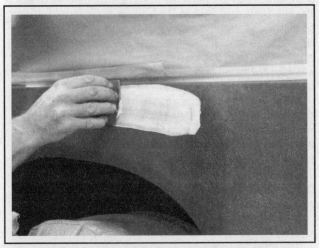

11 The primer will help reveal imperfections or scratches. Fill these with glazing compound. Follow the label instructions and sand it with 360 or 400-grit sandpaper until it's smooth. Repeat the glazing, sanding and respraying until the primer reveals a perfectly smooth surface

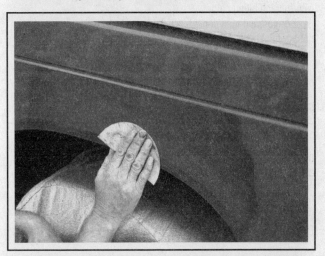

12 Finish sand the primer with very fine sandpaper (400 or 600-grit) to remove the primer overspray. Clean the area with water and allow it to dry. Use a tack rag to remove any dust, then apply the finish coat. Don't attempt to rub out or wax the repair area until the paint has dried completely (at least two weeks)

21 If the back side of the dent is accessible, it can be hammered out gently from behind using a soft-face hammer. While doing this, hold a block of wood firmly against the opposite side of the metal to absorb the hammer blows and prevent the metal from being stretched.

22 If the dent is in a section of the body which has double layers, or some other factor makes it inaccessible from behind, a different technique is required. Drill several small holes through the metal inside the damaged area, particularly in the deeper sections. Screw long, self tapping screws into the holes just enough for them to get a good grip in the metal. Now the dent can be pulled out by pulling on the protruding heads of the screws with locking pliers.

23 The next stage of repair is the removal of paint from the damaged area and from an inch or so of the surrounding metal. This is easily done with a wire brush or sanding disk in a drill motor, although it can be done just as effectively by hand with sandpaper. To complete the preparation for filling, score the surface of the bare metal with a screwdriver or the tang of a file or drill small holes in the affected area. This will provide a good grip for the filler material. To complete the repair, see the Section on filling and painting.

REPAIR OF RUST HOLES OR GASHES

24 Remove all paint from the affected area and from an inch or so of the surrounding metal using a sanding disk or wire brush mounted in a drill motor. If these are not available, a few sheets of sandpaper will do the job just as effectively.

25 With the paint removed, you will be able to determine the severity of the corrosion and decide whether to replace the whole panel, if possible, or repair the affected area. New body panels are not as expensive as most people think and it is often quicker to install a new panel than to repair large areas of rust.

26 Remove all trim pieces from the affected area except those which will act as a guide to the original shape of the damaged body, such as headlight shells, etc. Using metal snips or a hacksaw blade, remove all loose metal and any other metal that is badly affected by rust. Hammer the edges of the hole in to create a slight depression for the filler material.

27 Wire brush the affected area to remove the powdery rust from the surface of the metal. If the back of the rusted area is accessible, treat it with rust inhibiting paint.

28 Before filling is done, block the hole in some way. This can be done with sheet metal riveted or screwed into place, or by stuffing the hole with wire mesh.

29 Once the hole is blocked off, the affected area can be filled and painted. See the following subsection on filling and painting.

FILLING AND PAINTING

30 Many types of body fillers are available, but generally speaking, body repair kits which contain filler paste and a tube of resin hardener are best for this type of repair work. A wide, flexible plastic or nylon applicator will be necessary for imparting a smooth and contoured finish to the surface of the filler material. Mix up a small amount of filler on a clean piece of wood or cardboard (use the hardener sparingly). Follow the manufacturer's instructions on the package, otherwise the filler will set incorrectly.

31 Using the applicator, apply the filler paste to the prepared area. Draw the applicator across the surface of the filler to achieve the desired contour and to level the filler surface. As soon as a contour that approximates the original one is achieved, stop working the paste. If you continue, the paste will begin to stick to the applicator. Continue to add thin layers of paste at 20-minute intervals until the level of the filler is just above the surrounding metal.

32 Once the filler has hardened, the excess can be removed with a body file. From then on, progressively finer grades of sandpaper should be used, starting with a 180-grit paper and finishing with 600-grit wet-or-dry paper. Always wrap the sandpaper around a flat rubber or wooden block, otherwise the surface of the filler will not be completely flat. During the sanding of the filler surface, the wet-or-dry paper should be periodically rinsed in water. This will ensure that a very smooth finish is produced in the final stage.

33 At this point, the repair area should be surrounded by a ring of bare metal, which in turn should be encircled by the finely feathered edge of good paint. Rinse the repair area with clean water until all of the dust produced by the sanding operation is gone.

34 Spray the entire area with a light coat of primer. This will reveal any imperfections in the surface of the filler. Repair the imperfections with fresh filler paste or glaze filler and once more smooth the surface with sandpaper. Repeat this spray-and-repair procedure until you are satisfied that the surface of the filler and the feathered edge of the paint are perfect. Rinse the area with clean water and allow it to dry completely.

35 The repair area is now ready for painting. Spray painting must be carried out in a warm, dry, windless and dust free atmosphere. These conditions can be created if you have access to a large indoor work area, but if you are forced to work in the open, you will have to pick the day very carefully. If you are working indoors, dousing the floor in the work area with water will help settle the dust which would otherwise be in the air. If the repair area is confined to one body panel, mask off the surrounding panels. This will help minimize the effects of a slight mismatch in paint color. Trim pieces such as chrome strips, door handles, etc., will also need to be masked off or removed. Use masking tape and several thickness of newspaper for the masking operations.

36 Before spraying, shake the paint can thoroughly, then spray a test area until the spray painting technique is mastered. Cover the repair area with a thick coat of primer. The thickness should be built up using several thin layers of primer rather than one thick one. Using 600-grit wet-or-dry sandpaper, rub down the surface of the primer until it is very smooth. While doing this, the work area should be thoroughly rinsed with water and the wet-or-dry sandpaper periodically rinsed as well. Allow the primer to dry before spraying additional coats.

37 Spray on the top coat, again building up the thickness by using several thin layers of paint. Begin spraying in the center of the repair area and then, using a circular motion, work out until the whole repair area and about two inches of the surrounding original paint is covered. Remove all masking material 10 to 15 minutes after spraying on the final coat of paint. Allow the new paint at least two weeks to harden, then use a very fine rubbing compound to blend the edges of the new paint into the existing paint. Finally, apply a coat of wax.

6 Body repair - major damage

1   Major damage must be repaired by an auto body shop specifically equipped to perform unibody repairs. These shops have the specialized equipment required to do the job properly.

2   If the damage is extensive, the body must be checked for proper alignment or the vehicle's handling characteristics may be adversely affected and other components may wear at an accelerated rate.

3   Due to the fact that some of the major body components (hood, fenders, doors, etc.) are separate and replaceable units, any seriously damaged components should be replaced rather than repaired. Sometimes the components can be found in a wrecking yard that specializes in used vehicle components, often at considerable savings over the cost of new parts.

7 Hinges and locks - maintenance

Once every 3000 miles, or every three months, the hinges and latch assemblies on the doors, hood and trunk (or liftgate) should be given a few drops of light oil or lock lubricant. The door latch strikers should also be lubricated with a thin coat of grease to reduce wear and ensure free movement. Lubricate the door and trunk (or liftgate) locks with spray-on graphite lubricant.

8 Windshield and fixed glass - replacement

Replacement of the windshield and fixed glass requires the use of special fast-setting adhesive/caulk materials and some specialized tools and techniques. These operations should be left to a dealer service department or a shop specializing in glass work.

9 Hood - removal, installation and adjustment

➡**Note: The hood is somewhat awkward to remove and install, at least two people should perform this procedure.**

REMOVAL AND INSTALLATION

▶ **Refer to illustrations 9.3 and 9.4**

1   Open the hood, then place blankets or pads over the fenders and cowl area of the body. This will protect the body and paint as the hood is lifted off.

2   Disconnect any cables or wires that will interfere with removal. Disconnect the windshield washer tubing near the right-side hinge.

3   Make marks around the hood hinge to ensure proper alignment during installation (see illustration).

4   Have an assistant support one side of the hood. Remove the clips from each end of the support, then detach the supports from the hood. Take turns removing the hinge-to-hood bolts and lift off the hood (see illustration).

5   Installation is the reverse of removal. Align the hinge bolts with the marks made in Step 3.

ADJUSTMENT

▶ **Refer to illustrations 9.9 and 9.10**

6   Fore-and-aft and side-to-side adjustment of the hood is done by

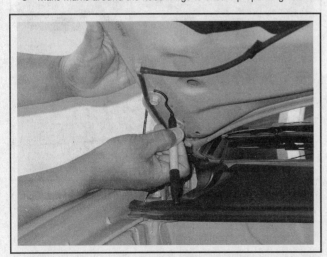

9.3 Draw alignment marks around the hood hinges to ensure proper alignment of the hood when it's reinstalled

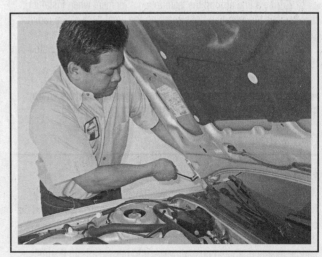

9.4 Support the hood with your shoulder while removing the hood bolts

9.9 To adjust the hood latch horizontally or vertically, loosen these bolts

9.10 To adjust the vertical height of the leading edge of the hood so that it's flush with the fenders, turn each edge cushion (arrow indicates one) clockwise (to lower the hood) or counterclockwise (to raise the hood)

moving the hinge plate slot after loosening the bolts or nuts.

➡ Note: The factory bolts are "centering" type that will not allow adjustment. To adjust the hood in relation to the hinges, these bolts must be replaced with standard bolts with flat washers and lock washers.

7 Mark around the entire hinge plate so you can determine the amount of movement.

8 Loosen the bolts and move the hood into correct alignment. Move it only a little at a time. Tighten the hinge bolts and carefully lower the hood to check the position.

9 If necessary after installation, the entire hood latch assembly can be adjusted up-and-down as well as from side-to-side on the radiator support so the hood closes securely and flush with the fenders. Scribe a line or mark around the hood latch mounting bolts to provide a reference point, then loosen them and reposition the latch assembly, as necessary (see illustration). Following adjustment, retighten the mounting bolts.

10 Finally, adjust the hood bumpers on the radiator support so the hood, when closed, is flush with the fenders (see illustration).

11 The hood latch assembly, as well as the hinges, should be periodically lubricated with white, lithium-base grease to prevent binding and wear.

10 Hood latch and release cable - removal and installation

LATCH

▶ Refer to illustration 10.2

1 Scribe a line around the latch to aid alignment when installing, then remove the retaining bolts securing the hood latch to the radiator support (see illustration 9.9). Remove the latch.

2 Disconnect the hood release cable by disengaging the cable from the latch (see illustration).

3 Installation is the reverse of removal.

➡ Note: Adjust the latch so the hood engages securely when closed and the hood bumpers are slightly compressed.

CABLE

▶ Refer to illustration 10.4

4 Working in the passenger compartment, lift the hood release handle lever upward, then pull down on the cable housing end and disengage the cable from the hood release lever handle (see illustration). If the handle lever needs to be replaced simply pull outward on the handle retaining tab and push downward to release it from the instrument panel.

5 Attach a piece of thin wire or string to the end of the cable.

10.2 Pry out the cable retainer from the backside of the hood latch assembly, then disengage the cable

6 Working in the engine compartment, disconnect the hood release cable from the latch as described in Steps 1 and 2. Unclip all the cable retaining clips on the radiator support and the inner fenderwell.

7 Pull the cable forward into the engine compartment until you can

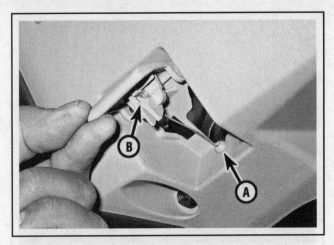

10.4 Lift upward on the handle and pull the cable housing end (A) from the base of the handle, then detach the cable end (B) from the lever

see the wire or string, then remove the wire or string from the old cable and fasten it to the new cable.

8   With the new cable attached to the wire or string, pull the wire or string back through the firewall until the new cable reaches the inside handle.

9   Working in the passenger compartment, install the new cable into the hood release lever, making sure the cable housing fits snugly into the notch in the handle bracket.

➡**Note: Pull on the cable with your fingers from the passenger compartment until the cable stop seats in the grommet on the firewall.**

10 The remainder of the installation is the reverse of removal.

11 Bumpers - removal and installation

FRONT BUMPER

▶ **Refer to illustrations 11.3, 11.5a and 11.5b**

1   Apply the parking brake, raise the vehicle and support it securely on jackstands.

2   Working below the vehicle, remove the lower splash shields.

3   Remove the inner fender splash shields (see illustration).

4   Detach the fasteners securing the top of the grille (see illustrations 13.1 and 13.2).

5   Detach the fasteners securing the sides and bottom of the bumper cover. Pull the cover outward slightly and disconnect the connectors from the fog lights, if equipped. Remove the cover from the vehicle (see illustrations).

6   Installation is the reverse of removal. Make sure the tabs (if equipped) on the back of the bumper cover fit into the corresponding clips on the body before attaching the bolts and screws. An assistant would be helpful at this point.

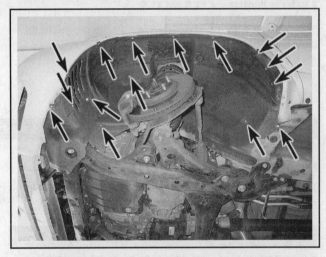

11.3 Detach the main portion of the inner fenderwell, secured by bolts, screws and plastic clips

11.5a Inside the inner fenderwell, remove the screw securing the bumper cover to the fender

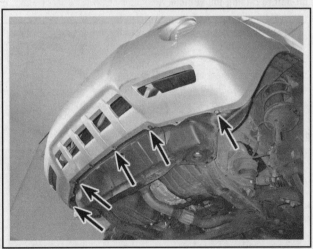

11.5b Remove the fasteners securing the bottom of the bumper cover

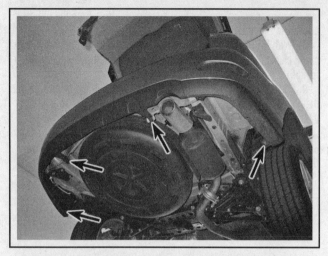

11.8 Remove the fasteners securing the bottom of the bumper cover

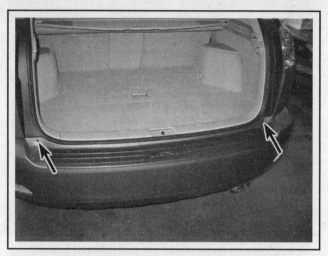

11.9 Open the liftgate and remove the two upper bumper cover fasteners

REAR BUMPER

▶ **Refer to illustrations 11.8 and 11.9**

7 Working in the rear wheelwell, detach the plastic clips and the upper bolt securing the edge of the bumper cover.

8 Detach the fasteners securing the bottom of the bumper cover (see illustration).

9 Open the liftgate and remove the fasteners securing the inside edge of the bumper cover (see illustration). Pull the bumper cover out and away from the vehicle.

10 Installation is the reverse of removal.

12 Front fender - removal and installation

▶ **Refer to illustrations 12.5, 12.7, 12.8 and 12.9**

1 Loosen the front wheel lug nuts. Raise the vehicle, support it securely on jackstands and remove the front wheel.

※※ WARNING:

If the vehicle is equipped with an electronically modulated air suspension, make sure that the height control switch is turned off before raising the vehicle.

2 Open the hood.

3 Detach the inner fenderwell push pins, then remove the inner fender splash shields (see illustration 11.3).

4 Detach the front bumper cover from the fender to be removed (see Section 11).

5 Working inside the inner fenderwell, remove the fastener securing the fender to the body (see illustration).

6 If you're removing the passenger side fender, remove the radio antenna (see Chapter 12).

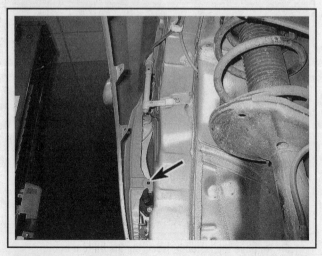

12.5 Remove the bolt in the inner fenderwell opening

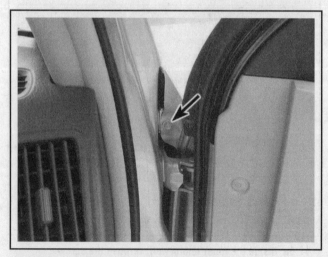

12.7 Remove the upper fender bolt with the door open

12.8 Remove the bolts along the lower edge of the fender

12.9 Remove the three bolts along the top of the fender

7   Open the front door and remove the upper fender-to-body bolt (see illustration).
8   Remove the fender-to-body lower bolts (see illustration).
9   Remove the fender upper mounting bolts (see illustration).
10 Lift off the fender. It's a good idea to have an assistant support

the fender while it's being moved away from the vehicle to prevent damage to the surrounding body panels.
11 Installation is the reverse of removal. Check the alignment of the fender to the hood and front edge of the door before final tightening of the fender fasteners.

13 Radiator grille - removal and installation

▶ **Refer to illustrations 13.1 and 13.2**

1   Open the hood. Remove fasteners along the top of the grille (see illustration).
2   Working from the back of the grille, use a pair of pliers and squeeze the plastic retaining clips at the upper corner of both sides of the grille (see illustration).

3   Pull the top of the grille out slightly and disengage the retaining clips at the bottom of the grille with a long screwdriver. The retaining clips can be disengaged by simply pressing downward on the tabs.
4   Once the retaining clips are disengaged, pull the grille out and remove it.
5   Installation is the reverse of removal.

13.1 Radiator grille upper fasteners

13.2 Disengage the two clips on the back side of the radiator grille

14 Cowl cover and vent tray - removal and installation

▸ **Refer to illustrations 14.2 and 14.3**

1 Remove the wiper arms (see Chapter 12).
2 Remove the push pin fasteners securing the cowl cover (see illustration).

➡**Note: Use a small screwdriver to pop the center button up on the plastic fasteners, but do not try to remove the center but-** tons. They stay in the fastener. Disengage the remaining clips and remove the cowl cover.

3 If the vent tray needs to be removed, first remove the wiper motor linkage assembly as described in Chapter 12, then remove the vent tray mounting bolts (see illustration).
4 Installation is the reverse of removal.

14.2 Remove the cowl fasteners

14.3 Remove the fasteners securing the cowl vent tray (right side shown, left side similar)

15 Door trim panels - removal and installation

❊❊ WARNING:

The models covered by this manual are equipped with Supplemental Restraint systems (SRS), more commonly known as airbags. Always disarm the airbag system before working in the vicinity of any airbag system component to avoid the possibility of accidental deployment of the airbag, which could cause personal injury (see Chapter 12).

❊❊ CAUTION:

Wear gloves when working inside the door openings to protect against cuts from sharp metal edges.

FRONT AND REAR DOORS

▸ **Refer to illustrations 15.2, 15.3a, 15.3b, 15.4, 15.5, 15.6, 15.7 and 15.8**

1 Disconnect the cable from the negative battery terminal (see Chapter 5, Section 1).
2 Remove the push-pin fastener (see illustration).
3 Remove the set-screw to remove the bezel around the inside door handle (see illustrations).
4 Using a trim removal tool, pry up the window switch plate and disconnect the electrical connectors (see illustration).
5 Using a trim removal tool, pry out the outside mirror trim plate (see illustration).

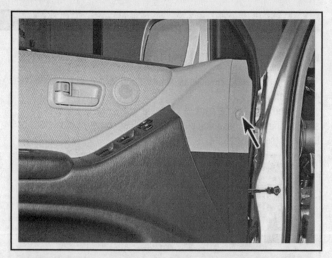

15.2 Remove the push-pin fastener from the door panel

6 Remove the door pull handle cover, then remove the screw underneath (see illustration).
7 Carefully pry the panel out until the clips disengage (see illustration). Work slowly and carefully around the outer edge of the trim panel until it's free. Unplug any wiring harness connectors and remove the panel.
8 For access to the door outside handle or the door window regulator inside the door, raise the window fully, then carefully peel back the plastic watershield (see illustration).
9 Installation is the reverse of removal.

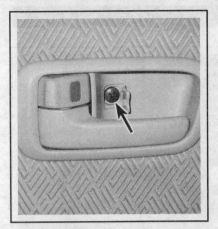

15.3a Remove the set screw securing the inside door handle . . .

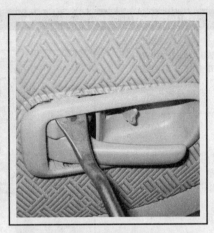

15.3b . . . then using a trim removal tool, carefully pry up the door handle bezel

15.4 Pry up the power window switch plate, then disconnect the electrical connectors

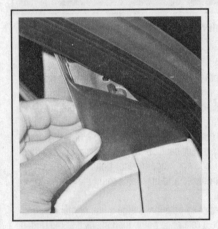

15.5 Pry off the mirror trim cover

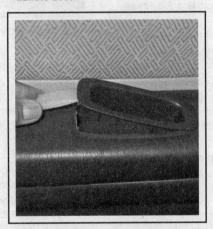

15.6 Pry out the trim cover on the door grip, then remove the door panel mounting screw

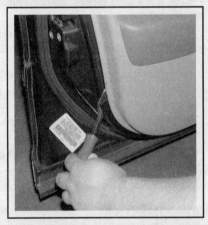

15.7 Carefully pry the clips free so the door trim panel can be removed

LIFTGATE

▶ **Refer to illustration 15.10**

10 Remove the assist strap mounting bolt (see illustration).

15.8 Starting in the upper corner, carefully peel back the plastic watershield for access to the inner door

11 Using a screwdriver or trim removal tool pry out the clips and remove the trim panel from the liftgate. Work slowly and carefully around the outer edge of the trim panel until it's free. Unplug any wiring harness connectors and remove the panel.

12 For access to other components inside the door, carefully peel back the plastic watershield.

13 Installation is the reverse of removal.

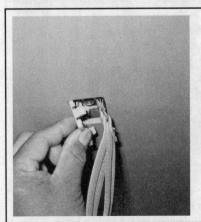

15.10 Remove the mounting fastener securing the door assist strap

16 Door - removal, installation and adjustment

➡Note: The door is heavy and somewhat awkward to remove and install - at least two people should perform this procedure.

REMOVAL AND INSTALLATION

▶ **Refer to illustrations 16.6 and 16.8**

1 Raise the window completely in the door and disconnect the cable from the negative battery terminal (see Chapter 5, Section 1).

2 Open the door all the way and support it from the ground on jacks or blocks covered with rags to prevent damaging the paint.

3 Remove the door trim panel and watershield as described in Section 15.

4 Disconnect all electrical connections, ground wires and harness retaining clips from the door.

➡Note: It is a good idea to label all connections to aid the reassembly process.

5 From the door side, detach the rubber conduit between the body and the door. Then pull the wiring harness through the conduit hole and remove it from the door.

6 Remove the door stop strut bolt (see illustration).

7 Mark around the door hinges with a pen or a scribe to facilitate realignment during reassembly.

8 With an assistant holding the door, remove the hinge-to-door bolts (see illustration) and lift the door off.

➡Note: Draw a reference line around the hinges before removing the bolts.

9 Installation is the reverse of removal.

ADJUSTMENT

▶ **Refer to illustration 16.13**

10 Having proper door-to-body alignment is a critical part of a well-functioning door assembly. First check the door hinge pins for excessive play. Fully open the door and lift up and down on the door without lifting the body. If a door has 1/16-inch or more excessive play, the

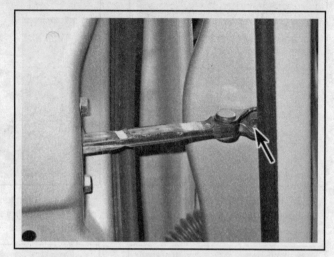

16.6 Remove the bolt retaining the door stop strut

hinges should be replaced.

11 Door-to-body alignment adjustments are made by loosening the hinge-to-body bolts or hinge-to-door bolts and moving the door. Proper body alignment is achieved when the top of the doors are parallel with the roof section, the front door is flush with the fender, the rear door is flush with the rear quarter panel and the bottom of the doors are aligned with the lower rocker panel. If these goals can't be reached by adjusting the hinge-to-body or hinge-to-door bolts, body alignment shims may have to be purchased and inserted behind the hinges to achieve correct alignment.

12 To adjust the door-closed position, scribe a line or mark around the striker plate to provide a reference point, then check that the door latch is contacting the center of the latch striker. If not, adjust the up and down position first.

13 Finally adjust the latch striker sideways position, so that the door panel is flush with the center pillar or rear quarter panel and provides positive engagement with the latch mechanism (see illustration).

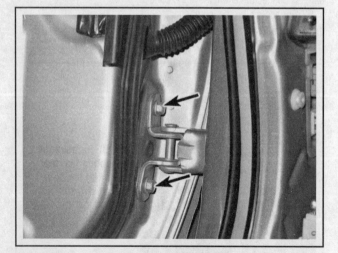

16.8 Remove the door hinge bolts with the door supported (arrows indicate bolts for the bottom hinge, top hinge similar)

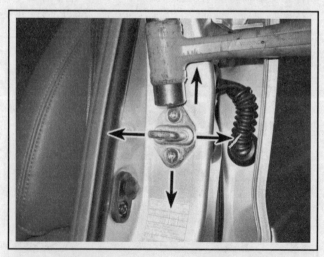

16.13 Adjust the door lock striker by loosening the mounting screws and gently tapping the striker in the desired direction

17 Door latch, lock cylinder and handles - removal and installation

✴✴ CAUTION:

Wear gloves when working inside the door openings to protect against cuts from sharp metal edges.

DOOR LATCH

▶ **Refer to illustrations 17.2 and 17.4**

1   Raise the window, then remove the door trim panel and watershield (see Section 15).
2   Working through the large access hole, disengage the rods from the handle and lock cylinder (see illustration). All door lock rods are attached by plastic clips. The plastic clips can be removed by unsnapping the portion engaging the connecting rod and then pulling the rod out of its locating hole.
3   Disconnect the electrical connectors at the latch. Disengage the handle-to-latch cables (see Steps 11 and 12).
4   Remove the screws securing the latch to the door (see illustration). Remove the latch assembly through the door opening.
5   Installation is the reverse of removal.

OUTSIDE HANDLE AND DOOR LOCK CYLINDER

▶ **Refer to illustrations 17.8a, 17.8b and 17.8c**

6   To remove the outside handle and lock cylinder assembly, raise the window and remove the door trim panel and watershield (see Section 15).

✴✴ CAUTION:

Take care not to scratch the paint on the outside of the door. Wide masking tape applied around the handle opening before beginning the procedure can help avoid scratches.

7   Working through the access hole, disengage the plastic clips that secure the outside door lock-to-latch rod and the outside door handle-to-latch rod (see illustration 17.2).
8   Remove the plug from the end of the door and remove the lock cylinder retaining screw. Withdraw the lock cylinder from the door and disconnect the electrical connector, if equipped. Unbolt the handle from the inside of the door and remove the handle and the handle frame (see illustrations).
9   Installation is the reverse of removal.

17.2 Disengage the rods from the handle and lock cylinder

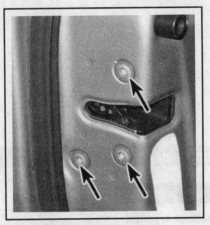

17.4 Remove the door latch mounting fasteners

17.8a Remove the plug from the end of the door to access the door lock cylinder retaining bolt

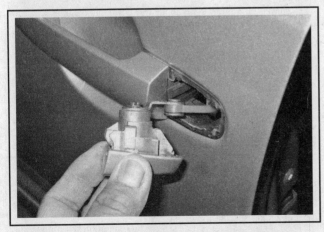

17.8b Slide the lock cylinder out

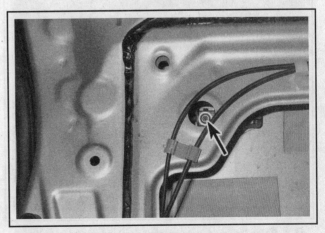

17.8c From inside the door opening, remove the door handle mounting bolt

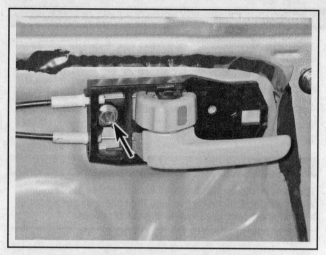

17.11 Remove the inside handle mounting fastener

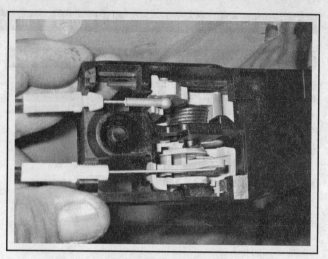

17.12 Detach the cables from the inside handle

INSIDE DOOR HANDLE

▶ **Refer to illustrations 17.11 and 17.12**

10 Remove the door trim panel (see Section 15).

11 Remove the handle retaining screw(s) and disengage the handle from the door (see illustration).

12 Disengage the handle-to-latch cables and remove the handle from the door (see illustration).

13 Installation is the reverse of removal.

18 Door window glass - removal and installation

✳✳ CAUTION:

Wear gloves when working inside the door openings to protect against cuts from sharp metal edges.

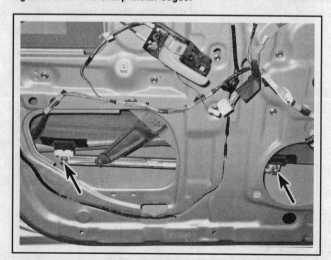

18.4 Raise the window to access the glass retaining bolts through the holes in the door frame

DOOR GLASS

▶ **Refer to illustration 18.4**

1 Remove the door trim panel and the plastic watershield (see Section 15).

2 Lower the window glass all the way down into the door.

3 Remove the door speaker (see Chapter 12).

4 Raise the window just enough to access the window retaining bolts through the holes in the door frame (see illustration).

5 Place a rag over the glass to help prevent scratching the glass and remove the two glass mounting bolts.

6 Remove the glass by pulling it up and out.

7 Installation is the reverse of removal.

LIFTGATE GLASS

8 Replacement of the back door glass requires the use of special fast-setting adhesive/caulk materials and some specialized tools and techniques. These operations should be left to a dealer service department or a shop specializing in glass work.

19 Door window glass regulator - removal and installation

♦ **Refer to illustration 19.4**

❋❋ CAUTION:

Wear gloves when working inside the door openings to protect against cuts from sharp metal edges.

1 Remove the door trim panel and the plastic watershield (see Section 15).

2 Remove the window glass (see Section 18).

3 Disconnect the electrical connector from the window regulator motor.

4 Loosen the temporary bolt, then remove the regulator/motor mounting bolts (see illustration).

5 Remove the regulator/motor assembly. Pull the equalizer arm and regulator assemblies through the service hole in the door frame to remove it.

6 Installation is the reverse of removal. Lubricate the rollers and wear points on the regulator with white grease before installation.

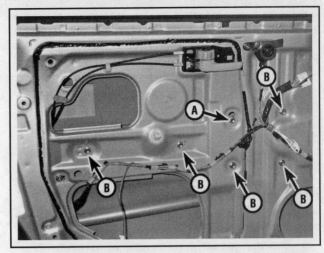

19.4 Loosen the temporary bolt (A), then remove the window regulator mounting bolts (B)

20 Mirrors - removal and installation

OUTSIDE MIRRORS

♦ **Refer to illustration 20.2**

1 Remove the door trim panel as described in Section 15.

2 Disconnect the electrical connector from the mirror, then remove the three mirror retaining bolts and detach the mirror from the vehicle (see illustration).

3 Installation is the reverse of removal.

INSIDE MIRROR

4 Disconnect the electrical connector from the mirror, if equipped.

5 On some models the mirror can be removed by carefully prying between the mirror mount and the notch in the base of the mirror stalk with a screwdriver tip covered with tape. There is a hairpin-type spring holding the mirror stalk in the base. Push the screwdriver in about 3/4-inch to release the spring. On some other models, the mirror can be removed by removing the set screw located at the base of the mirror stalk.

6 On models without a set screw, to install the mirror, reinsert the spring if it was removed earlier. Insert the mirror stalk's lug into the mount, pushing downward until the mirror is secured.

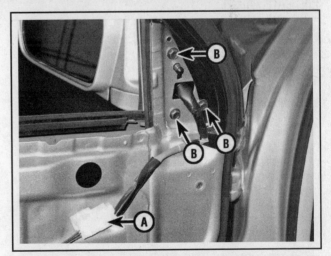

20.2 Disconnect the electrical connector (A), then remove the mirror mounting bolts (B)

7 If the mount plate itself has come off the windshield, adhesive kits are available at auto parts stores to resecure it. Follow the instructions included with the kit.

21 Liftgate - removal, installation and adjustment

→Note: The liftgate is heavy and somewhat awkward to remove and install - at least two people should perform this procedure.

REMOVAL AND INSTALLATION

1 Disconnect the cable from the negative battery terminal (see Chapter 5, Section 1).

2 Open the liftgate all the way and support it from the ground on jacks or blocks covered with rags to prevent damaging the paint.

3 Remove the liftgate trim panel and watershield as described in Section 15.

4 Disconnect all electrical connections, ground wires and harness retaining clips from the liftgate.

→Note: It is a good idea to label all connections to aid the reassembly process.

5 From the liftgate side, detach the rubber conduit between the body and the liftgate. Then pull the wiring harness through the conduit hole and remove it from the liftgate.

6 Remove the liftgate stop strut bolt(s).

7 Mark around the liftgate hinges with a pen or a scribe to facilitate realignment during reassembly.

8 With an assistant holding the liftgate, remove the hinge-to-liftgate bolts and lift the liftgate off.

→Note: Draw a reference line around the hinges before removing the bolts.

9 Installation is the reverse of removal.

ADJUSTMENT

10 Having proper liftgate-to-body alignment is a critical part of a well-functioning liftgate assembly. First check the liftgate hinge pins for excessive play. Fully open the liftgate and lift up and down on the liftgate without lifting the body. If a liftgate has 1/16-inch or more excessive play, the hinges should be replaced.

11 Liftgate-to-body alignment adjustments are made by loosening the hinge-to-body bolts or hinge-to-liftgate bolts and moving the liftgate. Proper body alignment is achieved when the top of the liftgate is parallel with the roof section and the sides of the liftgate are flush with the rear quarter panels and the bottom of the liftgate is aligned with the lower liftgate sill. If these goals can't be reached by adjusting the hinge-to-body or hinge-to-liftgate bolts, body alignment shims may have to be purchased and inserted behind the hinges to achieve correct alignment.

12 To adjust the liftgate-closed position, scribe a line or mark around the striker plate to provide a reference point, then check that the liftgate latch is contacting the center of the latch striker. If not, adjust the up and down position first.

13 Finally adjust the latch striker sideways position, so that the liftgate panel is flush with the rear quarter panel and provides positive engagement with the latch mechanism.

22 Liftgate latch, lock cylinder and handle - removal and installation

LIFTGATE LATCH

▶ Refer to illustration 22.3

1 Disconnect the cable from the negative battery terminal (see Chapter 5, Section 1).

2 Open the liftgate and remove the door trim panel and watershield as described in Section 15.

3 Working through the large access hole, disengage the outside liftgate handle-to-latch cable and the outside liftgate lock-to-latch rod (see illustration). Disconnect the electrical connector for the power lock.

4 All liftgate lock rods are attached by plastic clips. The plastic clips can be removed by unsnapping the portion engaging the connecting rod and then pulling the rod out of its locating hole.

5 Remove the fasteners securing the latch to the liftgate. Remove the latch assembly.

6 Installation is the reverse of removal.

LIFTGATE LOCK CYLINDER

▶ Refer to illustration 22.8

7 Open the liftgate and remove the door trim panel and watershield as described in Section 15.

8 Working through the large access hole, disengage the outside liftgate lock-to-latch rod (see illustration).

9 All liftgate lock rods are attached by plastic clips. The plastic

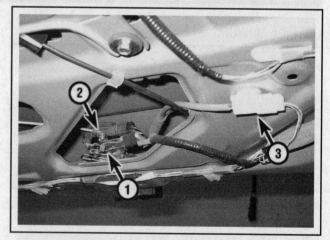

22.3 Liftgate latch details

1 Handle-to-latch cable
2 Liftgate lock-to-latch rod
3 Power lock electrical connector

clips can be removed by unsnapping the portion engaging the connecting rod and then pulling the rod out of its locating hole.

10 Remove the lock cylinder retaining fasteners. Remove the lock cylinder.

11 Installation is the reverse of removal.

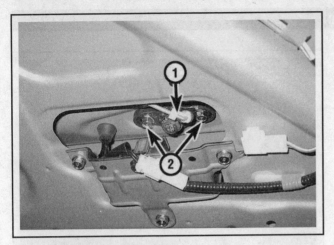

22.8 Liftgate lock cylinder details

1   Liftgate lock-to-latch rod
2   Lock cylinder mounting fasteners

LIFTGATE OUTSIDE HANDLE

♦ **Refer to illustration 22.13**

12 Open the liftgate and remove the door trim panel and watershield as described in Section 15.

22.13 Liftgate outside handle details

1   Liftgate handle-to-latch cable
2   Liftgate handle retaining fasteners

13 Working through the large access hole, disengage the outside liftgate handle-to-latch cable (see illustration).

14 Remove the handle retaining fasteners through the holes in the door frame and detach the handle from the liftgate.

15 Installation is the reverse of removal.

23 Center console - removal and installation

♦ **Refer to illustrations 23.2, 23.3 and 23.4**

❊❊ WARNING:

The models covered by this manual are equipped with Supplemental Restraint systems (SRS), more commonly known as airbags. Always disable the airbag system before working in the vicinity of any airbag system component to avoid the possibility of accidental deployment of the airbag, which could cause personal injury (see Chapter 12).

1   Disconnect the cable from the negative battery terminal (see Chapter 5, Section 1).

2   Using a trim stick, carefully disengage the clips securing the shifter bezel, then disconnect the electrical connectors from the back of the bezel (see illustration). Remove the shifter bezel.

3   Remove the trim covers at the front of the console (see illustration).

4   Remove the retaining screws and detach the rear half of the console from the vehicle (see illustration).

5   Installation is the reverse of removal.

23.2 Using a trim stick, carefully disengage the clips securing the shifter bezel

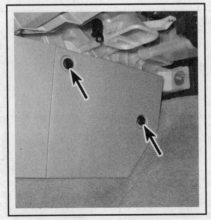

23.3 Remove the fasteners securing the trim covers at the front of the console

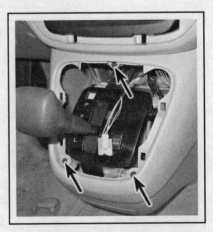

23.4 Remove the fasteners securing the rear half of the console

24 Dashboard trim panels - removal and installation

⁜ WARNING:

Models covered by this manual are equipped with a Supplemental Restraint System (SRS), more commonly known as airbags. Always disable the airbag system before working in the vicinity of any airbag system component to avoid the possibility of accidental deployment of the airbag, which could cause personal injury (see Chapter 12).

1 These panels provide access to various instrument panel mounting screws. Some of the covers use fasteners and others are easily pried off with a screwdriver or trim stick. If you're going to remove the instrument panel, remove all of the covers.

2 Disconnect the cable from the negative terminal of the battery (see Chapter 5, Section 1).

INSTRUMENT CLUSTER LOWER FINISH PANEL

▶ **Refer to illustration 24.4**

3 On models with tilt steering, lower the steering column as far down as it can go.

4 Using a trim stick, carefully pry the lower portion of the panel away from the instrument panel until the clips are released. Take care not to scratch the surrounding trim on the instrument panel (see illustration). Disconnect the electrical connectors for the switches mounted on the finish panel.

5 Installation is the reverse of the removal procedure. Make sure the clips are engaged properly before pushing the panel firmly into place.

INSTRUMENT CLUSTER BEZEL

▶ **Refer to illustration 24.7**

6 Remove the instrument cluster lower finish panel (see Steps 3 and 4).

7 Using a trim stick, carefully pry the bezel to release the clips, then remove the bezel (see illustration).

8 Installation is the reverse of the removal procedure. Make sure the clips are engaged properly before pushing the bezel firmly into place.

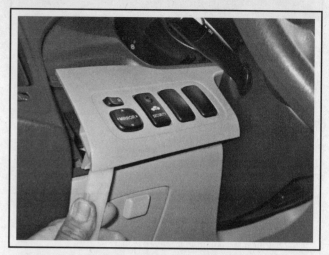

24.4 Carefully pry the lower portion of the panel away from the instrument panel until the clips are released

AUDIO UNIT AND AIR CONDITIONING CONTROL PANEL CENTER TRIM PANEL

▶ **Refer to illustration 24.9**

9 Using a trim stick, carefully pry the bezel to release the clips, then remove the panel (see illustration). Take care not to scratch the surrounding trim on the instrument panel.

10 Installation is the reverse of the removal procedure. Make sure the clips are engaged properly before pushing the panel firmly into place.

KNEE BOLSTER

▶ **Refer to illustrations 24.12 and 24.14**

11 Remove the instrument cluster lower finish panel (see Steps 3 and 4).

12 Remove the two fasteners securing the knee bolster cover (see illustration).

13 Pull the knee bolster out to disengage the clips behind it.

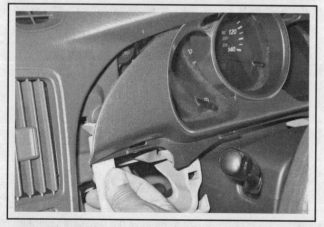

24.7 Carefully pry the bezel to release the clips, then remove the bezel

24.9 Carefully pry the bezel to release the clips, then remove the panel

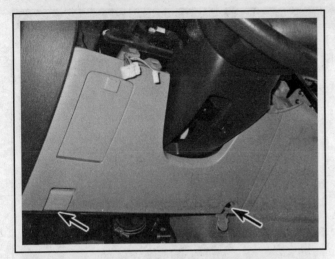

24.12 Remove the two fasteners securing the knee bolster cover

14 Remove the retaining bolts securing the knee bolster reinforcement, if needed for access to components under the dashboard (see illustration).

➡Note: On models equipped with a driver's knee airbag, refer to Chapter 12 for the removal procedure.

15 Installation is the reverse of the removal procedure.

GLOVE BOX DOOR

▶ **Refer to illustration 24.16**

16 Remove the two pins at the bottom of the glove box door and

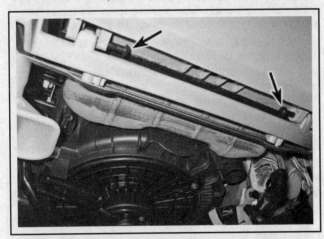

24.16 To remove the glove box door, remove the two pins at the bottom of the door

24.14 Remove the retaining bolts securing the knee bolster reinforcement

remove the door (see illustration).

17 Installation is the reverse of the removal procedure.

GLOVEBOX TRIM PANEL

▶ **Refer to illustration 24.19**

18 Remove the glove box door (see Step 16).

19 Remove the fasteners securing the trim panel, then remove the panel (see illustration).

20 Installation is the reverse of the removal procedure.

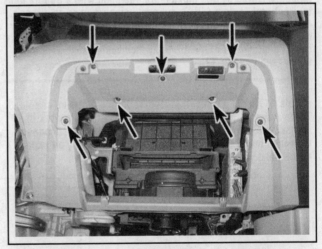

24.19 Remove the fasteners securing the trim panel, then remove the panel

25 Steering column covers - removal and installation

◆ Refer to illustration 25.2

✳✳ WARNING:

Models covered by this manual are equipped with a Supplemental Restraint System (SRS), more commonly known as airbags. Always disable the airbag system before working in the vicinity of any airbag system component to avoid the possibility of accidental deployment of the airbag, which could cause personal injury (see Chapter 12).

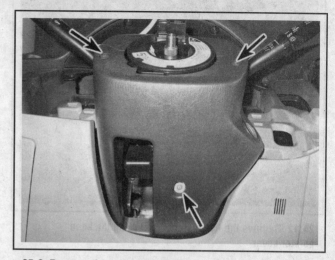

25.2 Remove the screws, then remove upper and lower covers

1 Disconnect the cable from the negative terminal of the battery (see Chapter 5, Section 1). On tilt steering columns, move the column to the lowest position. Refer to Chapter 10 and remove the steering wheel.

2 Remove the screws, then separate the halves and remove the upper and lower steering column covers (see illustration). On some models it may be necessary to remove the instrument cluster lower finish panel (see Section 24).

3 Installation is the reverse of the removal procedure.

26 Instrument panel - removal and installation

◆ Refer to illustrations 26.5, 26.9, 26.10, 26.12a, 26.12b, 26.13a, 26.13b and 26.13c

✳✳ WARNING:

Models covered by this manual are equipped with a Supplemental Restraint System (SRS), more commonly known as airbags. Always disable the airbag system before working in the vicinity of any airbag system component to avoid the possibility of accidental deployment of the airbag, which could cause personal injury (see Chapter 12).

➡Note 1: This is a difficult procedure for the home mechanic. There are many hidden fasteners, difficult angles to work in and many electrical connectors to tag and disconnect/connect. We recommend that this procedure be done only by an experienced do-it-yourselfer.

➡Note 2: During removal of the instrument panel, make careful notes of how each piece comes off, where it fits in relation to other pieces and what holds it in place. If you note how each part is installed before removing it, getting the instrument panel back together again will be much easier.

➡Note 3: It is not necessary, but it is suggested to remove both front seats to allow additional working space and lessen the chance of damage to the seats during this procedure.

1 Disconnect the cable from the negative battery terminal (see Chapter 5, Section 1).

2 Remove the dashboard trim panels (see Section 24) and the center floor console (see Section 23).

3 Remove the glove box (see Section 24).

4 Remove the instrument cluster (see Chapter 12).

5 Disconnect the electrical connector from the passenger's side airbag (see illustration).

26.5 Disconnect the electrical connector from the passenger's side airbag

26.9 Carefully detach the kick panels from each side by pulling them out and releasing the clips

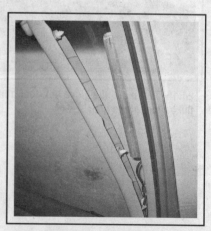

26.10 Remove the front pillar trim by carefully releasing the clips

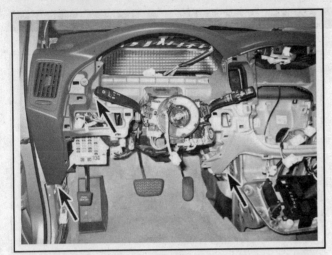

26.12a Remove all of the fasteners (bolts, screws and nuts) on the left . . .

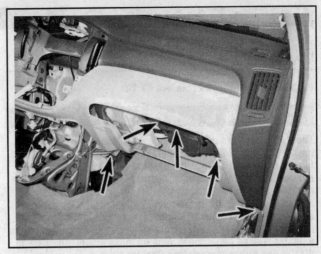

26.12b . . . and the right side of the instrument panel

6 Remove the audio unit and air conditioning control panel at the center of the dashboard (see Chapter 12).

7 Remove the driver's knee bolster and reinforcement panel (see Section 24).

8 Unscrew the bolts securing the steering column and lower it away from the instrument panel (see Chapter 10).

9 Remove the side kick panels (see illustration).

10 Remove the front pillar trim (see illustration).

11 A number of electrical connectors must be disconnected in order to remove the instrument panel. Most are designed so that they will only fit on the matching connector (male or female), but if there is any doubt, mark the connectors with masking tape and a marking pen before disconnecting them.

12 Remove all of the fasteners (bolts, screws and nuts) holding the instrument panel to the body (see illustrations). Once all are removed, lift the panel then pull it away from the windshield and take it out through the driver's door opening.

➡**Note: This is a two-person job.**

13 If you're also removing the instrument panel reinforcement tube, remove the fasteners securing the tube and take it out through the driver's door opening (see illustrations).

14 Installation is the reverse of removal.

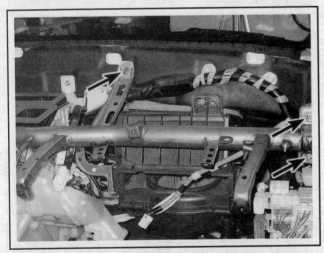

26.13a To remove the instrument panel reinforcement tube, remove the fasteners on the right side . . .

26.13b . . . and the fasteners on the left side . . .

26.13c . . . then remove the fasteners securing the center brace (right side shown, left side similar)

27 Seats - removal and installation

FRONT SEAT

▶ Refer to illustrations 27.2a and 27.2b

✳✳ WARNING 1:

The front seat belts on some models are equipped with pre-tensioners, which are pyrotechnic (explosive) devices designed to retract the seat belts in the event of a collision. On models equipped with pre-tensioners, do not remove the front seat belt retractor assemblies, and do not disconnect the electrical connectors leading to the assemblies. Problems with the pre-tensioners will turn on the SRS (airbag) warning light on the dash. If any pre-tensioner problems are suspected, take the vehicle to a dealer service department. Also on these models, be sure to disable the airbag system (see Chapter 12).

✳✳ WARNING 2:

On models with side-impact airbags, be sure to disarm the air-bag system before beginning this procedure (see Chapter 12).

1   Pry out the plastic covers to access the seat tracks and their mounting bolts.
2   Remove the retaining bolts (see illustrations).
3   Tilt the seat upward to access the underside, then disconnect any electrical connectors and lift the seat from the vehicle.
4   Installation is the reverse of removal.

27.2a Remove the front . . .

REAR SEAT

▶ Refer to illustrations 27.5, 27.6 and 27.7

5   Working at the front of the rear seats, remove the retaining bolts (see illustration).
6   Flip the seats backs down, then unclip the seat track covers (see illustration).
7   Remove the remaining seat retaining bolts (see illustration).
8   Installation is the reverse of removal.

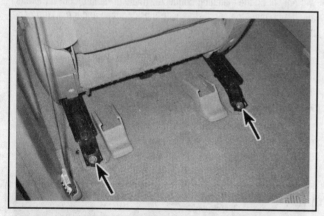

27.2b . . . and rear retaining bolts

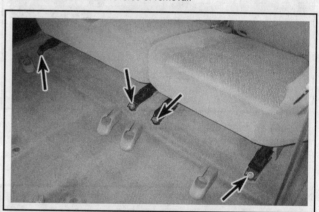

27.5 Remove the retaining bolts at the front of the rear seats

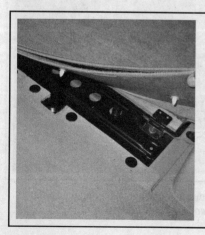

27.6 Unclip the seat track covers

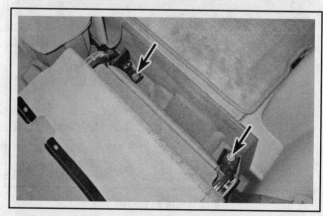

27.7 Remove the retaining bolts at the rear of the seats

Section

12

CHASSIS ELECTRICAL SYSTEM

1 General information

The electrical system is a 12-volt, negative ground type. Power for the lights and all electrical accessories is supplied by a lead/acid-type battery, which is charged by the alternator.

This Chapter covers repair and service procedures for the various electrical components not associated with the engine. Information on the battery, alternator and starter motor can be found in Chapter 5.

It should be noted that when portions of the electrical system are serviced, the cable should be disconnected from the negative battery terminal to prevent electrical shorts and/or fires.

2 Electrical troubleshooting - general information

◆ **Refer to illustrations 2.5a and 2.5b**

A typical electrical circuit consists of an electrical component, any switches, relays, motors, fuses, fusible links or circuit breakers related to that component and the wiring and connectors that link the component to both the battery and the chassis. To help you pinpoint an electrical circuit problem, wiring diagrams are included at the end of this Chapter.

Before tackling any troublesome electrical circuit, first study the appropriate wiring diagrams to get a complete understanding of what makes up that individual circuit. For instance, noting whether other components related to the circuit are operating correctly can often narrow down potential causes of trouble. If several components or circuits fail at one time, chances are the problem is in a fuse or ground connection, because several circuits are often routed through the same fuse and ground connections.

Electrical problems usually stem from simple causes, such as loose or corroded connections, a blown fuse, a melted fusible link or a failed relay. Visually inspect the condition of all fuses, wires and connections in a problem circuit before troubleshooting the circuit.

If test equipment and instruments are going to be utilized, use the diagrams to plan ahead of time where you will make the necessary connections in order to accurately pinpoint the trouble spot.

The basic tools needed for electrical troubleshooting include a circuit tester or voltmeter (a 12-volt bulb with a set of test leads can also be used), a continuity tester, which includes a bulb, battery and set of test leads, and a jumper wire, preferably with a circuit breaker incorporated, which can be used to bypass electrical components (see illustrations). Before attempting to locate a problem with test instruments, use the wiring diagram(s) to decide where to make the connections.

VOLTAGE CHECKS

◆ **Refer to illustration 2.6**

Voltage checks should be performed if a circuit is not functioning properly. Connect one lead of a circuit tester to either the negative battery terminal or a known good ground. Connect the other lead to a connector in the circuit being tested, preferably nearest to the battery or fuse (see illustration). If the bulb of the tester lights, voltage is present, which means that the part of the circuit between the connector and the battery is problem free. Continue checking the rest of the circuit in the same fashion. When you reach a point at which no voltage is present, the problem lies between that point and the last test point with voltage. Most of the time the problem can be traced to a loose connection.

➡**Note: Keep in mind that some circuits receive voltage only when the ignition key is in the Accessory or Run position.**

FINDING A SHORT

One method of finding shorts in a live circuit is to remove the fuse and connect a test light in place of the fuse terminals (fabricate two jumper wires with small spade terminals, plug the jumper wires into the fuse box and connect the test light). There should be voltage present in the circuit. Move the suspected wiring harness from side-to-side while watching the test light. If the bulb goes off, there is a short to ground somewhere in that area, probably where the insulation has rubbed through.

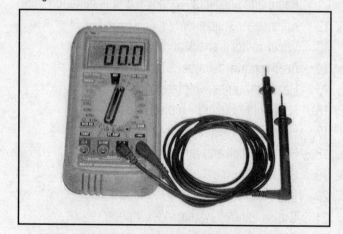

2.5a The most useful tool for electrical troubleshooting is a digital multimeter that can check volts, amps, and test continuity

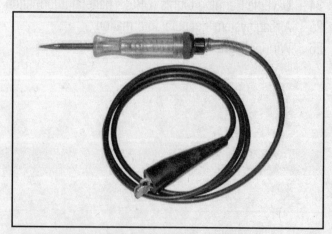

2.5b A simple test light is a very handy tool for testing voltage

2.6 In use, a basic test light's lead is clipped to a known good ground, then the pointed probe can test connectors, wires or electrical sockets - if the bulb lights, the circuit being tested has battery voltage

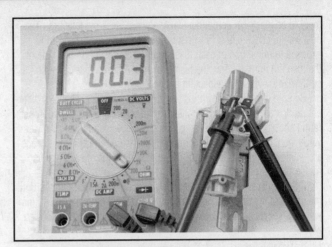

2.9 With a multimeter set to the ohm scale, resistance can be checked across two terminals - when checking for continuity, a low reading indicates continuity, a high reading or infinity indicates lack of continuity

GROUND CHECK

Perform a ground test to check whether a component is properly grounded. Disconnect the battery and connect one lead of a continuity tester or multimeter (set to the ohm scale), to a known good ground. Connect the other lead to the wire or ground connection being tested. If the resistance is low (less than 5 ohms), the ground is good. If the bulb on a self-powered test light does not go on, the ground is not good.

CONTINUITY CHECK

▶ Refer to illustration 2.9

A continuity check is done to determine if there are any breaks in a circuit - if it is passing electricity properly. With the circuit off (no power in the circuit), a self-powered continuity tester or multimeter can be used to check the circuit. Connect the test leads to both ends of the circuit (or to the "power" end and a good ground), and if the test light comes on the circuit is passing current properly (see illustration). If the resistance is low (less than 5 ohms), there is continuity; if the reading is 10,000 ohms or higher, there is a break somewhere in the circuit. The same procedure can be used to test a switch, by connecting the continuity tester to the switch terminals. With the switch turned On, the test light should come on (or low resistance should be indicated on a meter).

FINDING AN OPEN CIRCUIT

When diagnosing for possible open circuits, it is often difficult to locate them by sight because the connectors hide oxidation or terminal misalignment. Merely wiggling a connector on a sensor or in the wiring harness may correct the open circuit condition. Remember this when an open circuit is indicated when troubleshooting a circuit. Intermittent

problems may also be caused by oxidized or loose connections.

Electrical troubleshooting is simple if you keep in mind that all electrical circuits are basically electricity running from the battery, through the wires, switches, relays, fuses and fusible links to each electrical component (light bulb, motor, etc.) and to ground, from which it is passed back to the battery. Any electrical problem is an interruption in the flow of electricity to and from the battery.

CONNECTORS

Most electrical connections on these vehicles are made with multi-wire plastic connectors. The mating halves of many connectors are secured with locking clips molded into the plastic connector shells. The mating halves of large connectors, such as some of those under the instrument panel, are held together by a bolt through the center of the connector.

To separate a connector with locking clips, use a small screwdriver to pry the clips apart carefully, then separate the connector halves. Pull only on the shell, never pull on the wiring harness as you may damage the individual wires and terminals inside the connectors. Look at the connector closely before trying to separate the halves. Often the locking clips are engaged in a way that is not immediately clear. Additionally, many connectors have more than one set of clips.

Each pair of connector terminals has a male half and a female half. When you look at the end view of a connector in a diagram, be sure to understand whether the view shows the harness side or the component side of the connector. Connector halves are mirror images of each other, and a terminal shown on the right side end-view of one half will be on the left side end view of the other half.

3 Fuses and fusible links - general information

FUSES

▶ Refer to illustrations 3.1a, 3.1b and 3.3

The electrical circuits of the vehicle are protected by a combination of fuses, circuit breakers and fusible links. Fuse blocks are located

under the instrument panel and in the engine compartment (see illustrations).

Each of the fuses is designed to protect a specific circuit, and the various circuits are identified on the fuse panel cover.

Miniaturized fuses are employed in the fuse blocks. These compact fuses, with blade terminal design, allow fingertip removal and replacement. If an electrical component fails, always check the fuse first. The best way to check a fuse is with a test light. Check for power at the exposed terminal tips of each fuse. If power is present on one side of the fuse but not the other, the fuse is blown. A blown fuse can also be confirmed by visually inspecting it (see illustration).

Be sure to replace blown fuses with the correct type. Fuses of different ratings are physically interchangeable, but only fuses of the proper rating should be used. Replacing a fuse with one of a higher or lower value than specified is not recommended. Each electrical circuit needs a specific amount of protection. The amperage value of each fuse is molded into the fuse body.

If the replacement fuse immediately fails, don't replace it again until the cause of the problem is isolated and corrected. In most cases, this will be a short circuit in the wiring caused by a broken or deteriorated wire.

FUSIBLE LINKS

Some circuits are protected by fusible links. The links are used in circuits that are not ordinarily fused, such as the high-current side of the charging or starting circuits. Conventional inline fusible links, such as those used in the starter cable, are characterized by a bulge in the cable. Newer cartridge-type fusible links, which are similar in appearance to a large cartridge-type fuse, are located in their own fusible link block in the engine compartment fuse and relay box (see illustration

3.1a The engine compartment fuse and relay box is located at the left side of the engine compartment. There's a guide on the underside of the cover. The area between the arrows is the fusible link area, where cartridge-style fusible links are located

3.1a). After disconnecting the cable from the negative battery terminal, simply remove the fusible link and replace it with a unit of the same amperage.

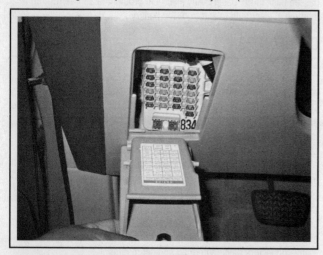

3.1b There's also a fuse box inside the vehicle, at the left end of the dash, behind a small access door, which has a fuse guide on it

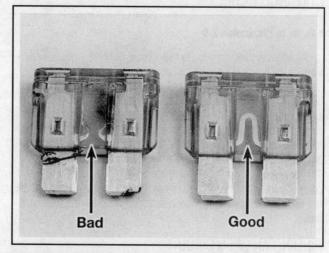

3.3 When a fuse blows, the element between the terminals melts

4 Circuit breakers - general information

Circuit breakers protect certain circuits, such as the power windows or heated seats. Depending on the vehicle's accessories, there might be circuit breakers in or near either of the fuse and relay boxes.

Because the circuit breakers reset automatically, an electrical overload in a circuit-breaker-protected system will cause the circuit to fail momentarily, then come back on. If the circuit does not come back on, check it immediately.

For a basic check, pull the circuit breaker up out of its socket on the

fuse panel, but just far enough to probe with a voltmeter. The breaker should still contact the sockets.

With the voltmeter negative lead on a good chassis ground, touch each end prong of the circuit breaker with the positive meter probe. There should be battery voltage at each end. If there is battery voltage only at one end, the circuit breaker must be replaced.

Some circuit breakers must be reset manually.

5 Relays - general information and testing

GENERAL INFORMATION

1   Several electrical accessories in the vehicle, such as the fuel injection system, horns, starter, and fog lamps use relays to transmit the electrical signal to the component. Relays use a low-current circuit (the control circuit) to open and close a high-current circuit (the power circuit). If the relay is defective, that component will not operate properly. Most relays are mounted in the engine compartment fuse/relay box, with some specialized relays located in the underhood box at the right fender. On some models, the ABS relays are located in a separate underhood relay box at the left side of the radiator. If a faulty relay is suspected, it can be removed and tested using the procedure below or by a dealer service department or a repair shop. Defective relays must be replaced as a unit. Identification of the circuit the relay controls is often marked on the top of the relay, but the decal or imprint inside the cover of the relay box should also indicate which circuits they control.

TESTING

▶ **Refer to illustrations 5.3a, 5.3b and 5.6**

2   Refer to the wiring diagrams for the circuit to determine the proper connections for the relay you're testing. If you can't determine the correct connection from the wiring diagrams, however, you may be able to determine the test connections from the information that follows.

3   There are four basic types of relays used on these models (see illustrations). Some are normally open type and some normally closed, while others include a circuit of each type.

4   On most relays, two of the terminals are the relay control circuit (they connect to the relay coil which, when energized, closes the large contacts to complete the circuit). The other terminals are the power circuit (they are connected together within the relay when the control-circuit coil is energized).

5   Some relays may be marked as an aid to help you determine which terminals make up the control circuit and which make up the power circuit. If the relay is not marked, refer to the wiring diagrams at the end of this Chapter to determine the proper hook-ups for the relay you're testing.

6   To test a relay, connect an ohmmeter across the two terminals of the power circuit, continuity should not be indicated (see illustration). Now connect a fused jumper wire between one of the two control circuit terminals and the positive battery terminal. Connect another jumper wire between the other control circuit terminal and ground. When the connections are made, the relay should click and continuity should be indicated on the meter. On some relays, polarity may be critical, so, if the relay doesn't click, try swapping the jumper wires on the control circuit terminals.

7   If the relay fails the above test, replace it.

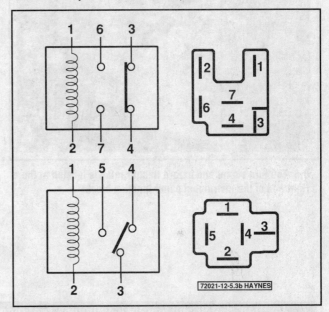

5.3b   These relays are normally closed types, where current flows though one circuit until the relay is energized, which interrupts that circuit and completes the second circuit

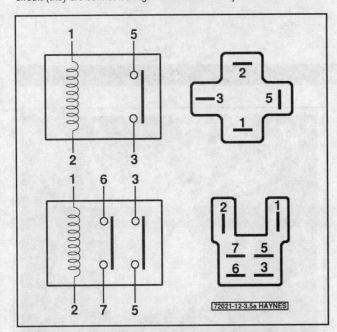

5.3a   These two relays are typical normally open types; the one above completes a single circuit (terminal 5 to terminal 3) when energized - the lower relay type completes two circuits (6 and 7, and 3 and 5) when energized

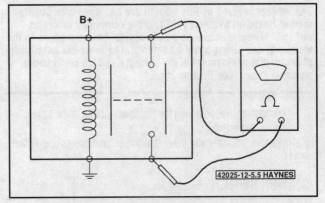

5.6   To test a typical four-terminal normally open relay, connect an ohmmeter to the two terminals of the power circuit - the meter should indicate continuity with the relay energized and no continuity with the relay not energized

6   Turn signal and hazard flasher relay - check and replacement

CHECK

WARNING:

The models covered by this manual are equipped with Supplemental Restraint Systems (SRS), more commonly known as airbags. Always disable the airbag system before working in the vicinity of any airbag system component to avoid the possibility of accidental deployment of the airbags, which could cause personal injury (see Section 25).

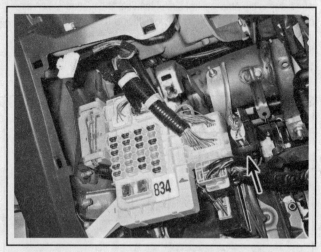

6.5 The turn signal and hazard flasher relay is located at the right end of the instrument panel junction block

1   When the turn signal and hazard flasher relay is functioning properly, you can hear an audible click when it's operating.

2   If the turn signals fail on one side or the other and the flasher unit does not make its characteristic clicking sound, or if a bulb on one side of the vehicle flashes much faster than normal but the bulb at the other end of the vehicle (on the same side) doesn't light at all, a turn signal bulb is probably faulty.

3   If both turn signals fail to blink, the problem might be a blown fuse, a faulty flasher unit, a defective switch or a loose or open connection. If a quick check of the fuse box indicates that the turn signal fuse has blown, check the wiring for a short before installing a new fuse.

REPLACEMENT

▶ **Refer to illustration 6.5**

➡ **Note: The turn signal and hazard flasher relay is located behind the knee bolster, on the right end of the instrument panel junction block.**

4   Remove the knee bolster (see Chapter 11).

5   Locate the turn signal and hazard flasher relay on the instrument panel junction block assembly (see illustration).

6   Disconnect the electrical connector from the flasher unit.

7   Remove the flasher unit from the instrument panel junction block assembly.

8   Make sure that the replacement unit is identical to the original. Compare the old one to the new one before installing it.

9   Installation is the reverse of removal.

7   Ignition switch and key lock cylinder - replacement

WARNING:

The models covered by this manual are equipped with Supplemental Restraint Systems (SRS), more commonly known as airbags. Always disable the airbag system before working in the vicinity of any airbag system component to avoid the possibility of accidental deployment of the airbag(s), which could cause personal injury (see Section 25).

1   Disconnect the cable from the negative terminal of the battery (see Chapter 5, Section 1).

2   Remove the upper and lower steering column covers (see Chapter 11).

IGNITION SWITCH

▶ **Refer to illustrations 7.3 and 7.4**

3   Disconnect the electrical connector from the ignition switch (see illustration).

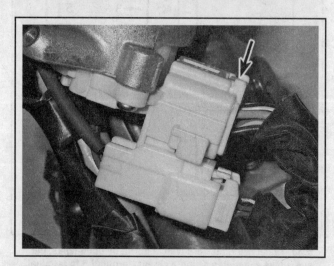

7.3 To disconnect the electrical connector from the ignition switch, depress this release tab and pull off the connector

CHASSIS ELECTRICAL SYSTEM **12-7**

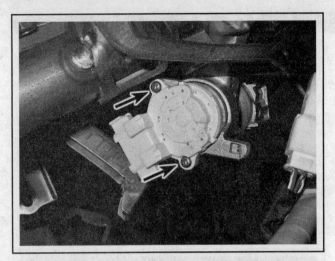

7.4 To detach the ignition switch from the key lock cylinder housing, remove these two screws

4   Remove the ignition switch retaining screws (see illustration) and remove the switch from the key lock cylinder housing.
5   Installation is the reverse of the removal.

KEY LOCK CYLINDER

▶ **Refer to illustrations 7.7, 7.8a and 7.8b**

6   Place the ignition key in the ACC position.
7   If the vehicle is equipped with an immobilizer system, discon-

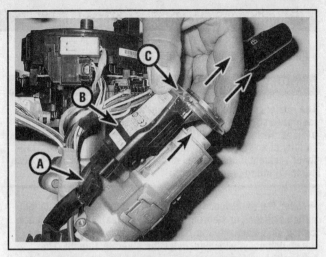

7.7 To remove the immobilizer module and the key lock cylinder illumination ring, disconnect the electrical connector (A), then slide off the immobilizer (B) and the illumination ring (C) from the key lock cylinder housing

nect the electrical connector from the immobilizer module, then slide the immobilizer module and the key lock cylinder illumination ring as a single assembly (see illustration).
8   Use an awl or punch to depress the key lock cylinder retaining pin and remove the key lock cylinder from the housing (see illustrations).
9   To install the lock cylinder, depress the retaining pin and guide the lock cylinder into the housing until the retaining pin extends itself back into the locating hole in the housing.
10   The remainder of installation is the reverse of removal.

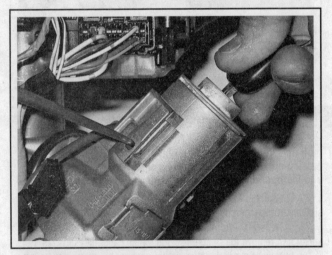

7.8a To remove the key lock cylinder from the lock cylinder housing, put the lock cylinder in the ACC position, insert an awl or punch into this small hole in the housing, push it in until it depresses the lock cylinder retaining pin . . .

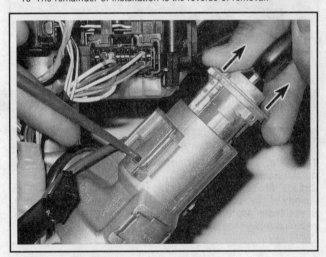

7.8b . . . and pull out the key lock cylinder

8   Multi-function switches - replacement

▸ Refer to illustrations 8.3a and 8.3b

❈❈ WARNING:

The models covered by this manual are equipped with Supplemental Restraint Systems (SRS), more commonly known as airbags. Always disable the airbag system before working in the vicinity of any airbag system components to avoid the possibility of accidental deployment of the airbag(s), which could cause personal injury (see Section 25).

➡Note: The multi-function switches (also referred to as combination switches or steering column switches) are two separate switch units connected to a central plastic housing known as the switch body, which encircles the steering column. The left multi-function switch controls the headlights and the turn signals; the right switch controls the windshield washer/wiper system. Either switch can be replaced separately.

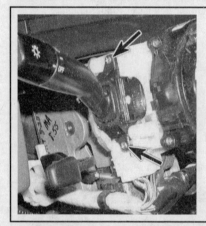

8.3a To detach either multi-function switch from the switch body on 1999 through 2003 Lexus models, remove these two screws

1   Disconnect the cable from the negative terminal of the battery (see Chapter 5, Section 1).

2   Remove the upper and lower steering column covers (see Chapter 11).

3   The multi-function switches are secured to the switch body one of two ways, depending on the year and model. On 1999 through 2003 Lexus models, each switch is secured to the switch body with a pair of screws. On 2004 and later Lexus models and on all Toyota models, the switches are secured to the switch body by retaining clips that must be depressed to disengage the switch (see illustrations).

4   Remove the multi-function switch and disconnect the electrical connectors from the switch.

5   Installation is the reverse of removal.

8.3b To detach either multi-function switch from the switch body on 2004 and later Lexus models and on all Toyota models, depress the clip with a screwdriver and slide out the switch

9   Instrument panel switches - replacement

❈❈ WARNING:

The models covered by this manual are equipped with Supplemental Restraint Systems (SRS), more commonly known as airbags. Always disable the airbag system before working in the vicinity of any airbag system component to avoid the possibility of accidental deployment of the airbag(s), which could cause personal injury (see Section 25).

SWITCHES LOCATED AT THE LOWER LEFT END OF THE INSTRUMENT PANEL

▸ Refer to illustrations 9.1, 9.2a, 9.2b, 9.3a and 9.3b

➡Note: Various switches are housed in the lower left end of the instrument panel, to the left of the steering column. On Lexus models, these switches are located on one or two small panels that are easily removed from the larger trim panel. On Toyota models, the switches are housed in a single larger panel that spans both sides of the steering column. Regardless of the size and shape of the panel, various switches are located here. They include (but are not limited to) the power mirror switch, the

theft deterrent system/engine immobilizer system switch, the outside mirror defogger and windshield wiper de-icer switch and the traction control system switch (2WD models).

1   Carefully pry the switch panel out of the instrument panel with a suitable trim panel tool (see illustration).

2   To replace the power mirror switch, disconnect the electrical connector and remove the switch from the switch panel (see illustrations).

3   To replace the theft deterrent/engine immobilizer system switch, disconnect the electrical connector and remove the switch from the switch panel (see illustrations).

4   The rest of the switches housed in the switch panel are replaced the same way as the two described above. To replace any of the other switches on this panel, refer to illustrations 9.2a, 9.2b, 9.3a and 9.3b.

5   Installation is the reverse of the removal procedure.

ELECTRICAL DEVICES ON THE CLIMATE CONTROL ASSEMBLY

▸ Refer to illustration 9.8

➡Note: On Toyota models, electrical devices such as the hazard warning switch, digital clock and passenger seatbelt reminder

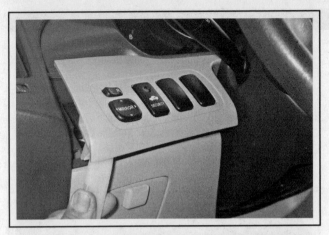

9.1 On 1999 through 2003 models, use a trim panel tool to pry off the switch panel from the instrument panel (Toyota Highlander shown, Lexus similar)

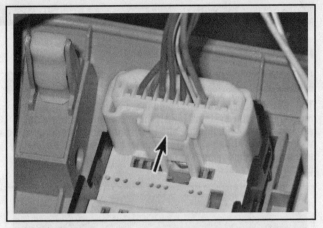

9.2a To disconnect the electrical connector from the power mirror switch, depress this release tab and pull out the connector

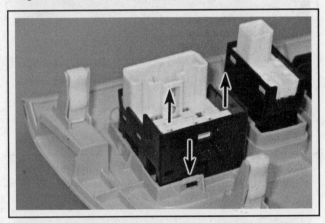

9.2b To remove the power mirror switch, carefully disengage the trim panel from the locking lugs on the switch housing (the other locking lug, on the other side of the switch housing, not visible in this photo) and pull the switch out of the trim panel

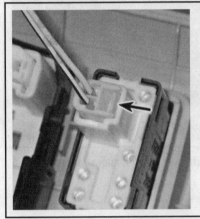

9.3a To disconnect the electrical connector from the theft deterrent/ engine immobilizer system switch, depress this release tab and pull off the connector

light protrude through the upper part of the center trim panel, but they're actually located on the heater and air conditioning control assembly.

6 Remove the center trim panel (see Chapter 11).

7 Remove the climate control assembly (see Chapter 3).

8 Remove the cover from the back of the climate control assembly (see illustration).

9 Remove the hazard warning switch, digital clock or passenger seatbelt reminder light from the climate control assembly.

10 Installation is the reverse of removal.

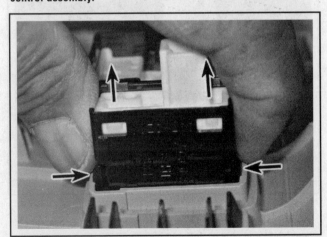

9.3b To remove the theft deterrent/engine immobilizer system switch, depress these release tabs on both sides of the switch and pull the switch out of the trim panel

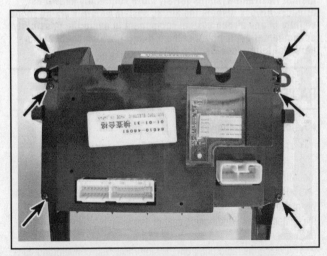

9.8 Remove these screws to detach the cover from the climate control assembly

10 Instrument cluster - removal and installation

▶ Refer to illustrations 10.3 and 10.4

❋❋ **WARNING:**

The models covered by this manual are equipped with Supplemental Restraint Systems (SRS), more commonly known as airbags. Always disable the airbag system before working in the vicinity of any airbag system component to avoid the possibility of accidental deployment of the airbag(s), which could cause personal injury (see Section 25).

1   Disconnect the cable from the negative battery terminal (see Chapter 5, Section 1).

2   Remove the instrument cluster trim panel (see Chapter 11).

3   Remove the cluster mounting screws (see illustration) and pull the instrument cluster towards the steering wheel.

4   Disconnect the electrical connectors from the backside of the cluster (see illustration).

5   Installation is the reverse of removal.

10.3 To detach the instrument cluster, remove these mounting screws

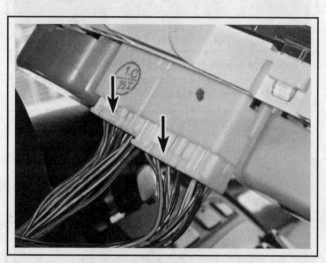

10.4 Pull out the cluster far enough to disconnect the electrical connectors from the backside of the cluster

11 Radio and speakers - removal and installation

❋❋ **WARNING:**

The models covered by this manual are equipped with Supplemental Restraint Systems (SRS), more commonly known as airbags. Always disable the airbag system before working in the vicinity of any airbag system component to avoid the possibility of accidental deployment of the airbag(s), which could cause personal injury (see Section 25).

RADIO

▶ Refer to illustration 11.4

1   Remove the center trim panel (see Chapter 11).

2   Remove the radio and heater/air conditioning control assembly mounting bolts (see illustration 10.4 in Chapter 3).

3   Pull out the radio and heater and air conditioning control assembly and disconnect the electrical connectors from the radio and from the heater and air conditioning control assembly (see illustration 10.5 in Chapter 3).

4   Detach the radio unit from the radio/heater/air conditioning control assembly mounting bracket (see illustration).

5   Installation is the reverse of removal.

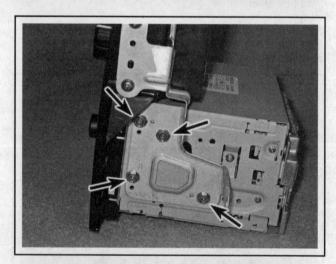

11.4 To detach the radio unit from the radio/heater/air conditioning mounting bracket, remove these bolts from each end (right end shown, left end identical)

11.7 To remove a door speaker, disconnect the electrical connector and remove the speaker mounting screws

SPEAKERS

Door speakers

▶ Refer to illustration 11.7

➡Note: All models have speakers in both front and rear doors. The photos accompanying this section depict front door speakers, but the procedure applies to rear door speakers as well, which are mounted in a fashion similar to the front door speakers.

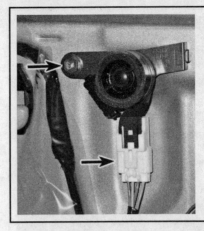

11.10 To remove a tweeter, disconnect the electrical connector and remove the mounting bolt

6   Remove the door trim panel (see Chapter 11).
7   Disconnect the electrical connector from the speaker, remove the speaker mounting screws (see illustration) and remove the speaker.
8   Installation is the reverse of removal.

Tweeters

▶ Refer to illustration 11.10

➡Note: The tweeters are located in the front doors.

9   Remove the door trim panel (see Chapter 11).
10   Disconnect the electrical connector, remove the mounting bolt (see illustration) and remove the tweeter.
11   Installation is the reverse of removal.

12 Antenna - replacement

CONVENTIONAL ANTENNA MAST

▶ **Refer to illustration 12.1**

1   Unscrew the antenna mast from the mounting base (see illustration).

Antenna mounting base

▶ **Refer to illustrations 12.3, 12.4, 12.5, 12.8 and 12.9**

2   Remove the antenna mast (see illustration 12.1).
3   Working inside the vehicle, remove the right A-pillar trim panel

(see illustration) and the right kick panel. (For help with removing either of these trim pieces, refer to Chapter 11.)
4   Disconnect the electrical connector that connects the cable from the antenna mounting base to the cable that goes through the dash (see illustration).
5   Remove the castellated nut from the top of the antenna mounting

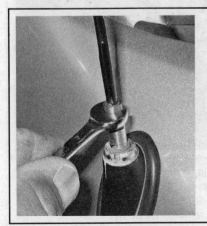

12.1 To remove the antenna mast from its mounting base, simply unscrew it

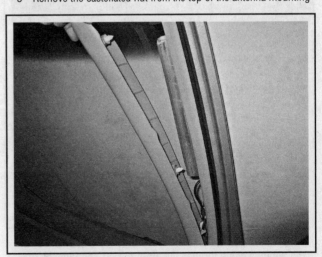

12.3 Use a trim removal tool or a slotted screwdriver to pry off the right A-pillar trim panel (if you're going to use a screwdriver, tape the tip to protect the plastic trim)

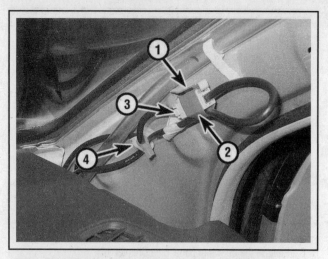

12.4 This electrical connector connects the cable from the antenna mounting base to the antenna cable that's routed through the dash. To disconnect it, detach it from the mounting bracket (1), cut the tape (2) wrapped around it, then depress the release tab (3) and disconnect the connector. Then pull the antenna mounting base cable out of the clip (4)

base (see illustration).

6   Loosen the right front wheel lug nuts, raise the front of the vehicle and place it securely on jackstands, then remove the right front wheel.

※※ WARNING:

If the vehicle is equipped with an electronically modulated air suspension, make sure that the height control switch is turned off before raising the vehicle.

---

7   Remove the right wheelhousing splash shield (see Chapter 11).

8   Locate the lower part of the antenna mounting base assembly (see illustration). Carefully pull the antenna mounting base cable

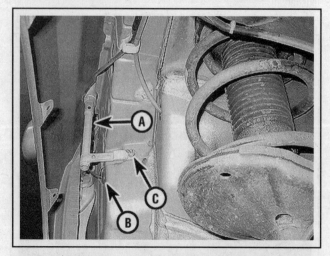

12.8 To remove the antenna mounting base (A), snake out the antenna cable through the grommet (B) and remove mounting bracket nut (C). When installing the mounting base assembly, don't forget to reattach the ground wire at the mounting nut

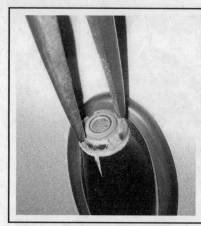

12.5 To detach the antenna mounting base from the fender, unscrew this castellated retaining nut. Special wrenches for this task are available at auto parts stores, but you can also use a pair of needle-nose pliers

through the grommet.

9   Using a flashlight, insert the antenna mounting base cable into the grommet and up through the right corner of the passenger compartment, between the end of the instrument panel and the A-pillar (see illustration).

10   Installation is otherwise the reverse of removal.

Antenna cable between the antenna cable and the radio

11   Remove the instrument panel assembly (see Chapter 11).

12   Study the routing of the antenna cable. Make a sketch if necessary.

13   Unclip the antenna cable from the backside of the instrument panel.

14   Installation is the reverse of removal. Make sure that the cable is routed correctly and that all clips are installed.

GRID-TYPE ANTENNA

15   Some of the vehicles covered by this manual are equipped with a wire grid-type antenna attached to the rear window glass. If you have this type of antenna on your vehicle, and if there's a problem with it, you can repair the antenna grid the same way that you'd repair the rear window defogger grid (see Section 14).

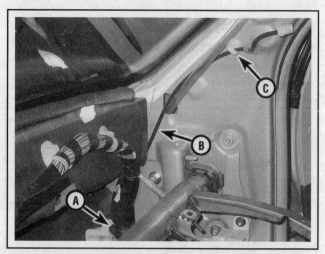

12.9 Using a flashlight, insert the antenna mounting base cable through the grommet (A), up between the end of the dash and the A-pillar (B), then through the clip on the A-pillar (C) (instrument panel removed for clarity)

13 Wiper motor - check and replacement

WIPER MOTOR CIRCUIT CHECK

➡Note: Refer to the wiring diagrams for the following checks. When checking for voltage, probe a grounded 12-volt test light to each terminal at a connector until it lights; this verifies voltage (power) at the terminal. If the following checks fail to locate the problem, have the system diagnosed by a dealer service department or other properly equipped repair facility.

1   If the wipers work slowly, make sure that the battery is in good condition and has a strong charge (see Chapter 5). If the battery is in good condition, remove the wiper motor (see below) and operate the wiper arms by hand. Check for binding linkage and pivots. Lubricate or repair the linkage or pivots as necessary. Reinstall the wiper motor. If the wipers still operate slowly, check for loose or corroded connections, especially the ground connection. If all connections look OK, replace the motor.

2   If the wipers fail to operate when activated, check the fuse in the driver's side interior fuse panel. If the fuse is OK, connect a jumper wire between the wiper motor's ground terminal and ground, then retest. If the motor works now, repair the ground connection. If the motor still doesn't work, turn the wiper switch to the HI position and check for voltage at the motor.

➡Note: The cowl cover will have to be removed (see Chapter 11) to access the wiper motor electrical connector.

3   If there's voltage at the connector, remove the motor and check it off the vehicle with fused jumper wires from the battery. If the motor now works, check for binding linkage (see Step 1). If the motor still doesn't work, replace it. If there's no voltage to the motor, check for voltage at the wiper control relays. If there's voltage at the wiper control relays and no voltage at the wiper motor, have the switch tested. If the switch is OK, the wiper control relay is probably bad. See Section 5 for relay testing.

4   If the interval (delay) function is inoperative, check the continuity of all the wiring between the switch and wiper control module.

5   If the wipers stop at the position they're in when the switch is turned off (fail to park), check for voltage at the park feed wire of the wiper motor connector when the wiper switch is OFF but the ignition is ON. If no voltage is present, check for an open circuit between the wiper motor and the fuse panel.

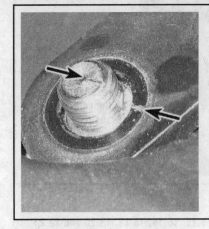

13.6a On Lexus models, the left (driver's) side wiper arm is a conventional single arm unit, secured by a retaining nut. Remove the nut and mark the relationship of the arm to the shaft

REPLACEMENT

Windshield wiper motor

Lexus models

▶ Refer to illustrations 13.6a, 13.6b, 13.6c, 13.8 and 13.9

6   Remove the wiper arm nuts and mark the relationship of the wiper arms to their shafts (see illustrations). Remove both wiper arms.

7   Remove the plastic cowl cover (see Chapter 11).

8   Disconnect the electrical connector from the wiper motor (see illustration).

9   Remove the windshield wiper motor and link assembly mounting bolts (see illustration) and remove the wiper motor and link assembly.

10 Using a screwdriver, carefully pry the link rod from the pivot pin on the motor's crank arm.

11 Remove the nut and washer that secures the crank arm to the motor shaft.

12 Mark the relationship of the crank arm to the motor shaft and remove the crank arm.

13 Remove the three motor mounting bolts and remove the motor from its mounting bracket on the link assembly.

14 Installation is the reverse of removal.

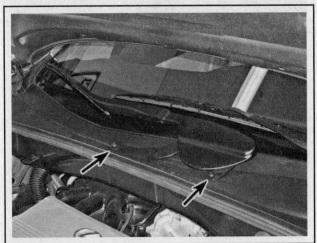

13.6b On Lexus models, the right (passenger) side wiper arm has two arms and nuts. Remove both nuts . . .

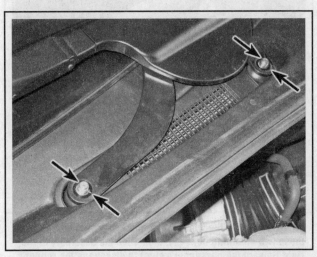

13.6c . . . and mark the relationship of the arms to the shafts

13.8 To disconnect the electrical connector from the windshield wiper motor, depress this release tab and pull off the connector (Lexus models)

Toyota models

▶ Refer to illustrations 13.15, 13.18, 13.19, 13.20 and 13.21

15 Remove the trim covers over the wiper arm mounting nuts (see illustration), then remove the wiper arm nuts.

16 Mark the position of each wiper arm to its shaft, then remove the arms (see illustration 13.6a).

17 Remove the plastic cowl cover (see Chapter 11).

13.15 Lift the end cap to access each wiper arm nut and remove the nut (Toyota models)

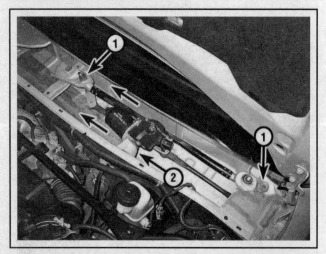

13.19 To detach the wiper motor and link assembly, remove these two mounting bolts (1), then slide the assembly to the right (toward the passenger's side) to disengage the rubber insulator from the semi-circular bracket (2) in the middle of the cowl

13.9 To detach the windshield wiper motor and link assembly from the cowl, remove these five bolts (Lexus models)

18 Disconnect the electrical connector from the wiper motor (see illustration).

19 Remove the wiper motor and link assembly mounting bolts (see illustration) and remove the wiper motor and link assembly from the cowl area.

20 Use a screwdriver to pry the linkage rod from the crank arm pivot of the wiper motor (see illustration).

13.18 To disconnect the electrical connector from the wiper motor, depress this release tab and pull off the connector

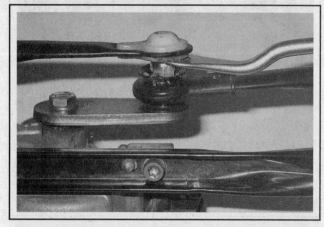

13.20 Use a trim panel tool to separate the link rod from the crank arm pivot

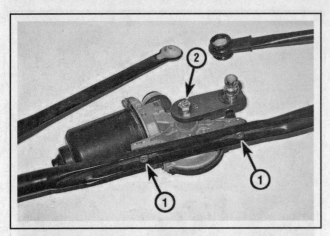

13.21 To detach the wiper motor from the link rod assembly, remove these two bolts (1). If you're replacing the motor, remove the crank arm nut (2)

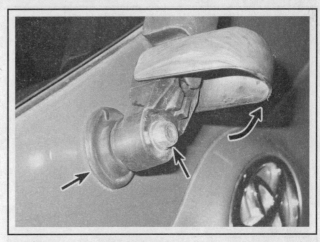

13.24 To remove the rear wiper arm, flip up the cap covering the retaining nut, remove the nut, mark the relationship of the wiper arm to the motor shaft, then remove the arm and the grommet

13.27 Disconnect the electrical connector from the rear wiper motor

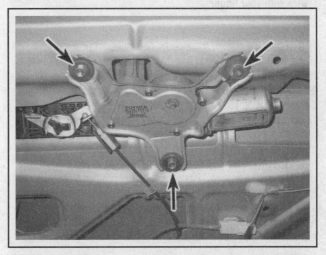

13.28 To detach the rear wiper motor from the hatch, remove these three bolts

21 Remove the crank arm nut (see illustration), mark the relationship of the crank arm to the motor shaft and remove the crank arm from the shaft.

22 Remove the wiper motor mounting bolts and separate the motor from the link rod assembly.

23 Installation is the reverse of removal.

Rear wiper motor

▶ **Refer to illustrations 13.24, 13.27 and 13.28**

24 Flip open the cap covering the rear wiper arm retaining nut (see illustration) and remove the nut.

25 Mark the relationship of the rear wiper arm to the motor shaft,

then remove the arm.

26 Remove the trim panel from the rear hatch (see Chapter 11).

27 Disconnect the electrical connector from the rear wiper motor (see illustration).

28 Remove the rear wiper motor mounting bolts (see illustration) and remove the motor.

29 Installation is the reverse of removal.

14 Rear window defogger - check and repair

1 The rear window defogger consists of a number of horizontal elements baked onto the glass surface.

2 Small breaks in the element can be repaired without removing the rear window.

CHECK

▶ **Refer to illustrations 14.4, 14.5 and 14.7**

3 Turn the ignition switch and defogger system switches to the ON

position. Using a voltmeter, place the positive probe against the defogger grid positive terminal and the negative probe against the ground terminal. If battery voltage is not indicated, check the fuse, defogger switch and related wiring. If voltage is indicated, but all or part of the defogger doesn't heat, proceed with the following tests.

4 When measuring voltage during the next two tests, wrap a piece of aluminum foil around the tip of the voltmeter positive probe and press the foil against the heating element with your finger (see illustration). Place the negative probe on the defogger grid ground terminal.

14.4 When measuring the voltage at the rear window defogger grid, wrap a piece of aluminum foil around the positive probe of the voltmeter and press the foil against the wire with your finger

14.5 To determine if a heating element has broken, check the voltage at the center of each element - if the voltage is 5 or 6-volts, the element is unbroken - if the voltage is 10 or 12-volts, the element is broken between the center and the ground side - if there is no voltage, the element is broken between the center and the positive side

5   Check the voltage at the center of each heating element (see illustration). If the voltage is 5 or 6-volts, the element is okay (there is no break). If the voltage is zero, the element is broken between the center of the element and the positive end. If the voltage is 10 to 12-volts the element is broken between the center of the element and ground. Check each heating element.

6   Connect the negative lead to a good body ground. The reading should stay the same. If it doesn't, the ground connection is bad.

7   To find the break, place the voltmeter negative probe against the defogger ground terminal. Place the voltmeter positive probe with the foil strip against the heating element at the positive terminal end and slide it toward the negative terminal end. The point at which the voltmeter deflects from several volts to zero is the point at which the heating element is broken (see illustration).

REPAIR

▶ **Refer to illustration 14.13**

8   Repair the break in the element using a repair kit specifically recommended for this purpose, available at most auto parts stores. Included in this kit is plastic conductive epoxy.

9   Prior to repairing a break, turn off the system and allow it to cool off for a few minutes.

10 Lightly buff the element area with fine steel wool, then clean it thoroughly with rubbing alcohol.

11 Use masking tape to mask off the area being repaired.

12 Thoroughly mix the epoxy, following the instructions provided with the repair kit.

13 Apply the epoxy material to the slit in the masking tape, overlapping the undamaged area about 3/4-inch on either end (see illustration).

14 Allow the repair to cure for 24 hours before removing the tape and using the system.

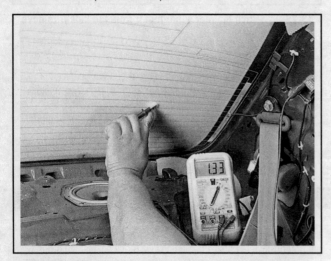

14.7 To find the break, place the voltmeter negative lead against the defogger ground terminal, place the voltmeter positive lead with the foil strip against the heating element at the positive terminal end and slide it toward the negative terminal end - the point at which the voltmeter reading changes abruptly is the point at which the element is broken

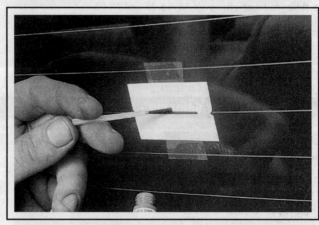

14.13 To use a defogger repair kit, apply masking tape to the inside of the window at the damaged area, then brush on the special conductive coating

15 Headlight bulbs - replacement

♦ **Refer to illustrations 15.2 and 15.3**

❉❉ WARNING:

Gas-filled bulbs are under pressure and may shatter if the surface is scratched or the bulb is dropped. Wear eye protection and handle the bulbs carefully, grasping only the base whenever possible. Do not touch the surface of the bulb with your fingers because the oil from your skin could cause it to overheat and fail prematurely. If you do touch the bulb surface, clean it with rubbing alcohol.

➡Note: The inner headlight bulbs are the high-beam bulbs. The outer bulbs are the low-beam bulbs (these bulbs are also illuminated when the high-beam bulbs are turned on).

1 If you're replacing a low-beam bulb on a Lexus model, disconnect the headlight harness connector from the short pigtail lead leading to the light bulb holder, then remove the rubber weather cover from the back of the headlight housing.

2 Disconnect the electrical connector from the bulb holder assembly (see illustration).

3 Grasp the bulb holder securely and rotate it counterclockwise to remove it from the housing (see illustration).

4 Without touching the bulb glass with your bare fingers, insert the new bulb assembly into the headlight housing. Twist it clockwise to lock it in place, then plug in the electrical connector.

5 If you replaced the low-beam bulb on a Lexus model, reinstall the round rubber cover over the low-beam bulb.

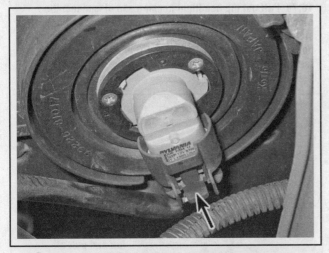

15.2 To disconnect the electrical connector from either headlight bulb holder, depress this release tab and pull off the connector

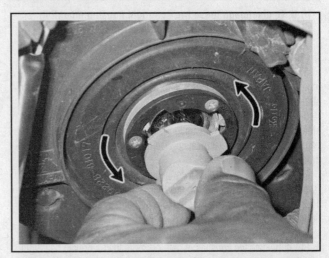

15.3 To remove the headlight bulb holder from the headlight housing, rotate the bulb holder counterclockwise and pull it out (this step also applies to the fog lamp bulb holder)

16 Headlight housing - replacement

♦ **Refer to illustrations 16.2, 16.6, 16.7a and 16.7b**

❉❉ WARNING:

These vehicles are equipped with gas-filled headlight bulbs that are under pressure and may shatter if the surface is damaged or the bulb is dropped. Wear eye protection and handle the bulbs carefully, grasping only the base whenever possible. Do not touch the surface of the bulb with your fingers because the oil from your skin could cause it to overheat and fail prematurely. If you do touch the bulb surface, clean it with rubbing alcohol.

1 Disconnect the cable from the negative battery terminal (see Chapter 5, Section 1).

2 Remove the upper headlight mounting bolts (see illustration).

3 Remove the grille and the bumper cover (see Chapter 11).

4 On 1999 through 2003 Lexus models, remove the two headlight housing mounting nuts from the outer edge of the headlight housing. On 2004 and later Lexus models, remove the headlight housing retaining bolt located at the outer lower corner of the housing.

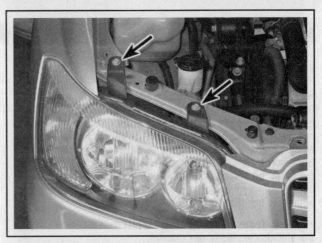

16.2 To detach the upper part of the headlight housing from the upper radiator crossmember, remove these two bolts (Toyota model shown, Lexus models similar)

16.6 To disengage these two headlight housing mounting clips, grasp the headlight firmly and carefully pull it out (Toyota model shown, Lexus models similar)

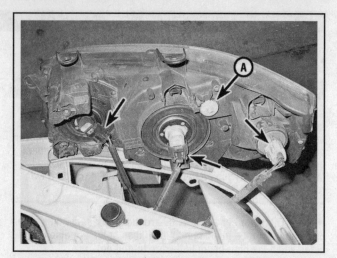

16.7a Pull out the headlight housing and disconnect the electrical connectors from the headlights and from the turn signal/sidemarker light. After you've installed the headlight housing, use the vertical adjuster (A) to adjust the headlights (see Section 17)

5   On Lexus models disengage the locator pin at the upper mounting bolt bracket. To disengage it, simply pull the mounting bracket straight up.

6   Disengage the two headlight housing clips located on the lower side of the housing (see illustration) and pull out the housing just far enough to access the electrical connectors on the backside of the headlight assembly.

7   Disconnect the electrical connectors from the headlight housing (see illustration). There are two headlight connectors (see illustration 15.2) and a third connector for the turn signal/parking and sidemarker light (see illustration).

8   If the headlight housing itself is damaged, you'll have to replace it. But if the damage consists of nothing more than a broken headlight housing mounting bracket, Lexus and Toyota sell bracket repair kits for the headlight housing mounting brackets.

9   Installation is the reverse of removal.

10   Be sure to check headlight adjustment when you're done (see Section 17).

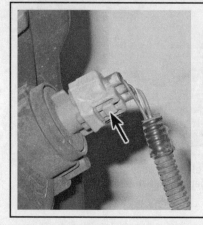

16.7b To disconnect the electrical connector from the front turn signal/parking and sidemarker light, depress this release tab and pull off the connector

17 Headlights - adjustment

▶ Refer to illustrations 17.2 and 17.5

➡ **Note 1: The headlights must be aimed correctly. If adjusted incorrectly they could blind the driver of an oncoming vehicle and cause a serious accident or seriously reduce your ability to see the road. The headlights should be checked for proper aim every 12 months and any time a new headlight is installed or front-end bodywork is performed. It should be emphasized that the following procedure is only an interim step, which will provide temporary adjustment until a properly equipped shop can adjust the headlights.**

➡ **Note 2: There are no horizontal adjustment screws on any of the models covered in this manual.**

1   There are several methods of adjusting the headlights. The simplest method requires masking tape, a blank wall and a level floor.

2   Position masking tape vertically on the wall in reference to the vehicle centerline and the centerlines of both headlights (see illustration).

3   Position a horizontal tape line in reference to the centerline of all the headlights.

➡ **Note: It may be easier to position the tape on the wall with the vehicle parked only a few inches away.**

4   Adjustment should be made with the vehicle parked 25 feet from the wall, sitting level, the gas tank half-full and no unusually heavy load in the vehicle.

5   Starting with the low beam adjustment, position the high intensity zone so it is two inches below the horizontal line. Make the adjustment by turning the vertical adjusting screw to raise or lower the beam (see illustration).

6   With the high beams on, the high intensity zone should be vertically centered with the exact center just below the horizontal line.

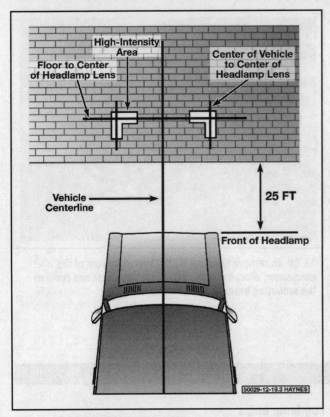

17.2 Headlight adjustment details

17.5 To turn the vertical adjuster wheel, insert a Phillips screwdriver into the gap at the top, engage the head of the screwdriver with the teeth on the adjuster wheel and turn the wheel (if you can't locate the adjuster wheel, refer to illustration 16.7a)

➡Note: It may not be possible to position the headlight aim exactly for both high and low beams. If a compromise must be made, keep in mind that the low beams are the most used and have the greatest effect on driver safety.

7   Have the headlights adjusted by a dealer service department or service station at the earliest opportunity.

18 Horn - check and replacement

✳✳ WARNING:

The models covered by this manual are equipped with Supplemental Restraint Systems (SRS), more commonly known as airbags. Always disable the airbag system before working in the vicinity of any airbag system components to avoid the possibility of accidental deployment of the airbag(s), which could cause personal injury (see Section 25).

CHECK

➡Note: Check the fuses before beginning electrical diagnosis.

1   Disconnect the electrical connector from the horn.
2   To test the horn, connect battery voltage to the horn terminal with a jumper wire. If the horn doesn't sound, replace it.
3   If the horn does sound, check for voltage at the terminal when the horn button is depressed. If there's voltage at the terminal, check for a bad ground at the horn.
4   If there's no voltage at the horn, check the relay (see Section 5).

5   If the relay is OK, check for voltage to the relay power and control circuits. If either of the circuits is not receiving voltage, inspect the wiring between the relay and the fuse panel.
6   If both relay circuits are receiving voltage, depress the horn button and check the circuit from the relay to the horn button for continuity to ground. If there's no continuity, check the circuit for an open. If there's no open circuit, replace the horn button.
7   If there's continuity to ground through the horn button, check for an open or short in the circuit from the relay to the horn.

REPLACEMENT

▶ Refer to illustrations 18.8a and 18.8b

➡Note: There are two horns. One is located in front of the condenser and the other is located below and in front of the engine compartment fuse and relay box.

8   Disconnect the electrical connector (see illustrations).
9   Remove the bracket bolt.
10   Installation is the reverse of removal.

18.8a To remove the horn that's located below and in front of the fuse and relay box, disconnect the electrical connector and remove the mounting bracket bolt

18.8b To remove the horn that's located in front of the condenser, disconnect the electrical connector and remove the mounting bracket bolt

19 Bulb replacement

EXTERIOR LIGHT BULBS

Front turn signal/parking and sidemarker light bulbs

▶ **Refer to illustrations 19.2 and 19.3**

➡**Note: The front turn signal/parking and sidemarker light bulbs are located in the headlight housing.**

1 Remove the headlight housing (see Section 16).
2 Remove the bulb holder for the front turn signal/parking and sidemarker light bulb from the headlight housing (see illustration).
3 Remove the bulb from the holder (see illustration).
4 Installation is the reverse of removal.

Fog lamp bulbs

5 Raise the front of the vehicle and place it securely on jackstands.
6 Disconnect the electrical connector from the fog lamp bulb holder.
7 Turn the fog lamp bulb holder counterclockwise and pull it out of the fog lamp housing.
8 Installation is the reverse of removal.

Center high-mounted brake light

▶ **Refer to illustrations 19.9, 19.10, 19.11 and 19.12**

➡**Note: The center high-mounted brake light bulb is located in the upper edge of the rear hatch.**

9 Open the rear hatch and remove the trim piece from the upper

19.2 To remove the bulb holder for the front turn signal/ parking and sidemarker light bulb, rotate it counterclockwise and pull it out of the headlight housing

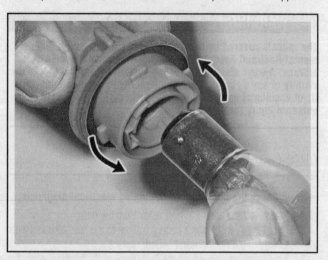

19.3 To remove the front turn signal/parking and sidemarker light bulb from the bulb holder, push it into the bulb holder, turn it counterclockwise and pull it out

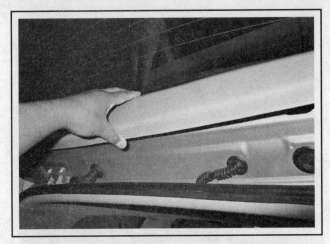

19.9 Remove this trim piece from the upper inside edge of the rear hatch to access the center high-mounted brake light bulb. To detach the trim piece, simply pull it off

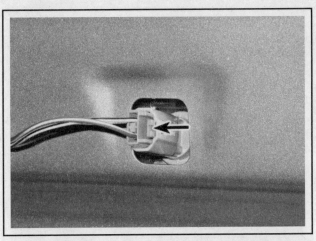

19.10 To disconnect the electrical connector from the center high-mounted brake light bulb holder, depress this release tab and pull off the connector

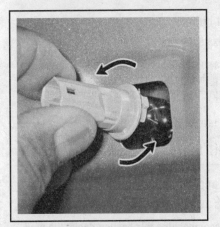

19.11 To remove the bulb holder from the center high-mounted brake light housing, rotate it counterclockwise and pull it out

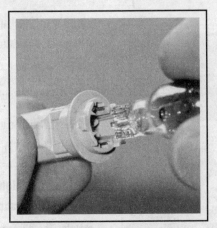

19.12 To remove the center high-mounted brake light bulb from the bulb holder, pull it straight out of the holder. After installing the new bulb, wipe it off with a clean cloth

19.14 Open this small panel to access the rear brake, turn signal, brake and back-up light bulbs

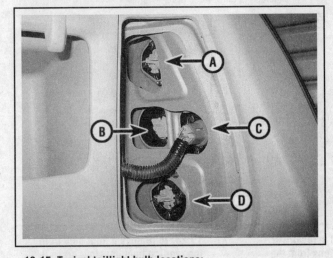

19.15 Typical taillight bulb locations:

A   Rear brake/taillight bulb   C   Brake/taillight bulb
B   Rear turn signal bulb   D   Back-up light bulb

inside edge of the hatch, between the two hinges (see illustration).

10 Disconnect the electrical connector from the center high-mounted brake light bulb holder (see illustration).

11 Remove the center high-mounted brake light bulb holder from the housing (see illustration).

12 Remove the center high-mounted brake light bulb from the holder (see illustration).

13 Installation is the reverse of removal.

Rear brake/tail, turn signal, brake and back-up light bulbs

▶ **Refer to illustrations 19.14, 19.15, 19.16 and 19.17**

14 Open the rear hatch. Remove the rear brake/tail, turn signal, brake and back-up light access panel (see illustration).

15 Locate the bulb that you want to replace (see illustration).

16 Remove the bulb holder from the taillight housing (see illustration).

17 Remove the bulb from its holder (see illustration).

18 Installation is the reverse of removal.

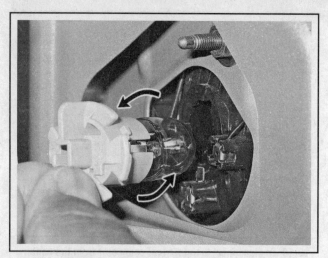

19.16 To remove a taillight bulb holder from the taillight housing, rotate it counterclockwise and pull it out of the taillight housing

19.17 To remove a taillight bulb from its holder, pull it straight out of the holder. After installing the bulb, wipe it off

License plate light bulbs

▶ Refer to illustrations 19.19 and 19.20

19 Remove the license plate light housing screws (see illustration) and remove the license plate light housing.

20 Remove the bulb from the license plate light housing (see illustration).

21 Installation is the reverse of removal.

INTERIOR LIGHT BULBS

Map reading light and dome light bulbs

▶ Refer to illustrations 19.22 and 19.23

➡Note: The photos accompanying this section depict the map reading light, but they apply to the dome light as well.

22 Remove the lens from the map reading light housing (see illustration).

23 Remove the bulb from the terminals (see illustration).

24 Installation is the reverse of removal.

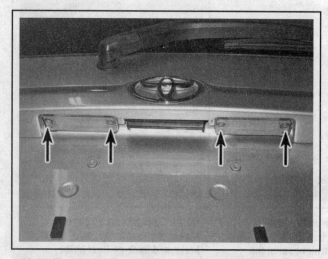

19.19 To remove either license plate light housing, remove the two screws from the lens

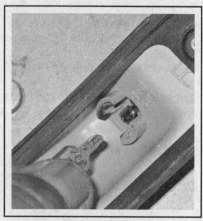

19.20 To remove a bulb from a license plate light housing, pull it straight out

19.22 Using a small screwdriver, carefully pry off the lens from the map reading light housing (this procedure applies to the dome light as well)

19.23 To remove a light bulb from the reading light housing, grasp it firmly and pull it straight down from the terminal clips. If necessary, pry only on the metal ends of the bulb, not the glass

20 Electric side view mirrors - description

1   Most electric rear view mirrors use two motors to move the glass; one for up and down adjustments and one for left-right adjustments.

2   The control switch has a selector portion that sends voltage to the left or right side mirror. With the ignition ON but the engine OFF, roll down the windows and operate the mirror control switch through all functions (left-right and up-down) for both the left and right side mirrors.

3   Listen carefully for the sound of the electric motors running in the mirrors.

4   If the motors can be heard but the mirror glass doesn't move, there's a problem with the drive mechanism inside the mirror. Remove and disassemble the mirror to locate the problem.

5   If the mirrors do not operate and no sound comes from the mirrors, check the fuse (see Chapter 1).

6   If the fuse is OK, remove the mirror control switch from the dashboard. Have the switch continuity checked by a dealership service department or other qualified automobile repair facility.

7   Test the ground connections. Refer to the wiring diagrams at the end of this Chapter.

8   If the mirror still doesn't work, remove the mirror and check the wires at the mirror for voltage.

9   If there's not voltage in each switch position, check the circuit between the mirror and control switch for opens and shorts.

10   If there's voltage, remove the mirror and test it off the vehicle with jumper wires. Replace the mirror if it fails this test.

21 Cruise control system - description

1999 THROUGH 2003 MODELS

1   The cruise control system maintains vehicle speed with a vacuum actuated servo located in the engine compartment, which is connected to the throttle body by a cable. The system consists of the cruise control unit, brake switch, control switches, vacuum hose and vehicle speed sensor. Some features of the system require special testers and diagnostic procedures that are beyond the scope of this manual. Listed below are some general procedures that may be used to locate common problems.

2   Locate and check the fuse (see Section 3).

3   Check the brake light switch (see Chapter 9).

4   Visually inspect the control cable between the actuator assembly and the accelerator pedal for free movement, replace if necessary.

5   Test drive the vehicle to determine if the cruise control is now working. If it isn't, take it to a dealer for further diagnosis.

2004 AND LATER MODELS

6   These models have an electrically controlled throttle body - there is no accelerator cable. The accelerator pedal communicates with the throttle body through the Powertrain Control Module (PCM). (See Chapters 4 and 6 for more information about the electronic throttle control system.) The PCM also controls the cruise control system, which is now an integral function of the electronic throttle control system. If the system malfunctions, take it to a dealer service department or other qualified repair shop for further diagnosis.

22 Power window system - description

1   The power window system operates electric motors, mounted in the doors, which lower and raise the windows. The system consists of the control switches, relays, the motors, regulators, glass mechanisms and associated wiring.

2   The power windows can be lowered and raised from the master control switch by the driver or by remote switches located at the individual windows. Each window has a separate motor that is reversible. The position of the control switch determines the polarity and therefore the direction of operation.

3   The circuit is protected by a fuse and a circuit breaker. Each motor is also equipped with an internal circuit breaker; this prevents one stuck window from disabling the whole system.

4   The power window system will only operate when the ignition switch is ON. In addition, many models have a window lockout switch at the master control switch that, when activated, disables the switches at the rear windows and, sometimes, the switch at the passenger's window also. Always check these items before troubleshooting a window problem.

5   These procedures are general in nature, so if you can't find the problem using them, take the vehicle to a dealer service department or other properly equipped repair facility.

6   If the power windows won't operate, always check the fuse and circuit breaker first.

7   If only the rear windows are inoperative, or if the windows only operate from the master control switch, check the rear window lockout switch for continuity in the unlocked position. Replace it if it doesn't have continuity.

8   Check the wiring between the switches and fuse panel for continuity. Repair the wiring, if necessary.

9   If only one window is inoperative from the master control switch, try the other control switch at the window.

➡**Note: This doesn't apply to the driver's door window.**

10   If the same window works from one switch, but not the other, check the switch for continuity. Have the switch checked at a dealer service department or other qualified automobile repair facility.

11   If the switch tests OK, check for a short or open in the circuit between the affected switch and the window motor.

12   If one window is inoperative from both switches, remove the trim panel from the affected door and check for voltage at the switch and at the motor while the switch is operated.

13   If voltage is reaching the motor, disconnect the glass from the

regulator (see Chapter 11). Move the window up and down by hand while checking for binding and damage. Also check for binding and damage to the regulator. If the regulator is not damaged and the window moves up and down smoothly, replace the motor. If there's binding or damage, lubricate, repair or replace parts, as necessary.

14 If voltage isn't reaching the motor, check the wiring in the circuit for continuity between the switches and motors. You'll need to consult the wiring diagram for the vehicle. If the circuit is equipped with a relay, check that the relay is grounded properly and receiving voltage.

15 Test the windows after you are done to confirm proper repairs.

23 Power door lock system - description

1 A power door lock system operates the door lock actuators mounted in each door. The system consists of the switches, actuators, a control unit and associated wiring. Diagnosis can usually be limited to simple checks of the wiring connections and actuators for minor faults that can be easily repaired.

2 Power door lock systems are operated by bi-directional solenoids located in the doors. The lock switches have two operating positions: Lock and Unlock. When activated, the switch sends a ground signal to the door lock control unit to lock or unlock the doors. Depending on which way the switch is activated, the control unit reverses polarity to the solenoids, allowing the two sides of the circuit to be used alternately as the feed (positive) and ground side.

3 Some vehicles may have an anti-theft system incorporated into the power locks. If you are unable to locate the trouble using the following general Steps, consult a dealer service department or other qualified repair shop.

4 Always check the circuit protection first. Some vehicles use a combination of circuit breakers and fuses.

5 Operate the door lock switches in both directions (Lock and Unlock) with the engine off. Listen for the click of the solenoids operating.

6 Test the switches for continuity. Remove the switches and have them checked by a dealer service department or other qualified automobile repair facility.

7 Check the wiring between the switches, control unit and solenoids for continuity. Repair the wiring if there's no continuity.

8 Check for a bad ground at the switches or the control unit.

9 If all but one lock solenoid operate, remove the trim panel from the affected door (see Chapter 11) and check for voltage at the solenoid while the lock switch is operated. One of the wires should have voltage in the Lock position; the other should have voltage in the Unlock position.

10 If the inoperative solenoid is receiving voltage, replace the solenoid.

11 If the inoperative solenoid isn't receiving voltage, check the relay for an open or short in the wire between the lock solenoid and the control unit.

➡ Note: It's common for wires to break in the section of harness that goes between the body and door because opening and closing the door fatigues and eventually breaks the wires.

24 Daytime Running Lights (DRL) - general information

The Daytime Running Lights (DRL) system used on Canadian models illuminates the headlights whenever the engine is running. The only exception is with the engine running and the parking brake engaged. Once the parking brake is released, the lights will remain on as long as the ignition switch is on, even if the parking brake is later applied.

The DRL system supplies reduced power to the headlights so they won't be too bright for daytime use, while prolonging headlight life.

25 Airbag system - general information

All models are equipped with a Supplemental Restraint System (SRS), more commonly known as the airbag system. The airbag system is designed to protect the driver and the front seat passenger from serious injury in the event of a head-on or frontal collision. It consists of the impact sensors, a driver's airbag module in the center of the steering wheel, a passenger's airbag module in the glove box area of the instrument panel and a sensing/diagnostic module mounted in the center of the floorpan, ahead of the center console. Some models are also equipped with side-impact airbags, side-curtain airbags and a driver's knee airbag.

AIRBAG MODULES

Driver's airbag

The airbag inflator module contains a housing incorporating the cushion (airbag) and inflator unit, mounted in the center of the steering wheel The inflator assembly is mounted on the back of the housing over a hole through which gas is expelled, inflating the bag almost instantaneously when an electrical signal is sent from the system. A spiral cable assembly on the steering column under the steering wheel carries this signal to the module. This spiral cable assembly can transmit an electrical signal regardless of steering wheel position. The igniter in the airbag converts the electrical signal to heat and ignites the powder, which inflates the bag.

Driver's knee airbag

On some 2004 and later models there is a second driver's airbag located under the steering column, just ahead of the knee bolster. Its only purpose is to protect the driver's knees from serious damage in the event of a collision. If you're not sure whether you have a drive's knee airbag, refer to your owner's manual.

Passenger's airbag

The airbag is mounted inside the right end of the instrument panel, in the vicinity of the glove box compartment. It's similar in design to the driver's airbag, except that it's larger than the steering wheel unit. The passenger airbag is mounted between the instrument panel reinforcement bar and the underside of the instrument panel. The trim cover (on the side of the instrument panel that faces toward the passenger) is textured and colored to match the instrument panel and has a molded seam that splits open when the bag inflates.

Side impact airbags

Extra protection is provided by side-impact airbags on some models. These are smaller devices, which are located on the outer sides of the seat backs, and deploy in the event of a severe side-impact collision.

Side curtain airbags

In addition to the side-impact airbags, extra side-impact protection is also provided by side-curtain airbags on some models. These are long airbags that, in the event of a side impact, come out of the headliner at each side of the car and come down between the side windows and the seats. They are designed to protect the heads of both front seat and rear seat passengers.

SENSING AND DIAGNOSTIC MODULE

The sensing and diagnostic module supplies the current to the airbag system in the event of the collision, even if battery power is cut off. It checks this system every time the vehicle is started, causing the "AIR BAG" light to go on then off, if the system is operating properly. If there is a fault in the system, the light will go on and stay on, flash, or the dash will make a beeping sound. If this happens, the vehicle should be taken to your dealer immediately for service.

DISARMING THE SYSTEM AND OTHER PRECAUTIONS

✳✳ WARNING:

Failure to follow these precautions could result in accidental deployment of the airbag and personal injury.

Whenever working in the vicinity of the steering wheel, steering column or any of the other SRS system components, the system must be disarmed.

To disarm the airbag system:

a) *Point the wheels straight ahead and turn the key to the Lock position.*

b) *Disconnect the cable from the negative battery terminal.*
c) *Wait at least two minutes for the back-up power supply to be depleted.*

Whenever handling an airbag module:

Always keep the airbag opening (the trim side) pointed away from your body. Never place the airbag module on a bench of other surface with the airbag opening facing the surface. Always place the airbag module in a safe location with the airbag opening facing up.

Never measure the resistance of any SRS component. An ohmmeter has a built-in battery supply that could accidentally deploy the airbag.

Never use electrical welding equipment on a vehicle equipped with an airbag without first disconnecting the electrical connector for each airbag.

Never dispose of a live airbag module. Return it to a dealer service department or other qualified repair shop for safe deployment and disposal.

COMPONENT REMOVAL AND INSTALLATION

Driver's side airbag module and spiral cable

Refer to Chapter 10, *Steering wheel - removal and installation*, for the driver's side airbag module and spiral cable removal and installation procedures.

Driver's knee airbag

1 Disarm the airbag system (see above).
2 Remove the knee bolster (see Chapter 11).
3 Disconnect the big yellow electrical connector for the driver's knee airbag.
4 Remove the four driver's knee airbag mounting bolts and remove the airbag module.
5 Installation is the reverse of removal.

Other airbag modules

We don't recommend removing any of the other airbag modules. These jobs are best left to a professional. Should you ever have to remove the instrument panel, you'll have to disconnect the electrical connector for the passenger airbag module and remove one nut and bolt that secures the passenger airbag module to the instrument panel reinforcement bracket, but you don't have to remove the passenger airbag module to remove the instrument panel.

26 Wiring diagrams - general information

Since it isn't possible to include all wiring diagrams for every year covered by this manual, the following diagrams are those that are typical and most commonly needed.

Prior to troubleshooting any circuits, check the fuse and circuit breakers (if equipped) to make sure they're in good condition. Make sure the battery is properly charged and check the cable connections (see Chapter 1).

When checking a circuit, make sure that all connectors are clean, with no broken or loose terminals. When unplugging a connector, do not pull on the wires - pull only on the connector housings.

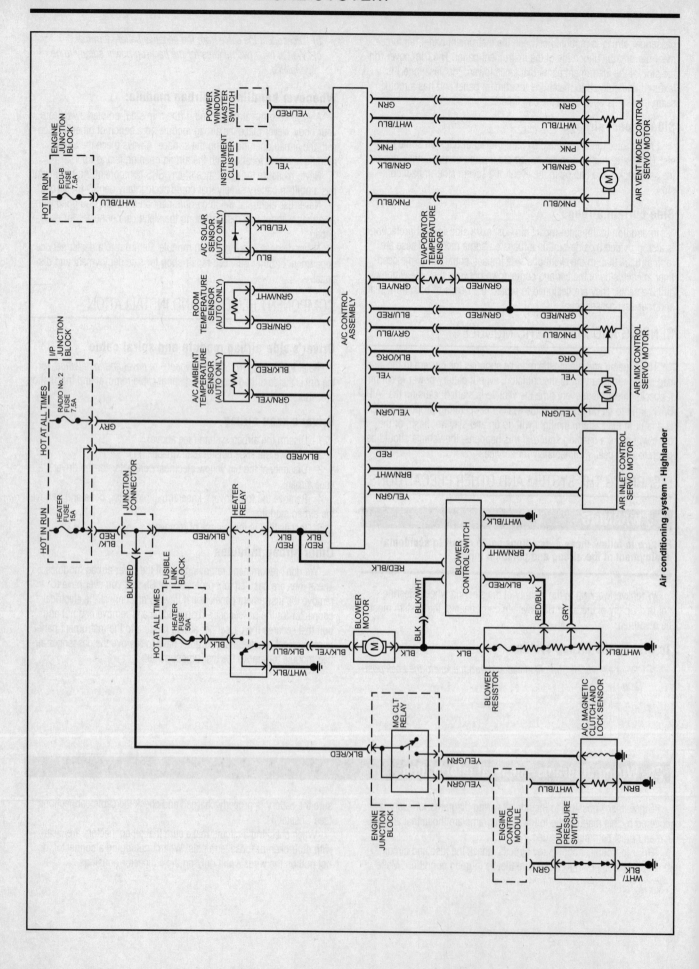

Air conditioning system - Highlander

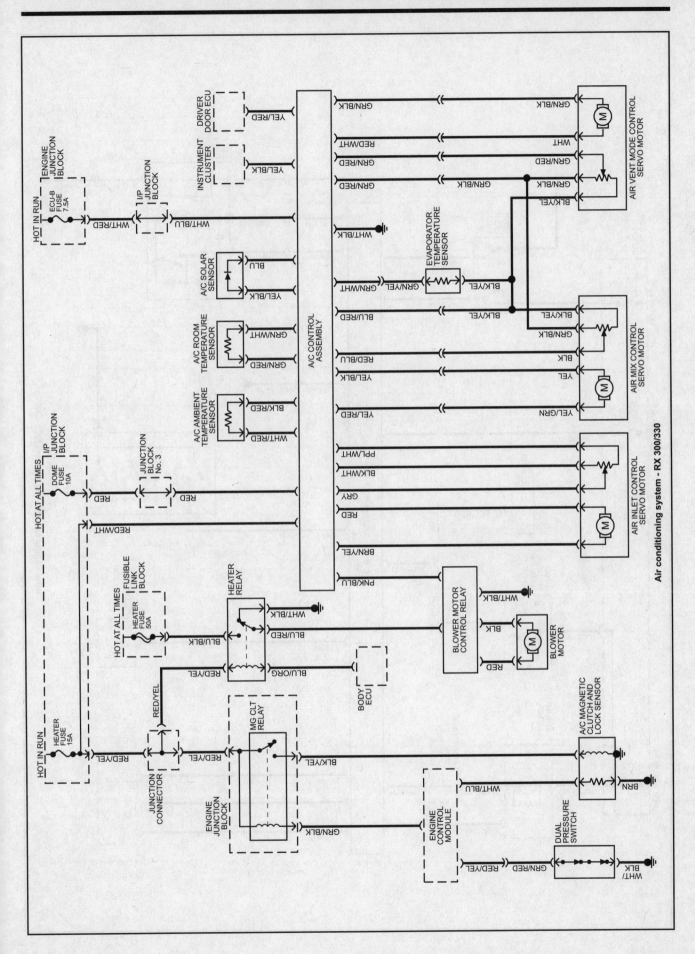

Air conditioning system - RX 300/330

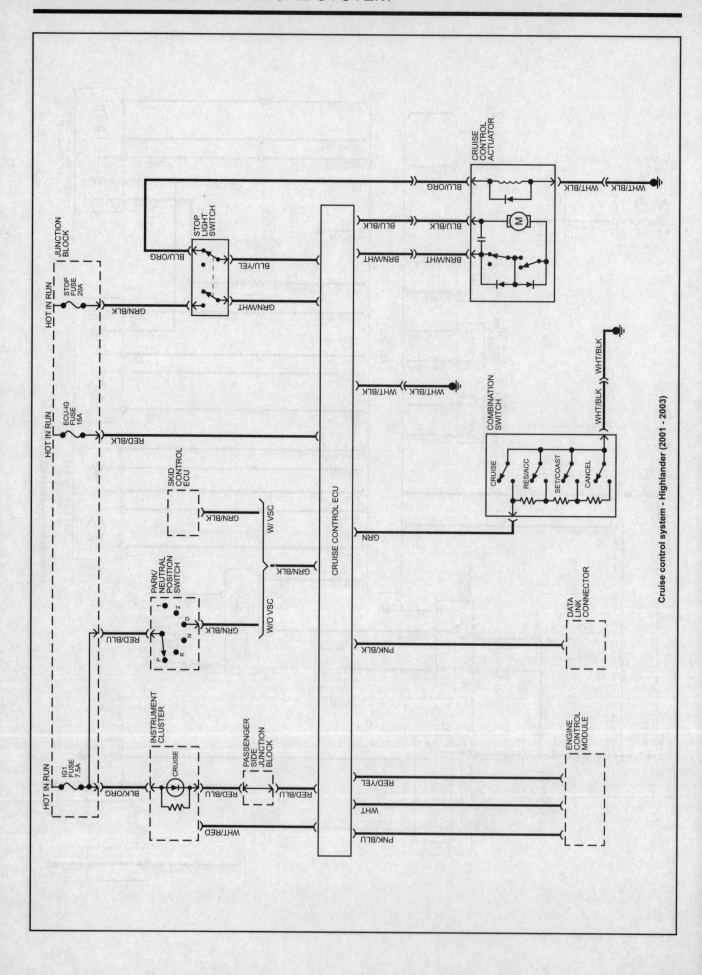

Cruise control system - Highlander (2001 - 2003)

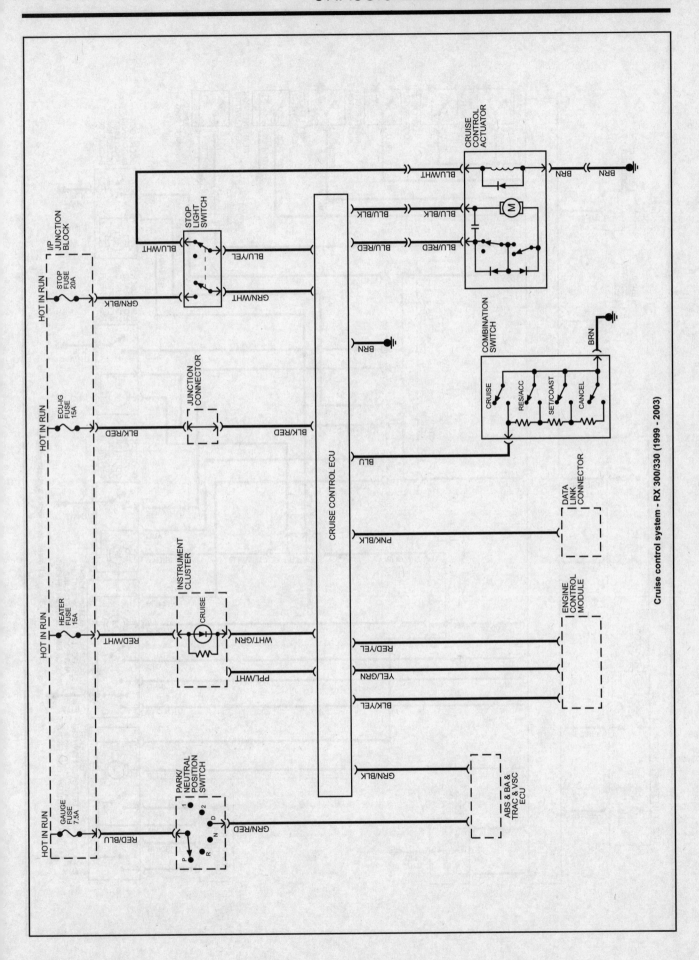

Cruise control system - RX 300/330 (1999 - 2003)

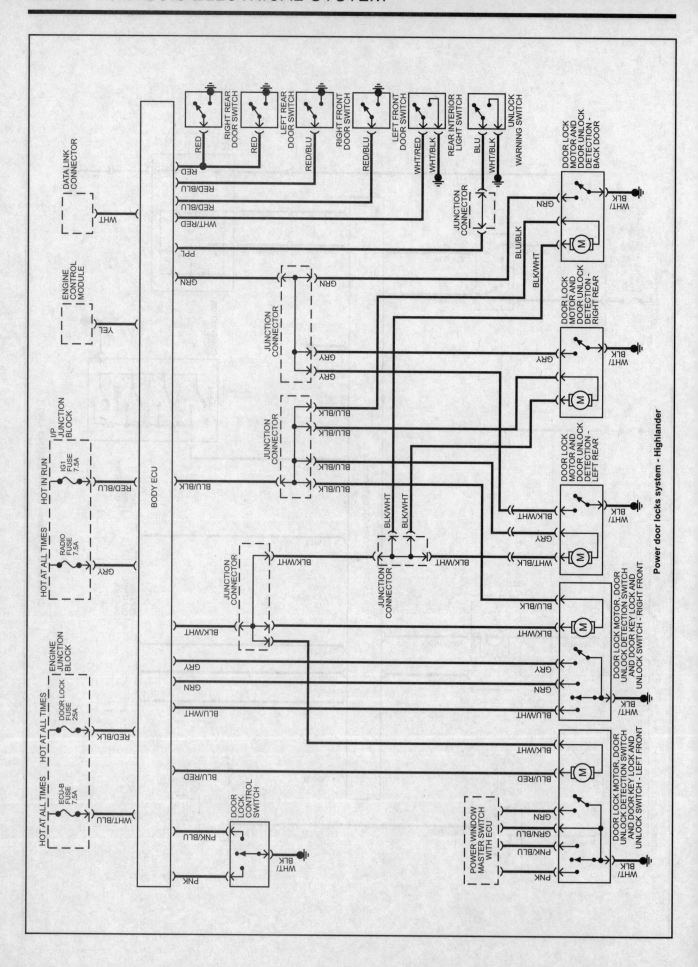

Power door locks system - Highlander

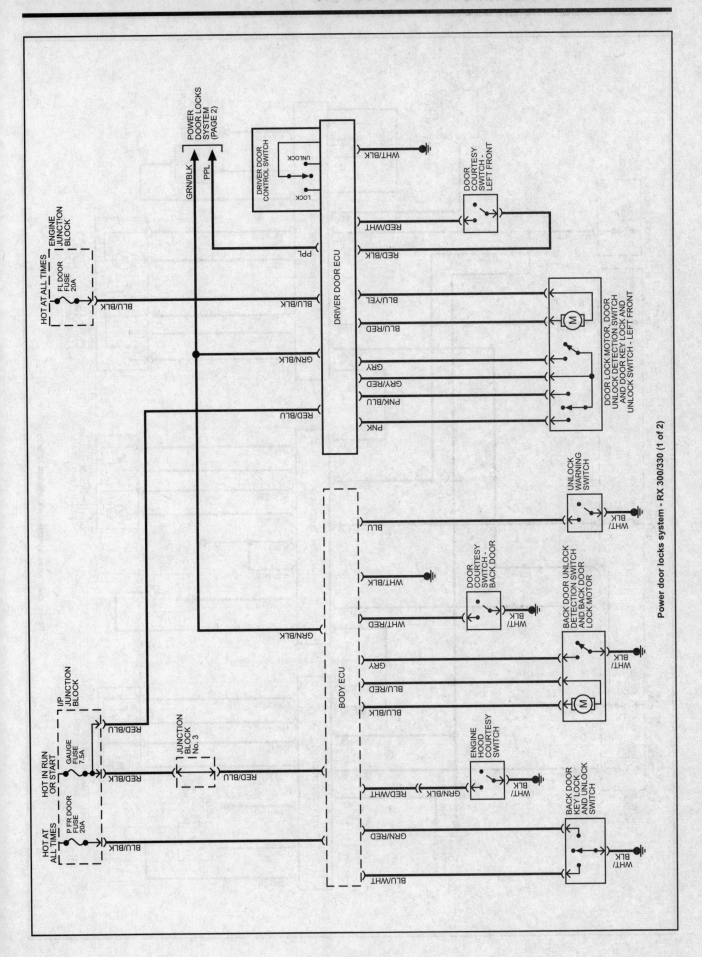

Power door locks system - RX 300/330 (1 of 2)

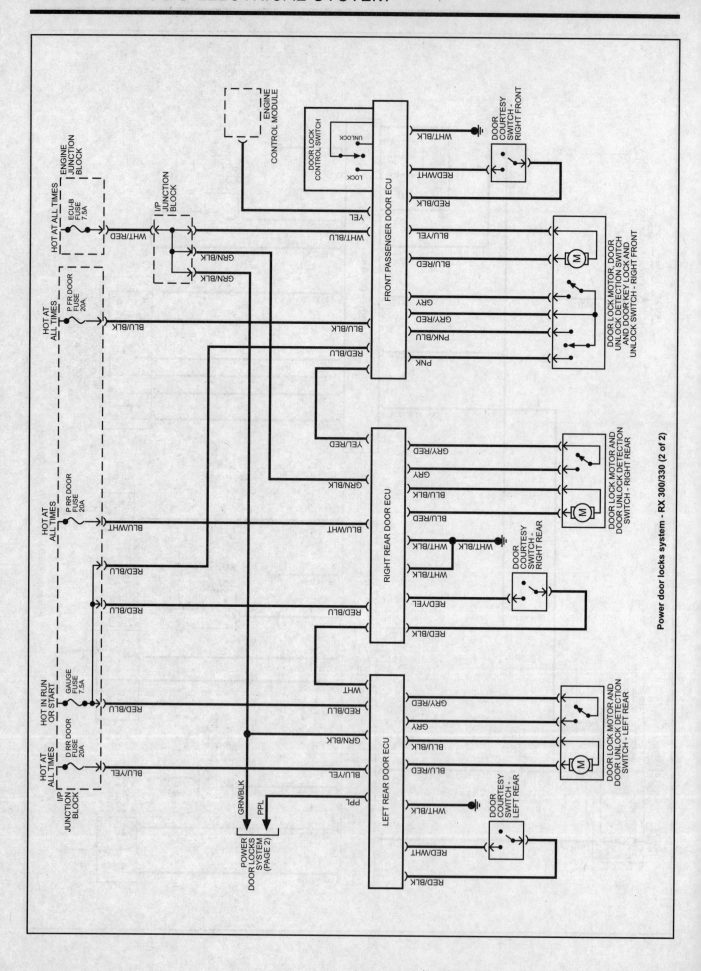

Power door locks system - RX 300/330 (2 of 2)

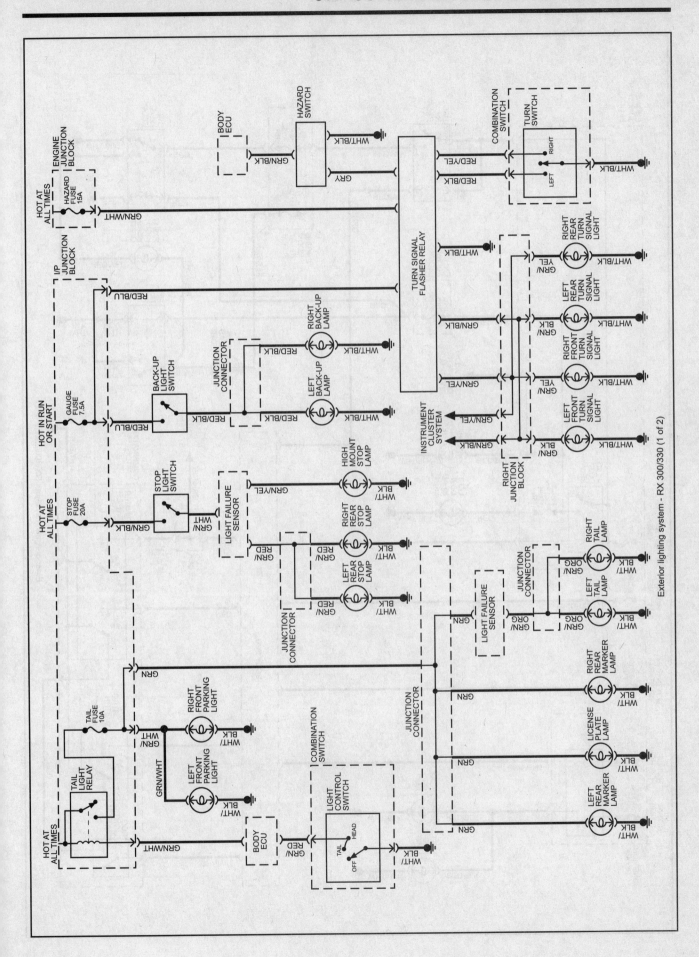

Exterior lighting system - RX 300/330 (1 of 2)

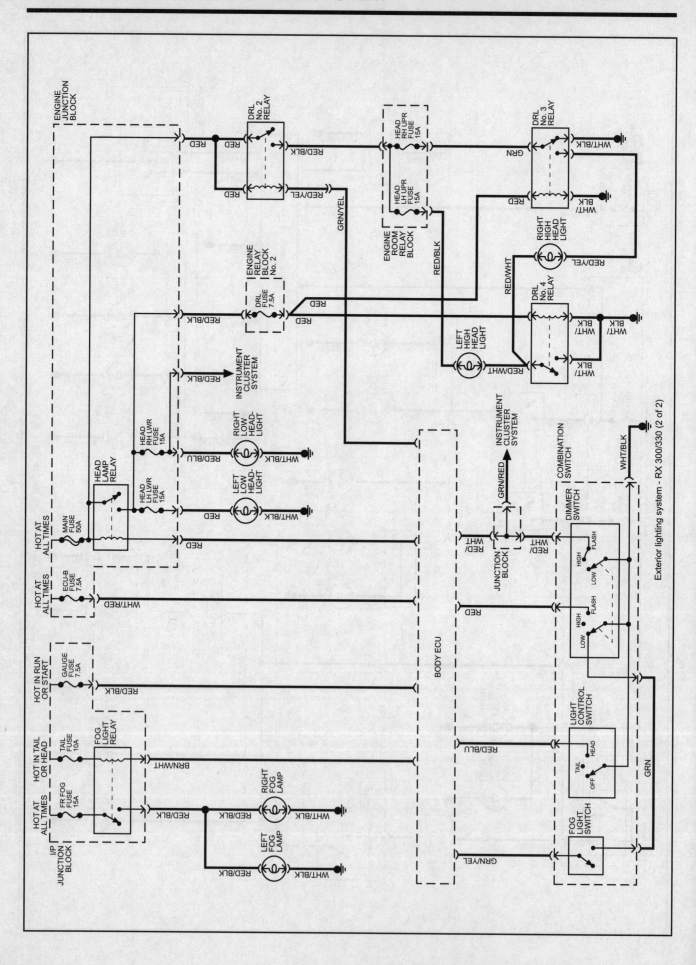

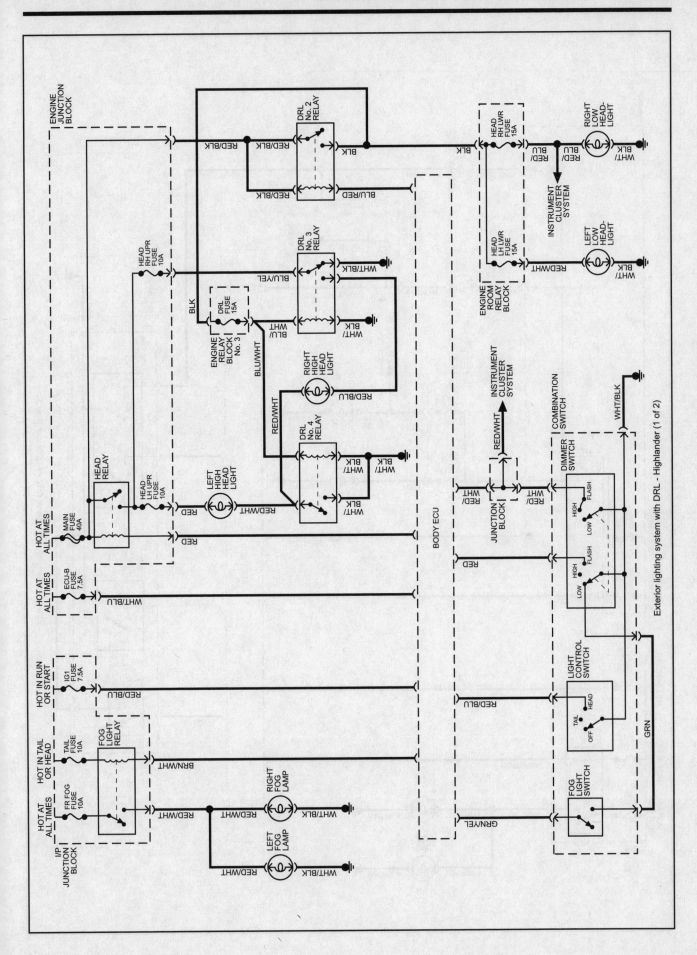

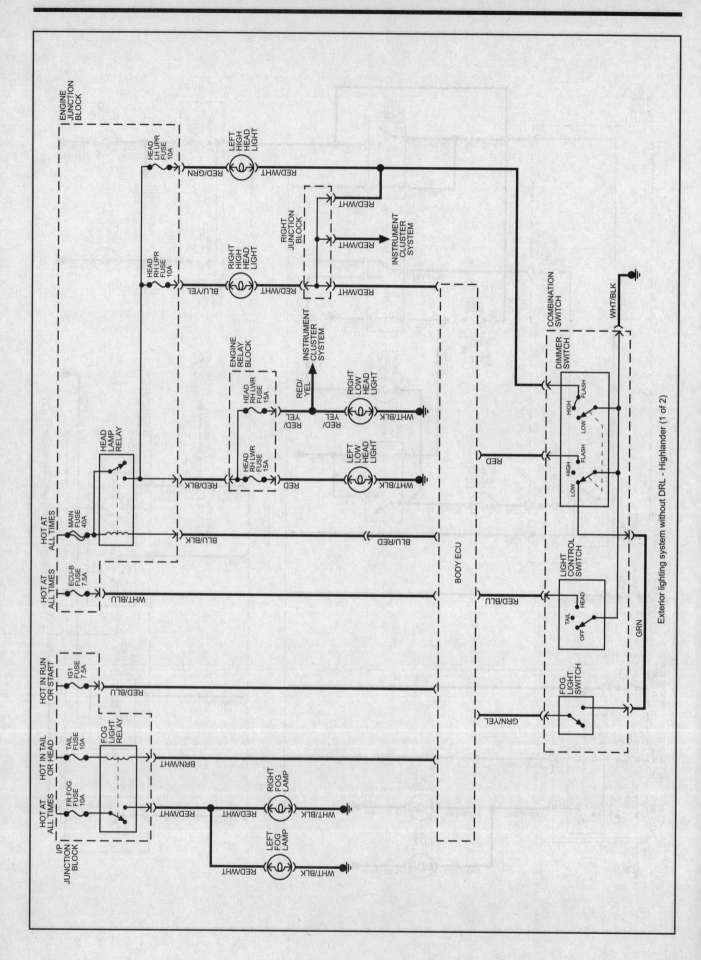

Exterior lighting system without DRL - Highlander (1 of 2)

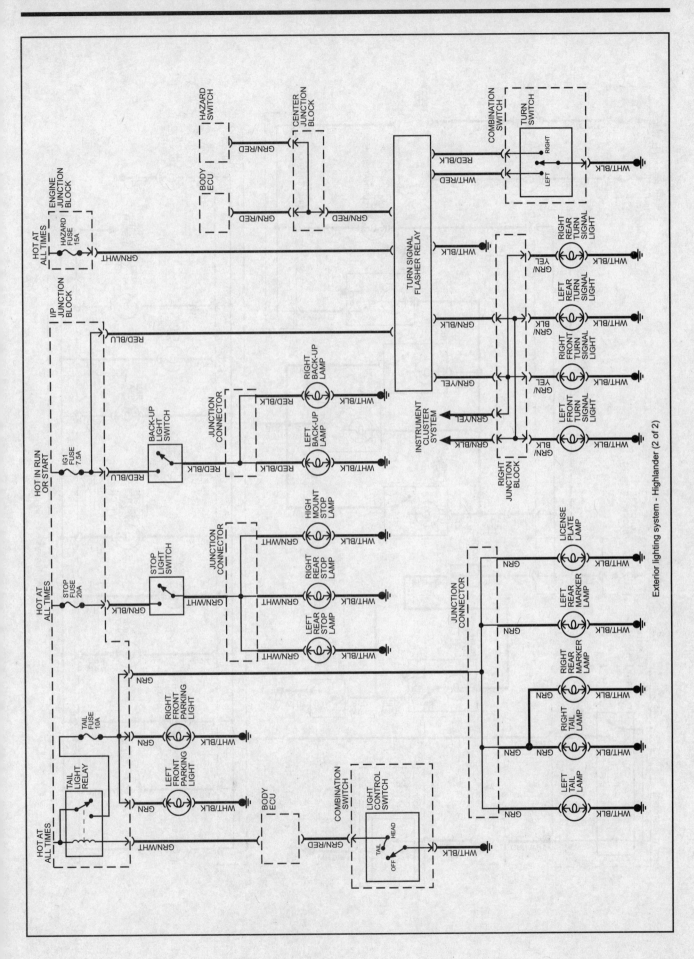

Exterior lighting system - Highlander (2 of 2)

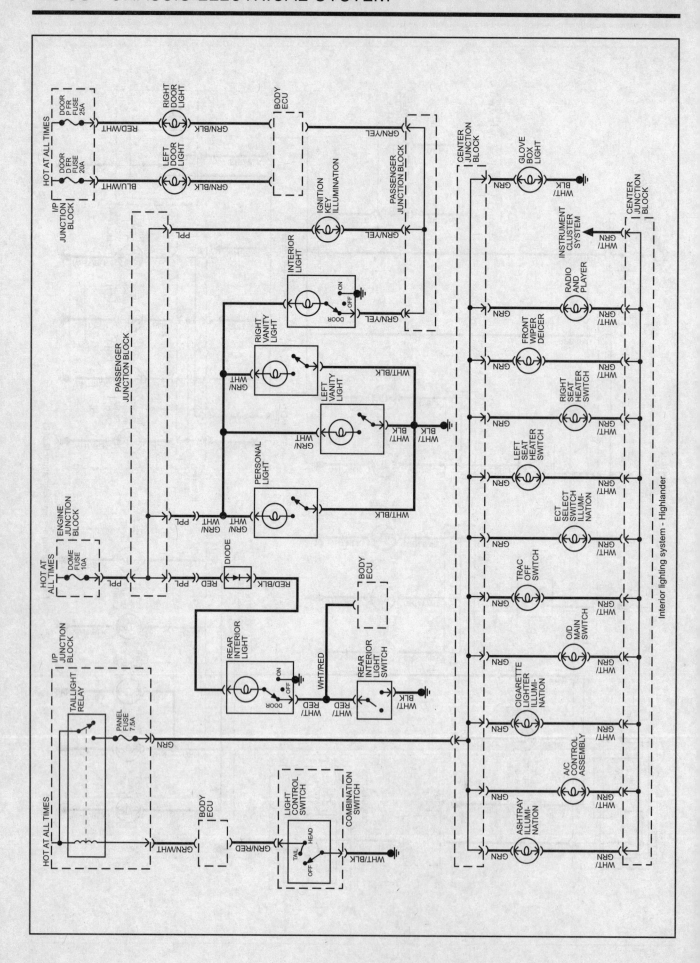

Interior lighting system - Highlander

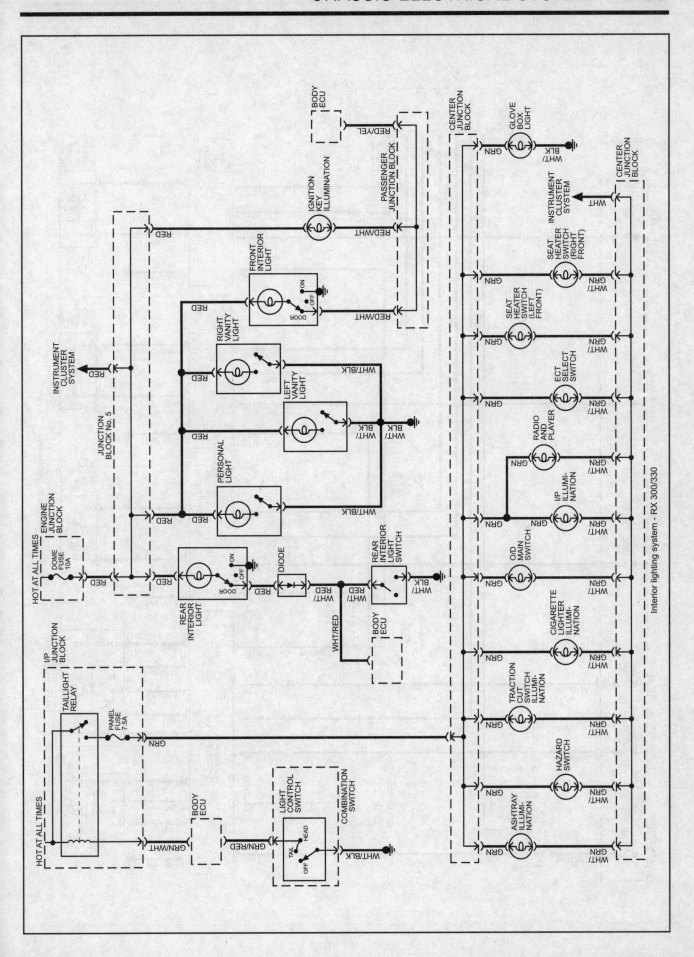

Interior lighting system - RX 300/330

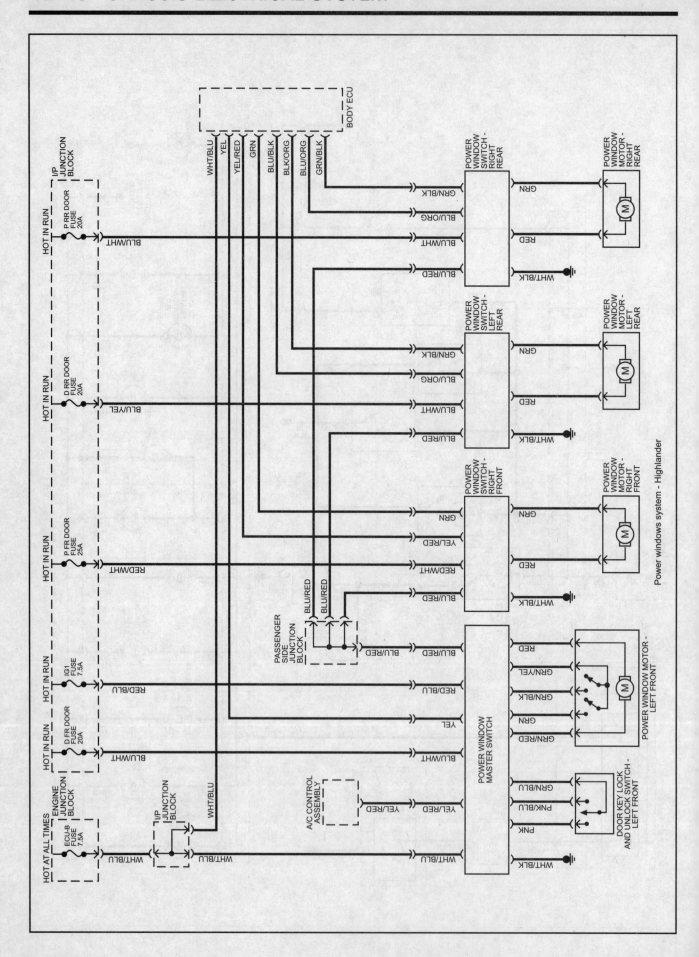

Power windows system - Highlander

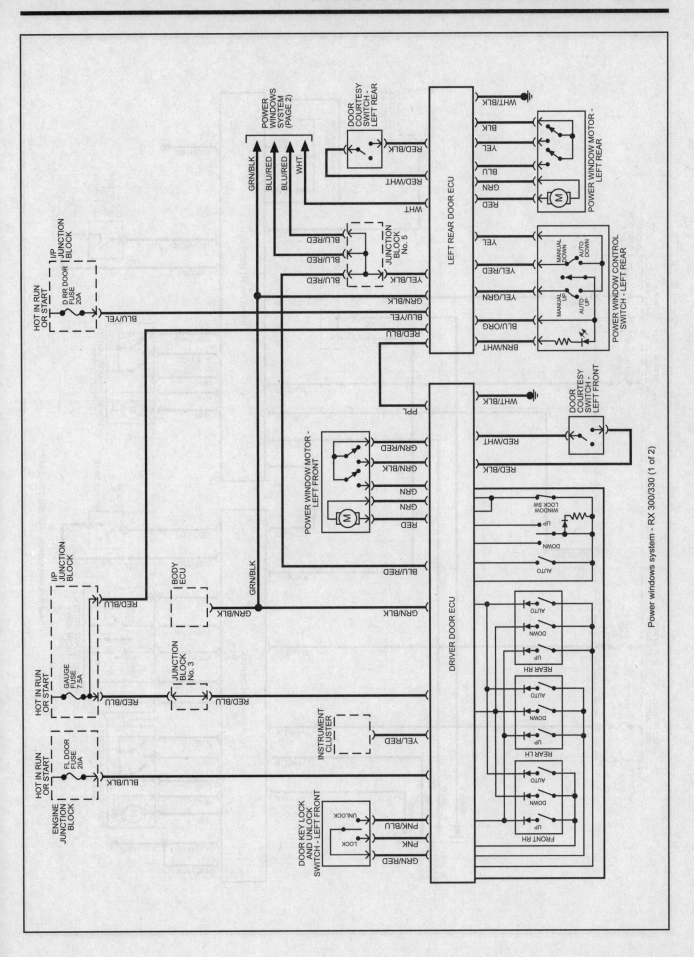

Power windows system - RX 300/330 (1 of 2)

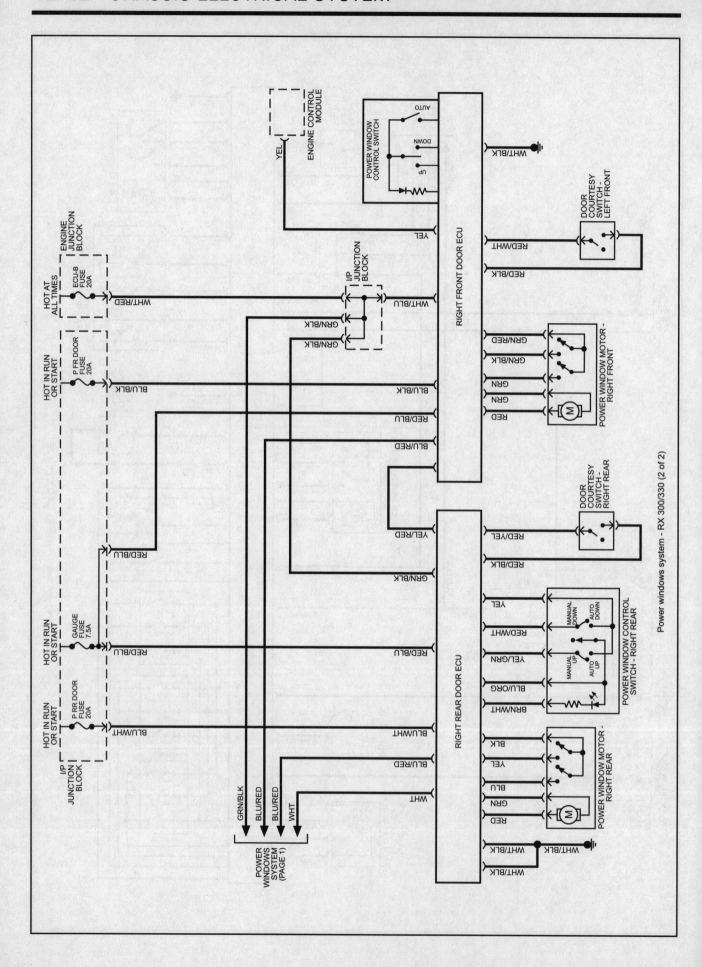

Power windows system - RX 300/330 (2 of 2)

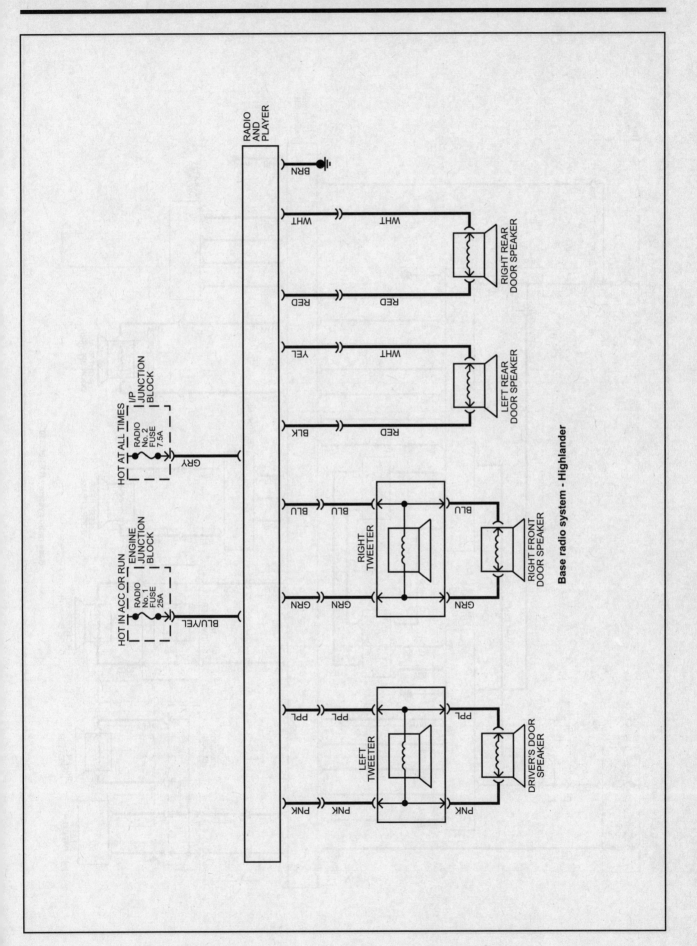

Base radio system - Highlander

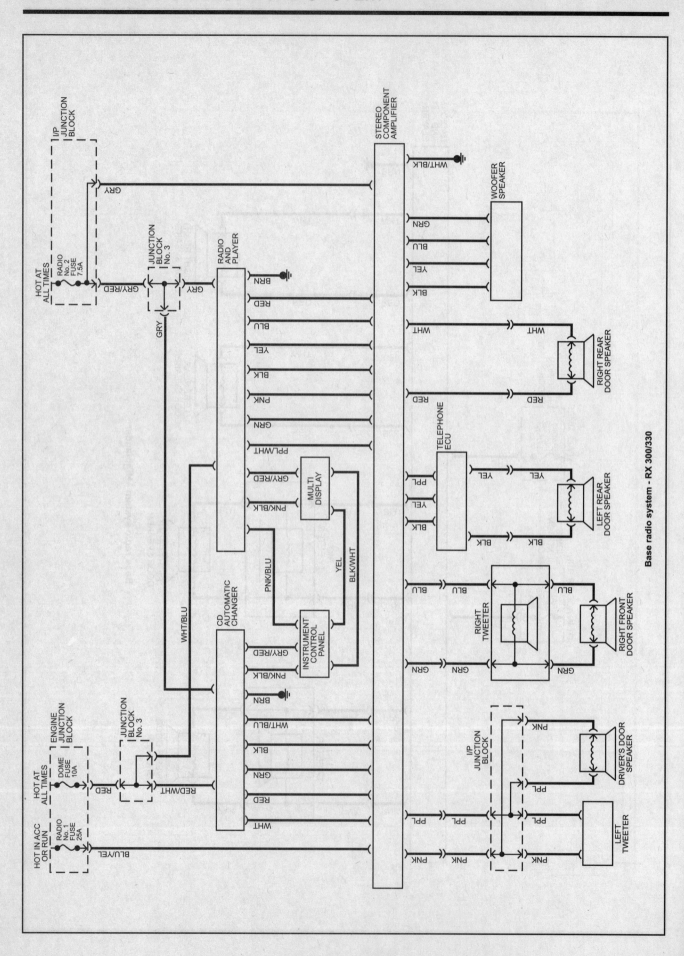

Base radio system - RX 300/330

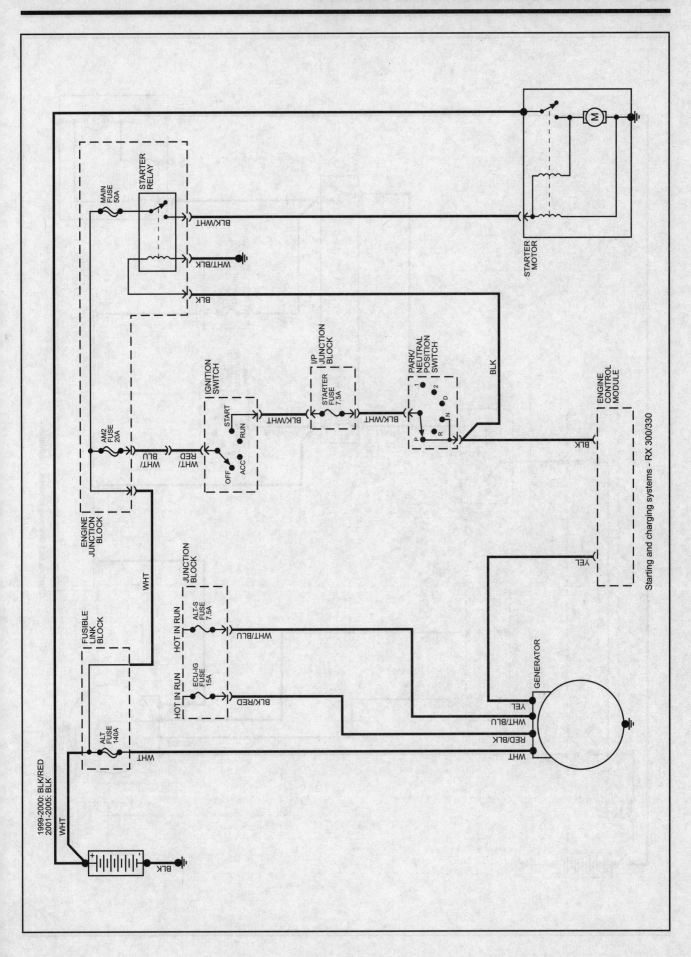

Starting and charging systems - RX 300/330

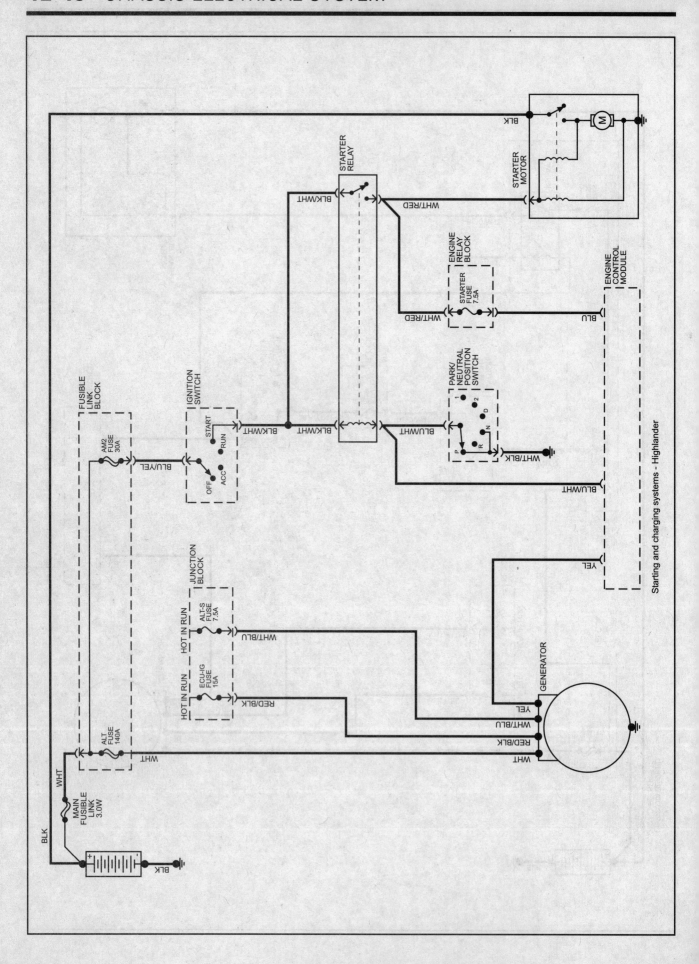

Starting and charging systems - Highlander

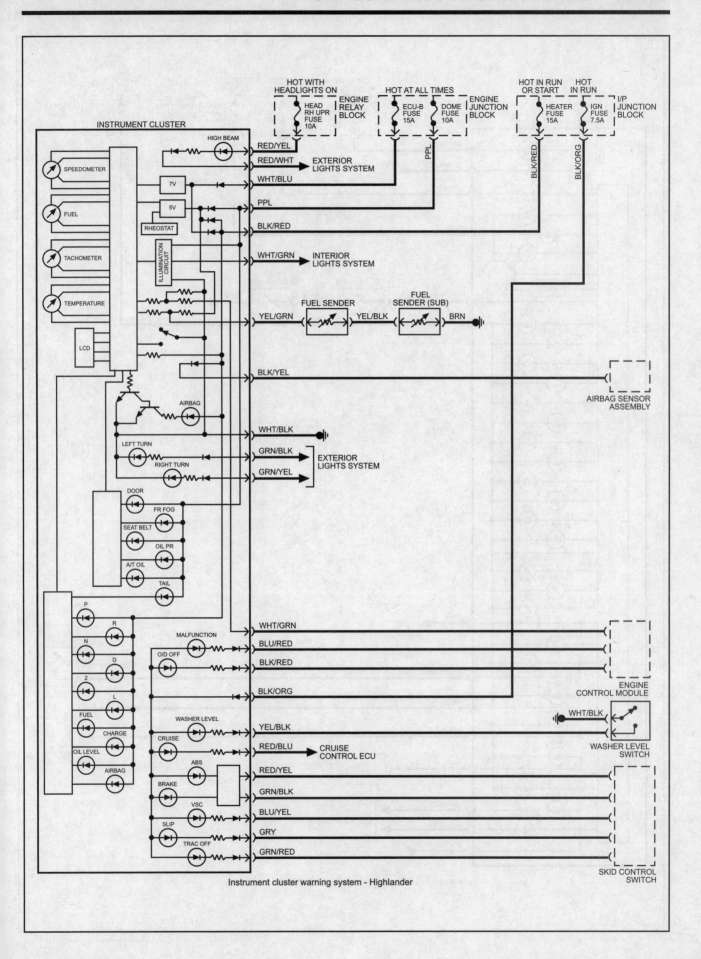

Instrument cluster warning system - Highlander

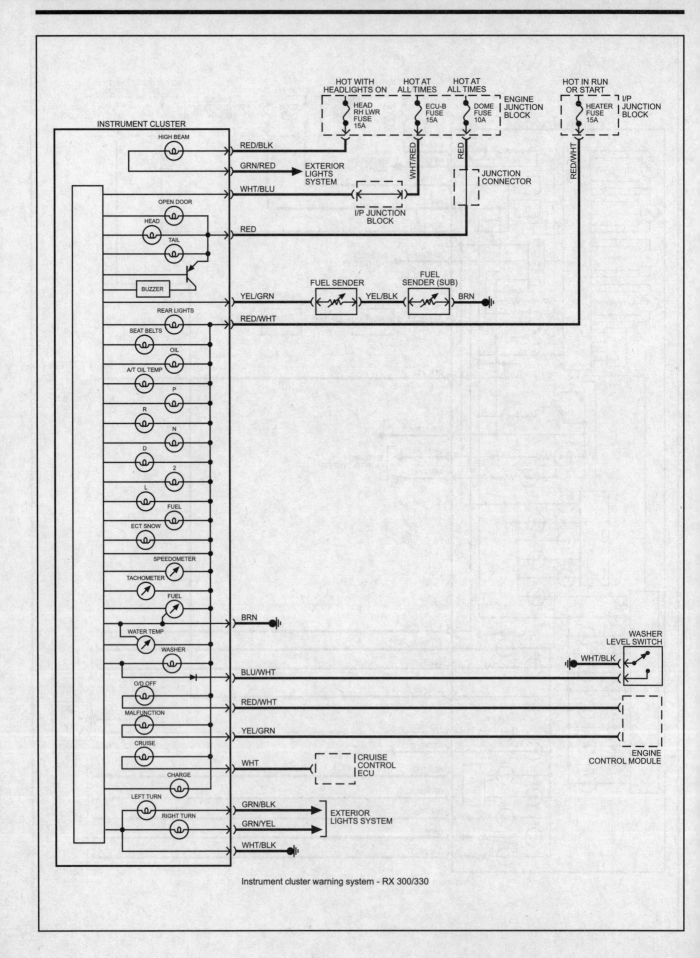

Instrument cluster warning system - RX 300/330

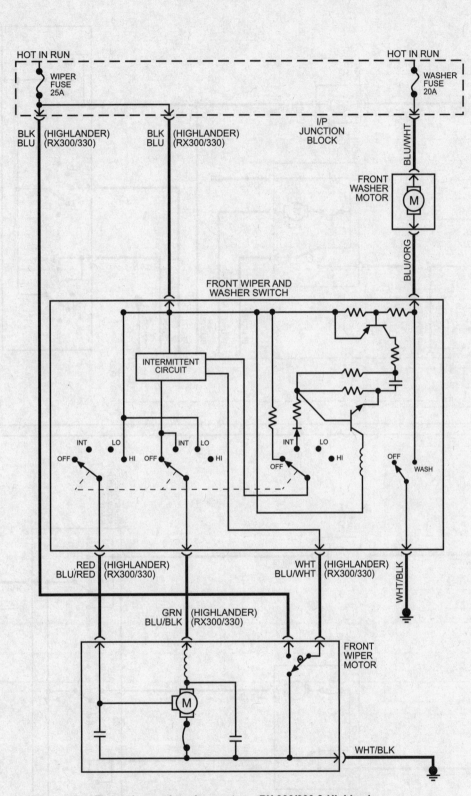

Front wiper and washer system - RX 300/330 & Highlander

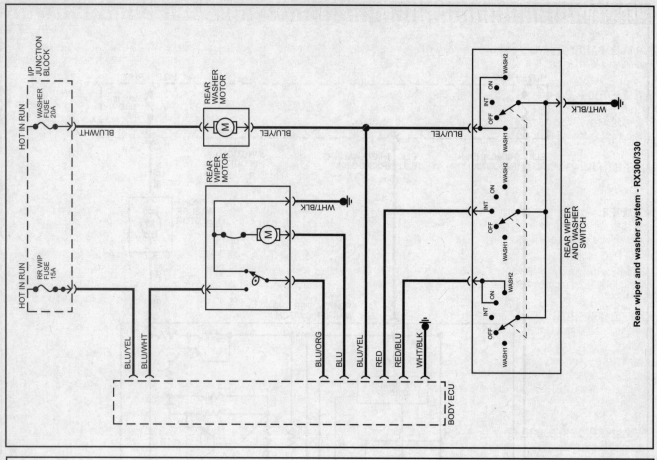

Rear wiper and washer system - RX300/330

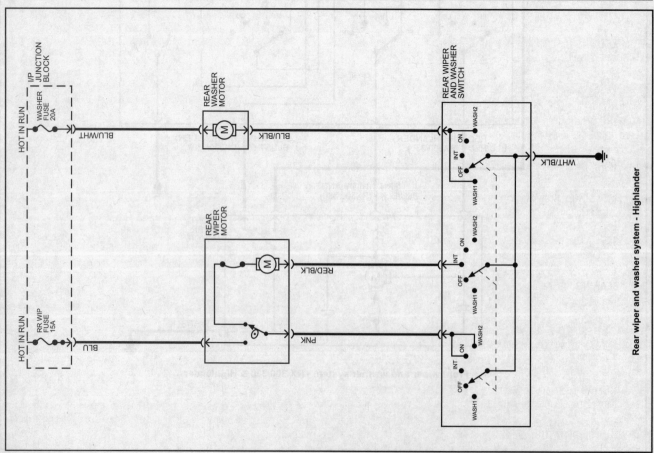

Rear wiper and washer system - Highlander

GLOSSARY

AIR/FUEL RATIO: The ratio of air-to-gasoline by weight in the fuel mixture drawn into the engine.

AIR INJECTION: One method of reducing harmful exhaust emissions by injecting air into each of the exhaust ports of an engine. The fresh air entering the hot exhaust manifold causes any remaining fuel to be burned before it can exit the tailpipe.

ALTERNATOR: A device used for converting mechanical energy into electrical energy.

AMMETER: An instrument, calibrated in amperes, used to measure the flow of an electrical current in a circuit. Ammeters are always connected in series with the circuit being tested.

AMPERE: The rate of flow of electrical current present when one volt of electrical pressure is applied against one ohm of electrical resistance.

ANALOG COMPUTER: Any microprocessor that uses similar (analogous) electrical signals to make its calculations.

ARMATURE: A laminated, soft iron core wrapped by a wire that converts electrical energy to mechanical energy as in a motor or relay. When rotated in a magnetic field, it changes mechanical energy into electrical energy as in a generator.

ATMOSPHERIC PRESSURE: The pressure on the Earth's surface caused by the weight of the air in the atmosphere. At sea level, this pressure is 14.7 psi at 32°F (101 kPa at 0°C).

ATOMIZATION: The breaking down of a liquid into a fine mist that can be suspended in air.

AXIAL PLAY: Movement parallel to a shaft or bearing bore.

BACKFIRE: The sudden combustion of gases in the intake or exhaust system that results in a loud explosion.

BACKLASH: The clearance or play between two parts, such as meshed gears.

BACKPRESSURE: Restrictions in the exhaust system that slow the exit of exhaust gases from the combustion chamber.

BAKELITE: A heat resistant, plastic insulator material commonly used in printed circuit boards and transistorized components.

BALL BEARING: A bearing made up of hardened inner and outer races between which hardened steel balls roll.

BALLAST RESISTOR: A resistor in the primary ignition circuit that lowers voltage after the engine is started to reduce wear on ignition components.

BEARING: A friction reducing, supportive device usually located between a stationary part and a moving part.

BIMETAL TEMPERATURE SENSOR: Any sensor or switch made of two dissimilar types of metal that bend when heated or cooled due to the different expansion rates of the alloys. These types of sensors usually function as an on/off switch.

BLOWBY: Combustion gases, composed of water vapor and unburned fuel, that leak past the piston rings into the crankcase during normal engine operation. These gases are removed by the PCV system to prevent the buildup of harmful acids in the crankcase.

BRAKE PAD: A brake shoe and lining assembly used with disc brakes.

BRAKE SHOE: The backing for the brake lining. The term is, however, usually applied to the assembly of the brake backing and lining.

BUSHING: A liner, usually removable, for a bearing; an anti-friction liner used in place of a bearing.

CALIPER: A hydraulically activated device in a disc brake system, which is mounted straddling the brake rotor (disc). The caliper contains at least one piston and two brake pads. Hydraulic pressure on the piston(s) forces the pads against the rotor.

CAMSHAFT: A shaft in the engine on which are the lobes (cams) which operate the valves. The camshaft is driven by the crankshaft, via a belt, chain or gears, at one half the crankshaft speed.

CAPACITOR: A device which stores an electrical charge.

CARBON MONOXIDE (CO): A colorless, odorless gas given off as a normal byproduct of combustion. It is poisonous and extremely dangerous in confined areas, building up slowly to toxic levels without warning if adequate ventilation is not available.

CARBURETOR: A device, usually mounted on the intake manifold of an engine, which mixes the air and fuel in the proper proportion to allow even combustion.

CATALYTIC CONVERTER: A device installed in the exhaust system, like a muffler, that converts harmful byproducts of combustion into carbon dioxide and water vapor by means of a heat-producing chemical reaction.

CENTRIFUGAL ADVANCE: A mechanical method of advancing the spark timing by using flyweights in the distributor that react to centrifugal force generated by the distributor shaft rotation.

CHECK VALVE: Any one-way valve installed to permit the flow of air, fuel or vacuum in one direction only.

CHOKE: A device, usually a moveable valve, placed in the intake path of a carburetor to restrict the flow of air.

CIRCUIT: Any unbroken path through which an electrical current can flow. Also used to describe fuel flow in some instances.

CIRCUIT BREAKER: A switch which protects an electrical circuit from overload by opening the circuit when the current flow exceeds a predetermined level. Some circuit breakers must be reset manually, while most reset automatically.

COIL (IGNITION): A transformer in the ignition circuit which steps up the voltage provided to the spark plugs.

COMBINATION MANIFOLD: An assembly which includes both the intake and exhaust manifolds in one casting.

COMBINATION VALVE: A device used in some fuel systems that routes fuel vapors to a charcoal storage canister instead of venting them into the atmosphere. The valve relieves fuel tank pressure and allows fresh air into the tank as the fuel level drops to prevent a vapor lock situation.

COMPRESSION RATIO: The comparison of the total volume of the cylinder and combustion chamber with the piston at BDC and the piston at TDC.

CONDENSER: 1. An electrical device which acts to store an electrical charge, preventing voltage surges. 2. A radiator-like device in the air conditioning system in which refrigerant gas condenses into a liquid, giving off heat.

CONDUCTOR: Any material through which an electrical current can be transmitted easily.

CONTINUITY: Continuous or complete circuit. Can be checked with an ohmmeter.

COUNTERSHAFT: An intermediate shaft which is rotated by a mainshaft and transmits, in turn, that rotation to a working part.

CRANKCASE: The lower part of an engine in which the crankshaft and related parts operate.

CRANKSHAFT: The main driving shaft of an engine which receives reciprocating motion from the pistons and converts it to rotary motion.

CYLINDER: In an engine, the round hole in the engine block in which the piston(s) ride.

CYLINDER BLOCK: The main structural member of an engine in which is found the cylinders, crankshaft and other principal parts.

CYLINDER HEAD: The detachable portion of the engine, usually fastened to the top of the cylinder block and containing all or most of the combustion chambers. On overhead valve engines, it contains the valves and their operating parts. On overhead cam engines, it contains the camshaft as well.

DEAD CENTER: The extreme top or bottom of the piston stroke.

DETONATION: An unwanted explosion of the air/fuel mixture in the combustion chamber caused by excess heat and compression, advanced timing, or an overly lean mixture. Also referred to as "ping".

DIAPHRAGM: A thin, flexible wall separating two cavities, such as in a vacuum advance unit.

DIESELING: A condition in which hot spots in the combustion chamber cause the engine to run on after the key is turned off.

DIFFERENTIAL: A geared assembly which allows the transmission of motion between drive axles, giving one axle the ability to turn faster than the other.

DIODE: An electrical device that will allow current to flow in one direction only.

DISC BRAKE: A hydraulic braking assembly consisting of a brake disc, or rotor, mounted on an axle, and a caliper assembly containing, usually two brake pads which are activated by hydraulic pressure. The pads are forced against the sides of the disc, creating friction which slows the vehicle.

DISTRIBUTOR: A mechanically driven device on an engine which is responsible for electrically firing the spark plug at a predetermined point of the piston stroke.

DOWEL PIN: A pin, inserted in mating holes in two different parts allowing those parts to maintain a fixed relationship.

DRUM BRAKE: A braking system which consists of two brake shoes and one or two wheel cylinders, mounted on a fixed backing plate, and a brake drum, mounted on an axle, which revolves around the assembly.

DWELL: The rate, measured in degrees of shaft rotation, at which an electrical circuit cycles on and off.

ELECTRONIC CONTROL UNIT (ECU): Ignition module, module, amplifier or igniter. See Module for definition.

ELECTRONIC IGNITION: A system in which the timing and firing of the spark plugs is controlled by an electronic control unit, usually called a module. These systems have no points or condenser.

END-PLAY: The measured amount of axial movement in a shaft.

ENGINE: A device that converts heat into mechanical energy.

EXHAUST MANIFOLD: A set of cast passages or pipes which conduct exhaust gases from the engine.

FEELER GAUGE: A blade, usually metal, of precisely predetermined thickness, used to measure the clearance between two parts.

FIRING ORDER: The order in which combustion occurs in the cylinders of an engine. Also the order in which spark is distributed to the plugs by the distributor.

FLOODING: The presence of too much fuel in the intake manifold and combustion chamber which prevents the air/fuel mixture from firing, thereby causing a no-start situation.

FLYWHEEL: A disc shaped part bolted to the rear end of the crankshaft. Around the outer perimeter is affixed the ring gear. The starter drive engages the ring gear, turning the flywheel, which rotates the crankshaft, imparting the initial starting motion to the engine.

FOOT POUND (ft. lbs. or sometimes, ft.lb.): The amount of energy or work needed to raise an item weighing one pound, a distance of one foot.

FUSE: A protective device in a circuit which prevents circuit overload by breaking the circuit when a specific amperage is present. The device is constructed around a strip or wire of a lower amperage rating than the circuit it is designed to protect. When an amperage higher than that stamped on the fuse is present in the circuit, the strip or wire melts, opening the circuit.

GEAR RATIO: The ratio between the number of teeth on meshing gears.

GENERATOR: A device which converts mechanical energy into electrical energy.

HEAT RANGE: The measure of a spark plug's ability to dissipate heat from its firing end. The higher the heat range, the hotter the plug fires.

HUB: The center part of a wheel or gear.

HYDROCARBON (HC): Any chemical compound made up of hydrogen and carbon. A major pollutant formed by the engine as a byproduct of combustion.

HYDROMETER: An instrument used to measure the specific gravity of a solution.

INCH POUND (inch lbs.; sometimes in.lb. or in. lbs.): One twelfth of a foot pound.

INDUCTION: A means of transferring electrical energy in the form of a magnetic field. Principle used in the ignition coil to increase voltage.

INJECTOR: A device which receives metered fuel under relatively low pressure and is activated to inject the fuel into the engine under relatively high pressure at a predetermined time.

INPUT SHAFT: The shaft to which torque is applied, usually carrying the driving gear or gears.

INTAKE MANIFOLD: A casting of passages or pipes used to conduct air or a fuel/air mixture to the cylinders.

JOURNAL: The bearing surface within which a shaft operates.

KEY: A small block usually fitted in a notch between a shaft and a hub to prevent slippage of the two parts.

MANIFOLD: A casting of passages or set of pipes which connect the cylinders to an inlet or outlet source.

MANIFOLD VACUUM: Low pressure in an engine intake manifold formed just below the throttle plates. Manifold vacuum is highest at idle and drops under acceleration.

MASTER CYLINDER: The primary fluid pressurizing device in a hydraulic system. In automotive use, it is found in brake and hydraulic clutch systems and is pedal activated, either directly or, in a power brake system, through the power booster.

MODULE: Electronic control unit, amplifier or igniter of solid state or integrated design which controls the current flow in the ignition primary circuit based on input from the pick-up coil. When the module opens the primary circuit, high secondary voltage is induced in the coil.

NEEDLE BEARING: A bearing which consists of a number (usually a large number) of long, thin rollers.

OHM: (Ω) The unit used to measure the resistance of conductor-to-electrical flow. One ohm is the amount of resistance that limits current flow to one ampere in a circuit with one volt of pressure.

OHMMETER: An instrument used for measuring the resistance, in ohms, in an electrical circuit.

OUTPUT SHAFT: The shaft which transmits torque from a device, such as a transmission.

OVERDRIVE: A gear assembly which produces more shaft revolutions than that transmitted to it.

OVERHEAD CAMSHAFT (OHC): An engine configuration in which the camshaft is mounted on top of the cylinder head and operates the valve either directly or by means of rocker arms.

OVERHEAD VALVE (OHV): An engine configuration in which all of the valves are located in the cylinder head and the camshaft is located in the cylinder block. The camshaft operates the valves via lifters and pushrods.

OXIDES OF NITROGEN (NOx): Chemical compounds of nitrogen produced as a byproduct of combustion. They combine with hydrocarbons to produce smog.

OXYGEN SENSOR: Use with the feedback system to sense the presence of oxygen in the exhaust gas and signal the computer which can reference the voltage signal to an air/fuel ratio.

PINION: The smaller of two meshing gears.

PISTON RING: An open-ended ring with fits into a groove on the outer diameter of the piston. Its chief function is to form a seal between the piston and cylinder wall. Most automotive pistons have three rings: two for compression sealing; one for oil sealing.

PRELOAD: A predetermined load placed on a bearing during assembly or by adjustment.

PRIMARY CIRCUIT: the low voltage side of the ignition system which consists of the ignition switch, ballast resistor or resistance wire, bypass, coil, electronic control unit and pick-up coil as well as the connecting wires and harnesses.

PRESS FIT: The mating of two parts under pressure, due to the inner diameter of one being smaller than the outer diameter of the other, or vice versa; an interference fit.

RACE: The surface on the inner or outer ring of a bearing on which the balls, needles or rollers move.

REGULATOR: A device which maintains the amperage and/or voltage levels of a circuit at predetermined values.

RELAY: A switch which automatically opens and/or closes a circuit.

RESISTANCE: The opposition to the flow of current through a circuit or electrical device, and is measured in ohms. Resistance is equal to the voltage divided by the amperage.

RESISTOR: A device, usually made of wire, which offers a preset amount of resistance in an electrical circuit.

RING GEAR: The name given to a ring-shaped gear attached to a differential case, or affixed to a flywheel or as part of a planetary gear set.

ROLLER BEARING: A bearing made up of hardened inner and outer races between which hardened steel rollers move.

ROTOR: 1. The disc-shaped part of a disc brake assembly, upon which the brake pads bear; also called, brake disc. 2. The device mounted atop the distributor shaft, which passes current to the distributor cap tower contacts.

SECONDARY CIRCUIT: The high voltage side of the ignition system, usually above 20,000 volts. The secondary includes the ignition coil, coil wire, distributor cap and rotor, spark plug wires and spark plugs.

SENDING UNIT: A mechanical, electrical, hydraulic or electro-magnetic device which transmits information to a gauge.

SENSOR: Any device designed to measure engine operating conditions or ambient pressures and temperatures. Usually electronic in nature and designed to send a voltage signal to an on-board computer, some sensors may operate as a simple on/off switch or they may provide a variable voltage signal (like a potentiometer) as conditions or measured parameters change.

SHIM: Spacers of precise, predetermined thickness used between parts to establish a proper working relationship.

SLAVE CYLINDER: In automotive use, a device in the hydraulic clutch system which is activated by hydraulic force, disengaging the clutch.

SOLENOID: A coil used to produce a magnetic field, the effect of which is to produce work.

SPARK PLUG: A device screwed into the combustion chamber of a spark ignition engine. The basic construction is a conductive core inside of a ceramic insulator, mounted in an outer conductive base. An electrical charge from the spark plug wire travels along the conductive core and jumps a preset air gap to a grounding point or points at the end of the conductive base. The resultant spark ignites the fuel/air mixture in the combustion chamber.

SPLINES: Ridges machined or cast onto the outer diameter of a shaft or inner diameter of a bore to enable parts to mate without rotation.

TACHOMETER: A device used to measure the rotary speed of an engine, shaft, gear, etc., usually in rotations per minute.

THERMOSTAT: A valve, located in the cooling system of an engine, which is closed when cold and opens gradually in response to engine heating, controlling the temperature of the coolant and rate of coolant flow.

TOP DEAD CENTER (TDC): The point at which the piston reaches the top of its travel on the compression stroke.

TORQUE: The twisting force applied to an object.

TORQUE CONVERTER: A turbine used to transmit power from a driving member to a driven member via hydraulic action, providing changes in drive ratio and torque. In automotive use, it links the driveplate at the rear of the engine to the automatic transmission.

TRANSDUCER: A device used to change a force into an electrical signal.

TRANSISTOR: A semi-conductor component which can be actuated by a small voltage to perform an electrical switching function.

TUNE-UP: A regular maintenance function, usually associated with the replacement and adjustment of parts and components in the electrical and fuel systems of a vehicle for the purpose of attaining optimum performance.

TURBOCHARGER: An exhaust driven pump which compresses intake air and forces it into the combustion chambers at higher than atmospheric pressures. The increased air pressure allows more fuel to be burned and results in increased horsepower being produced.

VACUUM ADVANCE: A device which advances the ignition timing in response to increased engine vacuum.

VACUUM GAUGE: An instrument used to measure the presence of vacuum in a chamber.

VALVE: A device which control the pressure, direction of flow or rate of flow of a liquid or gas.

VALVE CLEARANCE: The measured gap between the end of the valve stem and the rocker arm, cam lobe or follower that activates the valve.

VISCOSITY: The rating of a liquid's internal resistance to flow.

VOLTMETER: An instrument used for measuring electrical force in units called volts. Voltmeters are always connected parallel with the circuit being tested.

WHEEL CYLINDER: Found in the automotive drum brake assembly, it is a device, actuated by hydraulic pressure, which, through internal pistons, pushes the brake shoes outward against the drums.

A

MASTER INDEX